D1788202

UNDERSTANDING INDUSTRIAL EXPERIMENTATION

Second Edition

Donald J. Wheeler

SPC Press, Inc.

Knoxville, Tennessee

5908 Toole Drive, Suite C
Knoxville, Tennessee, 37919 USA
(865) 584-5005

ISBN 0–945320–09–4

Seventh Printing

Contents

Glossary

$36.\overline{6}$	= 36.6666666666666... the bar above the last digit indicates that this value repeats an infinite number of times: p.101
2^{k-p}	a notation for an experimental design using 2^{k-p} treatment combinations to study k factors at 2 levels each: p.186
2^{k-p}_{II}	the 2^{k-p} notation with the Resolution of the design attached: p.220
α	the theoretical risk of a false alarm when making a decision: pp.57, 59
β	the theoretical risk of a missed signal when making a decision: p.57
μ	the mean of $f(x)$: (a parameter) pp.14, 15
ν	the symbol used to denote degrees of freedom for dispersion statistics: p.44
ν_1	the numerator degrees of freedom for a regular F-distribution: p.96
ν_2	the denominator degrees of freedom for a regular F-distribution: p.96
ν_a	the degrees of freedom found using formula on p.45
$\nu_{B(k,n)}$	the degrees of freedom found for k subgroups of size n in Table B, p.302: p.44
σ	the standard deviation of $f(x)$: (a parameter) pp.14, 15
σ^2	the variance of $f(x)$: (a parameter) p.15
τ	the target value for a product performance characteristic, Y: p.255
A_2	a constant for finding control chart limits: pp.55, 304
C	a theoretical contrast among subgroup means: (a parameter) p.112
$\hat{C}$	the estimated value for C: (a statistic) p.115
c_2	a bias correction factor given in Table A: pp.25, 301
c_4	a bias correction factor given in Table A: pp.25, 301
c_j	the jth coefficient for a contrast, C: p.112
d_2	a bias correction factor given in Table A: pp.20, 301
d_2^*	a bias correction factor given in Table B: pp.25, 302–303
d_3	a bias correction factor given in Table A: pp.22, 301
D_3	a constant for finding control chart limits: pp.55, 304
D_4	a constant for finding control chart limits: pp.55, 304
$E[\mathcal{L}(y)]$	the average loss due to departures from target: p.257
E(X)	the expected value for X: (a parameter) p.15
F	the symbol for a calculated F-Ratio: (a statistic) p.86
$F_{1-\alpha}(\nu_1,\nu_2)$	a percentile from the regular F-distribution: p.132
$f(x)$	the probability density function or probability distribution function: p.13
H	the multiplier for finding ANOM decision limits: p.59
H_L	the multiplier for the lower ANOMR decision limit: p.78

H_U	the multiplier for the upper ANOMR decision limit: p.78
HSD	Tukey's Honestly Significant Difference: p.75
k	the number of subgroups in a set of subgroups: pp.30, 59
K	the constant in the expression for the average loss due to deviation from target: p.256
$\hat{l}$	the estimated effect for a contrast, C: (a statistic) p. 121
L	the number of levels of a factor in ANOMR: p.77, also, the number of levels of a factor in a Plackett-Burman Design, p.205
$LCL_{\bar{X}}$	lower control limit for the Average Chart: pp.55, 304
$LDL_{\bar{R}}$	lower decision limit for ANOMR: p.78
$LDL_{\bar{X}}$	lower decision limit for ANOM: p.66
LSL	lower specification limit: p.254
$\mathcal{L}(y)$	the loss associated with the value y: p.254
m	the number of subgroups per level of a factor in ANOMR: p.77
Mod $F_{1-\alpha}$	a percentile from a Mod F distribution: p.147
Mod F $(j,p,k\text{-}1)$	a symbol used to denote both the Mod F distribution and a calculated Mod F-ratio (a statistic): p.146
MEAN(R)	the mean of the distribution of subgroup ranges: (a parameter) p.18
MEAN(X)	the mean of $f(x)$: (a parameter) p.15
MEAN($\bar{X}$)	the mean of the distribution of subgroup averages: (a parameter) p.18
MSB	the Mean Square Between: (a statistic) p. 86
MS(C)	the Mean Square for a Contrast: (a statistic) p.117
MSD(τ)	an alternative expression (yielding slightly diferent values) for $MSD_n(\tau)$: p.259
$MSD_n(\tau)$	the Mean Square Deviation about target: (a statistic) p.258
MSE	the Mean Square Error: (a statistic) p.144
MSE_p	the average of the p smallest of a set of SS(C) values: (a statistic) p.146
MSW	the Mean Square Within: (a statistic) pp.39, 86
n	the size of a single set of data or a subgroup: p.30
n_j	the size of the jth subgroup in a set of subgroups of unequal size: p.91
q	percentiles of the Studentized Range Distribution: pp.75, 313-315
R	the range of a set of numbers: (a statistic) p.10
$\bar{R}$	the average of several subgroup ranges: (a statistic) p.31
$\bar{\bar{R}}$	the average of all the subgroup ranges in ANOMR: (a statistic) p.77
$R_{\bar{X}}$	the range of a set of subgroup averages: (a statistic) p.33
SD(R)	the standard deviation of the distribution of subgroup ranges: (a parameter) p.18
SD($\bar{R}$)	the standard deviation of the average of several subgroup ranges: (a parameter) p.77
SD(X)	the standard deviation of $f(x)$: (a parameter) p.15
SD($\bar{X}$)	the standard deviation of the distribution of subgroup averages: (a parameter) p.18
s	the sample or subgroup standard deviation for a set of numbers: (a statistic) p.11

$\bar{s}$	the average of several subgroup standard deviations: (a statistic) p.31
$s_{\bar{X}}$	the standard deviation of a set of subgroup averages: (a statistic) p.32
$\overline{s^2}$	the average of the squares of several sample standard deviations: (a statistic) p.39
s_n	the root mean square deviation for a set of numbers: (a statistic) p.10
SS_j	the jth value in a set of SS(C) values when those values are arranged in ascending order: (a statistic) p.146
SSB	the Sum of Squares Between subgroups: (a statistic) p.91
SS(C)	the Sum of Squares for a Contrast: (a statistic) p.117
SS(CC)	the Sum of Squares for the Curvature Contrast: (a statistic) p.239
SST	the Total Sum of Squares: (a statistic) p.91
SSW	the Sum of Squares Within subgroups: (a statistic) p.91
UCL_R	upper control limit for the Range Chart: pp.55, 304
$UCL_{\bar{X}}$	upper control limit for the Average Chart: pp.55, 304
$UDL_{\bar{R}}$	upper decision limit for ANOMR: p.78
$UDL_{\bar{X}}$	upper decision limit for ANOM: p.66
USL	upper specification limit: p.254
V(X)	the variance of $f(x)$: (a parameter) p.15
X_i	the ith value in a set of numbers: pp.10,13
X_{ij}	the ith value in the jth subgroup in a set of subgrouped data: p.91
$\bar{X}$	the average of the X values: (a statistic) p.10
$\bar{X}_j$	the average of the jth subgroup in a set of subgrouped data: (a statistic) p.91
$\bar{\bar{X}}$	the Grand Average: (a statistic) for balanced data, this is the average of the subgroup averages, for unbalanced data this is the simple average of all of the individual values: pp.41, 91
$\tilde{X}$	the median of the X values: (a statistic) p.10
Y	a product performance characteristic: p. 255, and a response variable: p.56
y	a value of a product performance characteristic: p.255
Z	a symbol used to denote any one of several different quantities collectively known as Signal-to-Noise Ratios: p.280

Preface to Second Edition

One of the most difficult things that a teacher must do is to look at his subject from the perspective of the student. In a live classroom setting the students provide feedback which a teacher may use to achieve this perspective. In preparing a text an author has to depend upon experience and memory to try and achieve this perspective. While the first edition of this text grew out of the insights gained from over 12 years of teaching Design of Experiments and Analysis of Variance techniques to nonstatisticians, there were still some portions of the text which could have been clearer. Thus, the need for a new edition.

The changes in this edition are primarily aimed at making the text easier to understand. Toward this end Chapter One has been extensively revised. Some portions have been simplified, while other sections have been expanded in the interest of clarity. While some additional sections have been added throughout the remainer of the text, most of the revisions consist of expansions of the original material. The new topics included do not substantially alter the scope of this text: it is still intended to be a bridge to help engineers and scientists understand the principles and advantages of using experiments which are designed to facilitate the analysis of the data.

Finally, I must thank Dr. James Maynard for his many helpful suggestions. Most of these have been incorporated in this edition.

Don Wheeler
December 1989

Three Obstacles to Research

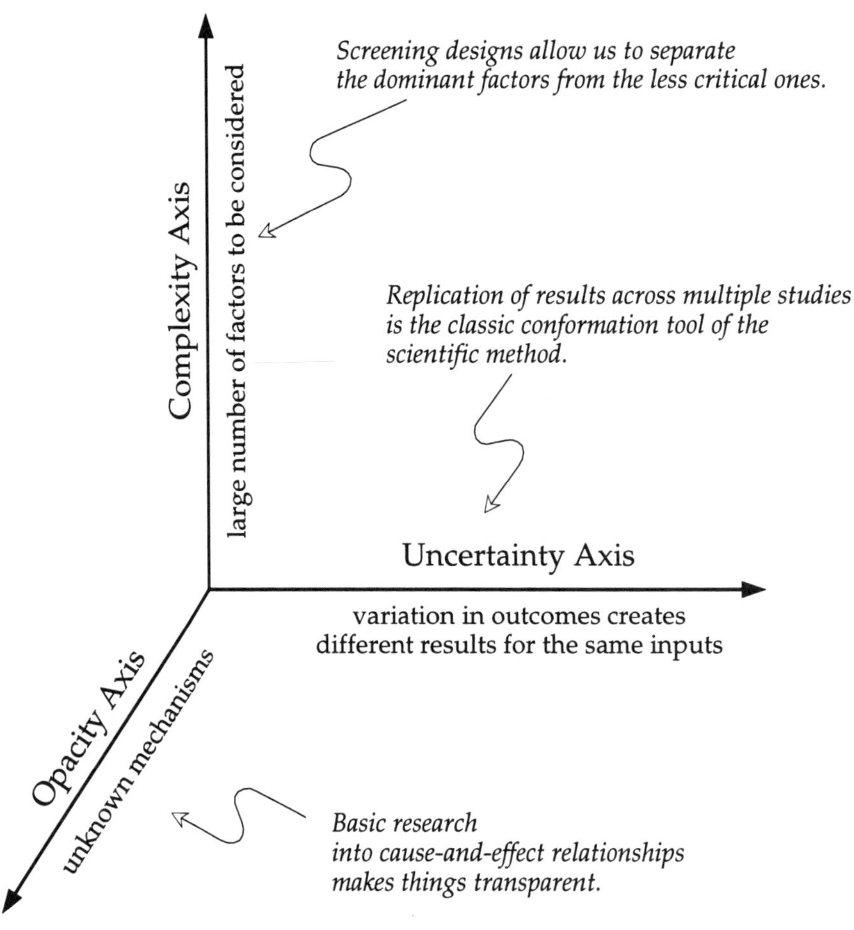

Myron Tribus

INTRODUCTION

In any R & D program there will usually be one or more points, somewhere between the theoretical concept and the production of a product, at which the engineer will need to collect and analyze some data. If this is done inefficiently, or if the results of the analysis are uninterpretable, additional data will usually be required, resulting in delays in development that can literally cost a company millions of dollars. It is the purpose of this book to provide the materials that engineers will need to carry out effective programs of industrial experimentation that will avoid both (1) the inefficient collection and analysis of data and (2) uninterpretable and uncommunicable results.

Throughout this book the emphasis will be upon the use of simple and robust techniques for exploratory studies, with the primary objective always being to obtain results that are interpretable, understandable, and easily communicated to others. Without such results, the most brilliant experiment, and the most sophisticated analysis, are useless. Thus, while some traditional tools are included, other, less traditional tools will also be suggested. When used in concert with each other, these various techniques comprise a very powerful approach to industrial experimentation.

Agricultural and Biomedical experiments provided the proving ground for the development of the traditional statistical tools commonly referred to as "Design of Experiments." Industrial experimentation differs from these traditional applications in two significant ways. First of all, in agricultural and biomedical experiments there is usually a multiplicity of experimental units (different fields, different offspring, different patients) while in industrial experimentation there will usually be only one experimental unit (the equipment in the plant or the lab). Secondly, in agricultural and biomedical experiments, confirmation of results may take years or decades to obtain, while in industrial experiments one can often confirm a finding tomorrow, next week, or at worst, next month. These two differences between industrial experimentation and the traditional applications of designed experimentation justify a nontraditional approach to the use of the tools.

In addition to these fundamental differences, there are other unique aspects of industrial experimentation. For example, the industrial experimenter is usually faced with a large array of independent variables (factors) whose effects upon a set of dependent variables (responses) are unknown. Simply sorting out the active factors from the inert factors will generally be a major step forward. With an objective of making a process function, the industrial experimenter will have to consider all relevant factors: simplifying assumptions will not be possible. Discovering new relationships and new theoretical knowledge may be a side benefit, but such knowledge is often not the primary objective of industrial research. Finally, the analysis techniques should be robust and easy to implement, while yielding results which are interpretable, communicable, and understandable.

Primary Objective is Exploration:
    Separate Active Factors from Inert Factors

Must Consider Large Numbers of Factors and Responses:
    Emphasis upon making process function
        rather than discovering new theories
    Simplifying assumptions are not possible

Analysis Must Be Appropriate:
    Understandable results required
    Results must be capable of being communicated
    Analysis should not require too much time or expertise
    Analysis should reliably detect the signals within the data

---

Uniqueness of Industrial Experimentation

Given these distinctions for industrial experimentation, the following strategy suggests itself. Initially one should identify all potential factors for the experiment. Any factor for which changes are feasible should be included at this stage. (When experimenting with an existing plant, the suggestions of the operators should definitely be included at this stage.) Next, some initial experiments should be performed which will allow the inert factors to be dropped from further consideration. Finally, those factors which appear to actively affect the response variables may be examined in greater detail by subsequent experiments. This would continue until the experimenter can identify some sort of "optimum operating conditions" for the process.

Identify Potential Factors:
    Include all factors for which changes are feasible

Screen Out Inert Factors:
    Delete factors that do not have a pronounced effect upon the response variable

Study Active Factors:
    Find the best levels for those factors which do have a pronounced effect upon the response variable

A Strategy for Industrial Experimentation

One should note that this strategy depends upon a sequence of experiments. (The only way to design the perfect experiment is to know the answers. If we don't know the answers, our experiments will be less than perfect, and subsequent experiments will be needed to refine our knowledge.)

Knowledge has always come from an iteration between ideas and data. Occasional flashes of inspiration may shorten some of these cycles, but the basic way in which we learn things is a cyclic process. Therefore, any strategy of experimentation which does not plan for these cycles is unrealistic.

W. A. Shewhart outlined this cyclic nature of experimentation in his book *Statistical Method from the Viewpoint of Quality Control* (Shewhart 1939, p.45) when he said that the dynamic scientific process of acquiring knowledge consists of three steps: making a hypothesis, carrying out an experiment, and testing the hypothesis, or, more concisely, Specification, Production, Inspection.

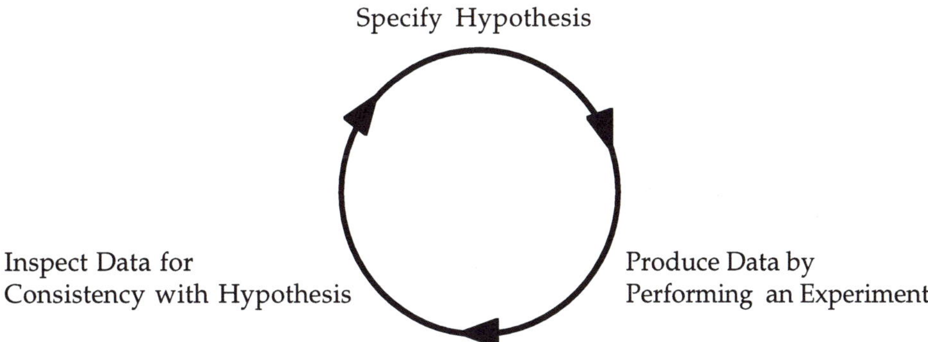

Figure 1: Shewhart's Cycle for the Scientific Process

A consequence of this sequential strategy for industrial experimentation is that we will not have to rely upon probabilistic arguments for the confirmation of results. The replication of results from one study to another will provide the most powerful confirmation possible. Moreover, the data from each experiment may be analyzed in an exploratory manner, with an emphasis on including all potentially active factors, rather than being analyzed in a conservative (i.e. exclusionary) manner. This shift in emphasis will have implications for the way the experiments are performed and the way the data are analyzed.

There is a subtle but important point in the previous paragraph. While we will not attempt to use probabilistic tools for the *confirmation* of results, we will still be faced with the problem of separating signals from the noise. Therefore we will utilize some of the traditional probabilistic procedures as *guidelines for separating the potential signals from the background noise*. Thus, while some of the analysis techniques included herein will look like the traditional techniques, they are being used and interpreted in a nontraditional manner.

Of course, with any probabilistic decision procedure there is always the question of specifying and interpreting the α-level for the procedure. Many contorted arguments have been advanced to interpret these α-levels in the context of a set of results. In spite of all of these arguments, the α-level remains a theoretical number which is calculated under a set of assumptions which are never fully satisfied in practice (it is the risk of a false alarm given data which are independently and normally distributed and which contain no signals—since data are never perfectly normally distributed, and data often do contain signals, this theoretical quantity does not have a clear interpretation in practice). So while the use of some traditional procedures will require the specification of an α-level, the number chosen will not be interpreted as implying any "confidence level" for the results. In practice, the α-level merely indicates how the line for separating potential signals from probable noise was obtained: large values of α will denote a more exploratory approach (one which will be more sensitive to signals which may be present), and a small α-level will indicate a more conservative approach. Thus, the α-level will tell you more about the experimenter's approach to the analysis than it will about the "significance level" or the "confidence level" for the results

Another major impediment to using the α-level to quantify the strength of the results is the fact that the results will always be extrapolated into the future. This point has been repeatedly made by Dr. Deming. The risk associated with the extrapolation from the data studied to future data is not covered by the α-level. Those effects which are found to be "significant at level α" are merely the best candidates for signals within the current data set. The real confirmation of whether they represent signals will be their persistence in subsequent experiments or operations. This nontrivial replication of results has been, and will continue to be, the traditional confirmatory tool of the scientific method. In order to establish any cause and effect relationship one must demonstrate the consistency of the effect across time and/or space. No probabilistic argument (using the data of one isolated study) can provide this type of evidence. So while we may use probability theory to obtain decision rules to help separate potential signals from probable noise, we are not attaching any real meaning to the α-level used. It is merely a theoretical value used as a guideline for practice.

All data must be interpreted in their context, and the sequential nature of industrial experimentation is part of this context. Since this sequential approach differs from the traditional approach, the user will need to have some understanding of the basic concepts in order to keep from becoming hopelessly confused. To help the reader avoid this confusion this text contains some discussions of the basic concepts and foundations. In these discussions, intuitive arguments have been used rather than rigorous proofs, and some results are given without derivation. (References have been provided for further reading.) It is hoped that these explanations will provide industrial practitioners with the understanding they will need to use the techniques effectively.

The basic structure of this book is shown in the figure entitled "Elements of Effective Industrial Experimentation" (see page 6). The capstone of the wall is the objective of this book, and the blocks in the wall are the elements discussed in this book. The relationships of the elements in this text are indicated by the arrows in the figure. Some elements and techniques support others, while some techniques operate parallel to, and independently of, others.

On the left side of the figure is one of the simplest and most understandable of all of the techniques, the Analysis of Means. In many cases this one tool will bridge the gap between raw data and understandable results.

In the middle of the figure is the Analysis of Variance and all of the special techniques that depend upon this basic analytic tool. While slightly more complex than the Analysis of Means, the Analysis of Variance will extend to special cases that the other technique will not handle.

On the right side of the figure is the concept that ties all the other techniques together into a unified approach to the problems of production: the Quadratic Loss Function for Product Performance. Using this concept, industrial experimentation may be said to have the objective of minimizing the transmission of variation into the product.

At the product design stage, the objective is to choose the design parameters in such a way that the product will be as insensitive as possible to variation in raw materials, variation in manufacturing conditions, and variation in the environment in which it is used. Industrial experimentation will provide the necessary inputs for making these choices. When this is done, the product will display a greater consistency, and will be easier to manufacture and use, than it would be otherwise.

At the process design stage, the objective is to choose the manufacturing conditions in such a way that the product will be as insensitive as possible to the variations in those manufacturing conditions. Again the objective is to achieve the greatest product consistency with the least manufacturing effort. In order to make these choices one may use either industrial experimentation or statistical process control techniques.

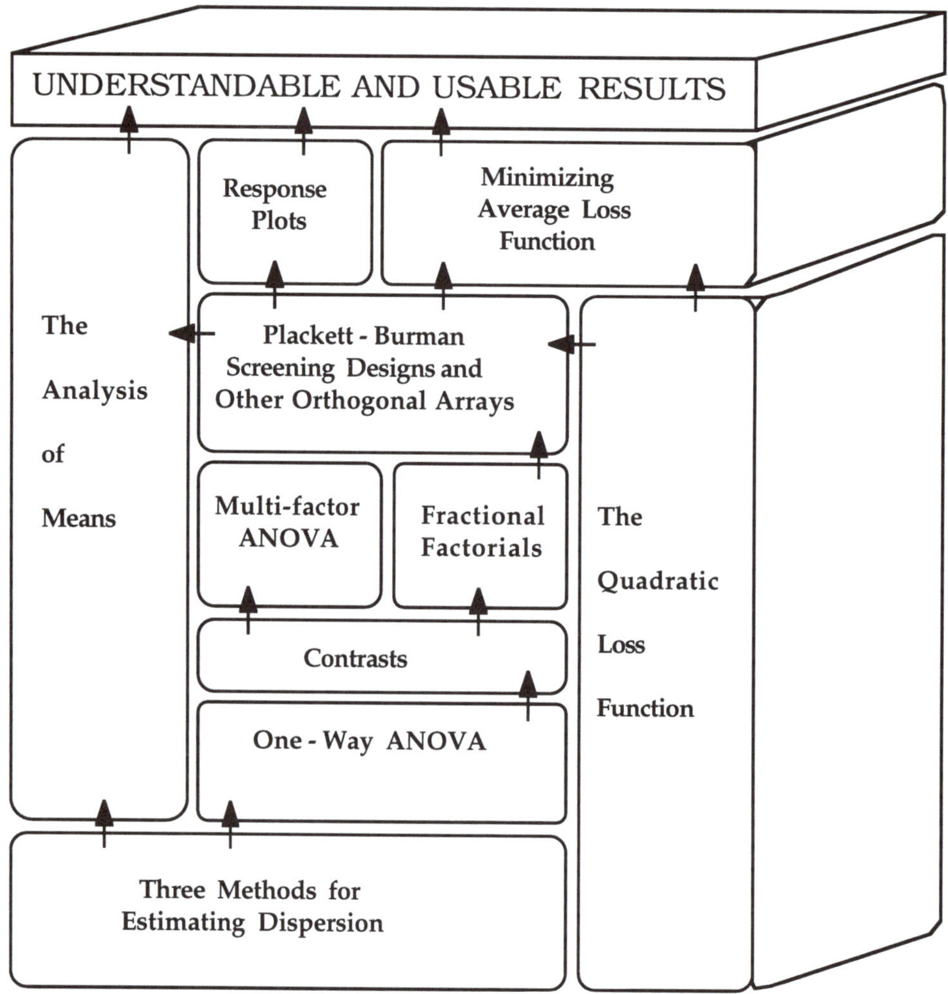

Figure 2: Elements of Effective Industrial Experimentation

Finally, in day-to-day manufacturing, the techniques of SPC are available to use in obtaining the maximum product consistency with the minimum effort. In short, once you have gone to the trouble of designing the product and the process, it is absurd to allow the process to operate with less consistency than it is capable of doing. Of course, the more robust the product design, the smaller the impact of process variation upon the product, which will result in a lower profile for the SPC effort on that process.

Thus, statistically aided manufacturing involves both industrial experimentation and statistical process control. The roles of these techniques are complimentary rather than contradictory.

Finally, one should note that statistically aided manufacturing has a different thrust from much of the automation being pushed today. Rather than adding remedies consisting of feedback and feed-forward loops, control devices, and complicated adjustment procedures, these statistical techniques are suited to discovering those factors which can be used to minimize the need for such complexity. Because complexity is the essence of Murphy's Law.

CHAPTER ONE

WORKING WITH VARIATION

Virtually all of the techniques presented in this book will involve some measure or estimate of variation. Since there are many different ways of measuring variation, and since some of these ways are interchangeable while others are not, a considerable amount of confusion exists among those seeking to use statistics in industry.

This chapter seeks to organize the issues involved in working with variation by making some categorical distinctions. Among these distinctions are:

(1) the difference between "statistics" and "parameters,"

(2) the difference between a "parameter value" and an "estimate of a parameter,"

(3) the different types of estimators, and

(4) the profound difference between the "three methods of estimating variaton."

Finally, these categorical distinctions are summarized by reference tables. Hopefully, these reference tables will help the reader avoid some of the more serious mistakes engendered by the confusion regarding variation.

Finally, the reader should not feel that all the details in this chapter must be mastered. If the broad outlines of the categorical distinctions are seen, and the reference tables can be used, then this chapter will have served its purpose. Therefore, you should read this chapter for the information it conveys without worrying over points that are not perfectly clear. They will usually become clear in the context of some application given later in the book.

1.1 Measures of Dispersion

Data are usually summarized by certain numbers that are properly called **statistics**. A statistic is simply a function of the data, such as the average or the range. When we want to describe the location of the center of some data, we will use one of the **Measures of Location**, such as the **average** or, occasionally, the **median**. When working with variation, we will use one of the **Measures of Dispersion**. Common measures of dispersion are the **range**, the **root mean square deviation**, and the **sample standard deviation**.

The **average, $\bar{X}$**, is the sum of the data divided by the number of values in the data. It is the balance point for the data (the center of mass), and is the most commonly used measure of location. Given the data {4, 5, 5, 4, 8 ,4, 3, 7}, the average is 40/8 = 5.0.

The **median, $\tilde{X}$**, is the fiftieth percentile for the data. If the data consist of an odd number of values, the median is the one "middle value" once the data have been arranged in numerical order. If the data consist of an even number of values, then the median is the average of the two "middle values" after the data have been arranged in numerical order. Given the data {4, 5, 5, 4, 8, 4, 3, 7} we first arrange them in ascending order: {3, 4, 4, 4, 5, 5, 7, 8}, find the two middle values to be 4 and 5, and compute the median to be (4 + 5)/2 = 4.5.

The **range, R,** is defined to be the difference between the maximum value and the minimum value. (When giving a range, we must also state the number of observations used in finding the range.) While the range is the simplest measure of dispersion to compute, it is primarily used with small data sets (data sets with fewer than 16 observations).

Given the data { 4, 5, 5, 4, 8, 4, 3, 7 }, the range would be R = (8 - 3) = 5, and this range would be said to be based upon $n = 8$ values.

The **root mean square deviation, s_n**, is, quite literally, defined by its name. This name, when read from right to left, lists the operations for finding s_n. The root mean square deviation is found by (1) obtaining the deviation of each datum from the average, (2) squaring these deviations, (3) averaging these squared deviations, and then (4) finding the square root of this average. The formula for this statistic is:

$$s_n = \sqrt{\frac{1}{n}\sum_{i=1}^{n}(X_i - \bar{X})^2}$$

For the data { 4, 5, 5, 4, 8, 4, 3, 7 }, $\bar{X}$ = 5.00, and $s_n = \sqrt{\frac{20}{8}} = \sqrt{2.5} = 1.5811$.

The root mean square deviation was the dispersion statistic preferred by W.A. Shewhart. (While Shewhart used the lower case Greek letter sigma, σ, to denote this statistic, modern notation will be used in this book.)

The **sample standard deviation,** *s*, is quite similar to the root mean square deviation. In fact, by changing the denominator from (*n*) to (*n–1*), the root mean square deviation becomes the sample standard deviation, *s*.

$$s = \sqrt{\frac{1}{n-1} \sum_{i=1}^{n} (X_i - \bar{X})^2}$$

For the data { 4, 5, 5, 4, 8, 4, 3, 7 }, $s = \sqrt{\frac{20}{7}} = \sqrt{2.8571} = 1.6903$.

The sample standard deviation was introduced by R. A. Fisher, and is commonly used in most of the subdisciplines of statistics. It represents a modification of the root mean square deviation. From a user's perspective there is little to recommend one of these dispersion statistics over the other. Both the sample standard deviation, *s*, and the root mean square deviation, s_n, are 100 percent efficient, and both are based on the same essential information. While different applications will generally specify one of these dispersion statistics, some applications will permit substitutions. More will be said about this later. At present the reader should merely become acquainted with these different names, symbols and formulas.

Since mathematical models used in probability theory are characterized by a *parameter* that is also called the standard deviation, the *statistic* above will be called the *sample standard deviation*. The adjective is added here merely in an attempt to avoid an ambiguity that students often find perplexing.

Thus, for the data set { 4, 5, 5, 4, 8, 4, 3, 7 }, the dispersion statistics given above have values ranging from 1.58 to 5.0. These different dispersion statistics yield different values because each statistic characterizes a slightly different aspect of the dispersion of the data.

EXERCISE 1.1: Calculation of Location and Dispersion Statistics:

The data are {3, 6, 6, 4, 5, 4, 6, 6}:

(a) find the average:

(b) find the median:

(c) find the range:

(d) find the root mean square deviation:

(e) find the sample standard deviation:

Unfortunately, both the name and the symbol used to denote the root mean square deviation may vary from book to book. To a lesser extent, the name and symbol for the sample standard deviation may change from place to place. This is especially true with the notations used on hand-held calculators and the code names shown on computer printouts. For this reason it is best to check the formula for any dispersion statistic which you encounter. (To make matters worse, some of the dispersion statistics floating around are distorted versions of the common statistics cited above.)

Likewise, because of the different values, we must be careful to insert the correct value whenever one of these statistics appears in a formula. While there are some substitutions that have very little impact upon the result, others have a profound impact. For this reason, the beginner should be careful about making any substitutions.

This proliferation of dispersion statistics, and the lack of a standard nomenclature, are the first two sources of confusion that the student of statistics encounters when working with variation.

1.2 The Notion of a Distribution

Assume that we have a process that is producing some product (either discrete items or bulk product), and assume that periodic checks are made upon some product characteristic. These checks will result in a sequence of values that can be denoted as

$$\{ X_1, X_2, \dots, X_i, \dots, X_n, \dots \}.$$

While the statistics computed from the first n values of this sequence will characterize the n values, the questions of interest are how well these statistics characterize the process during the time period covered by the n observations, and how well they predict future performance. This extrapolation from the data to the underlying process will be credible only when certain conditions exist.

These conditions were described by W.A. Shewhart when he defined a state of statistical control: *"A phenomenon will be said to be controlled when, through the use of past experience, we can predict, at least within limits, how the phenomenon may be expected to vary in the future."* Thus, the question of statistical control directly addresses the issue of this extrapolation from the data to the underlying process. If the process has been found to display "controlled variation" in the past, and if there is no evidence of "uncontrolled variation" in the present, then the extrapolation from the data to the underlying process will be credible. Moreover, *as long as the process continues to display controlled variation,* the statistics based upon the historical data will continue to characterize the production process. However, if the process is "out of control," there is no way to justify any extrapolation from the data to the underlying process. The statistics obtained from the data will not characterize the product being produced by the process.

Thus, the journey from a production process to a statistical distribution takes the following route.

If the production process displays a state of statistical control, then it will be characterized by measurements which display a stable, consistent and recurring pattern of variation. This stable pattern of variation will result in a histogram that, in effect, may be thought of as essentially unchanging over time. The histogram of this stable pattern of variation may then be approximated by some (continuous) function, $f(x)$.

The function $f(x)$ is called a **Probability Distribution Function** or a **Probability Density Function**, and may be described as follows. Let X denote the measurement of the product characteristic of interest. The function, $f(x)$, is defined so that the product $[f(x)][dx]$ approximates the proportion of measurements that will fall between the values of x and $x + dx$, where dx represents a small increment in x.

When a process displays statistical control, the measurements from the product stream . . .

$X_1, X_2, X_3, X_4, X_5, X_6, X_7, X_8, X_9, \ldots, X_n, \ldots$

will display a consistent and predictable amount of variation.

This implies that periodic histograms of the product measurements . . .

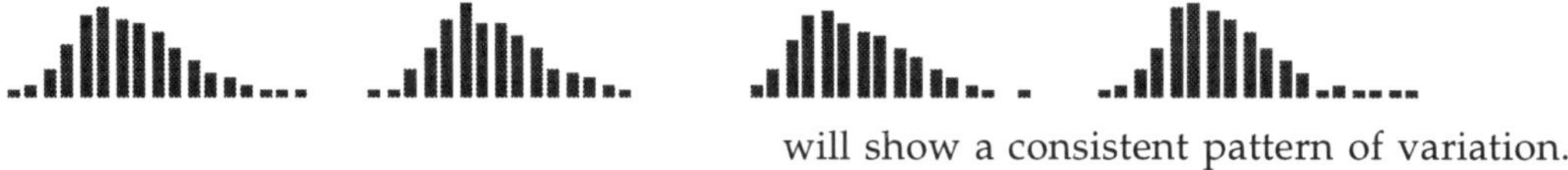

will show a consistent pattern of variation.

So that it is meaningful to approximate this pattern of variation . . .

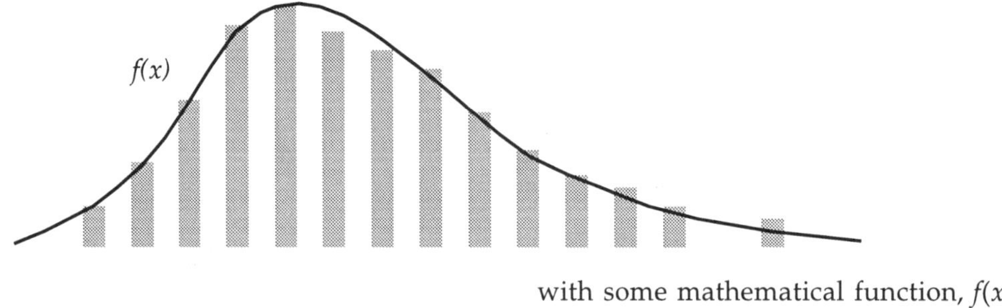

with some mathematical function, $f(x)$.

Which, in turn, may be characterized by parameters . . .

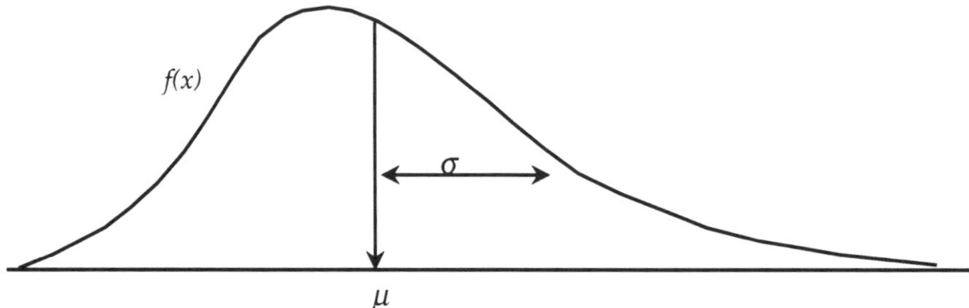

such as the mean, μ, and the standard deviation, σ.

Figure 1.1: The Notion of a Probability Distribution Function and Its Parameters

Of course, no real process is ever *exactly* stable in the sense which is implied in the picture above, but the notion of exact stability is useful in conceptualizing process stability and in building a theoretical model for working with the variables involved. (Dr. Henry Neave does an good job of discussing this in *The Deming Dimension*.) In particular, when the measurements obtained from a production process display statistical control one is often interested in characterizing the process location and the process dispersion. Since different samples will yield different averages and different ranges, it is helpful to use the **parameters** of the Probability Distribution Function, $f(x)$, to characterize both process location and process dispersion.

Process location may be characterized by the **mean** of $f(x)$. The mean of the distribution of X describes the balance point (or center of mass) of the distribution. This parameter is traditionally denoted by the symbols μ or E(X), and is defined as

$$\mu = E(X) = MEAN(X) = \int_{\text{all } x} x\, f(x)\, dx.$$

The symbol E(X) is usually called "the Expected Value of X." As shown, this text will also use the symbol MEAN(X) to denote the location parameter for the distribution of X.

Process dispersion may be characterized by the **variance** of $f(x)$. The variance of the distribution of X describes the rotational inertia of the distribution. This parameter is traditionally denoted by the symbols σ^2 or V(X), and is defined as

$$\sigma^2 = V(X) = \int_{\text{all } x} (x - \mu)^2 f(x)\, dx.$$

Another way of characterizing the dispersion of the process is to use the **standard deviation** of $f(x)$. The standard deviation of the distribution of X will be defined to be the square root of the variance of X. This parameter is generally denoted as σ or SD(X).

The usage of Greek letters, such as μ and σ, to denote parameters is a common convention in statistical notation. However, due to situations where some ambiguity might occur, this text will use the symbols MEAN(X), V(X), and SD(X) for the parameters of $f(x)$ interchangeably with the Greek letters.

In contrast to this, statistics (such as $\bar{X}$, R, and s) are generally denoted by Roman letters.

The integral equations above allow one to calculate the mean and variance of a distribution whenever the probability distribution function, $f(x)$, has a known form. In order to establish the form for $f(x)$ with any reasonable degree of accuracy it would be necessary to have several thousand observations collected while the process was in an *exact* state of statistical control. Since this will never happen in practice, one can never actually define $f(x)$. It is merely an abstraction.

Thus, the only way to use the integral equations given above to *calculate* values for the mean and variance of $f(x)$ is to *assume* a definite form for $f(x)$. Since values computed in this way would be no better than the assumption, it will generally be more satisfactory to *estimate* these parameters by using statistics obtained from the measurements generated by the process.

When a process displays statistical control, any sequence of data obtained from the process (in an objective manner) can be treated as a sample of values from the distribution of X, and the statistics computed from such data can be used to estimate parameters of the (abstract) Probability Distribution Function, $f(x)$.

When a process is out-of-statistical-control this whole structure falls apart. The failure to display statistical control means that the data will not show a stable and recurring pattern of variation. Therefore, we cannot begin to conceive of a meaningful Probability Distribution Function, $f(x)$, and the notion of the mean and variance of the distribution of X vanishes. Thus, while we may always calculate statistics from the data generated by an out-of-control process, these statistics cannot be used to estimate parameters because the process parameters are no longer well-defined. These statistics may be used to characterize the data upon which they are based, yet they cannot be used to extrapolate beyond these data simply because the context for such extrapolation has evaporated.

Thus, we come to the heart of the matter. The argument to this point is as follows:

(1) If data from a process display statistical control,
(2) then the notion of Probability Distribution Function, $f(x)$, is reasonable
(3) and the parameters of $f(x)$ will characterize process output;
(4) however, there will never be enough data to fully specify $f(x)$
(5) so we cannot actually compute the parameters μ and σ
(6) therefore we must use statistics to estimate parameters,
(7) and these estimates may be used to characterize process output.

Notice that the parameters provide a context for interpreting the statistics, and that this context is predicated upon the existence of a state of statistical control. This context allows the extrapolation beyond the product measured to the product not measured and to the underlying production process. Without this context, the statistics cannot be relied upon to do more than to characterize that product which was measured.

This is why it is always important to distinguish between ***statistics*** and ***parameters***.

Statistics are functions of the data. They can be (and often are) calculated even when the data are a meaningless collection of numbers.

Parameters are conceptualizations. They are descriptive constants for probability distribution functions. They exist only when the notion of a probability distribution function is well defined: that is, when the data display statistical control.

Since, in the real world, no process is ever in perfect statistical control, there is no way to ever compute "true values" for the parameters of a distribution. One must always make do with estimates, and sometimes these will be hard to obtain. In spite of this, the distinction between statistics and the notion of parameters of a distribution is both important and useful.

EXERCISE 1.2: Calculating E(X), V(X), SD(X) and Some Statistics:

Assume that the probability distribution function which describes a certain set of measurements has been found to be $f(x) = 1/2$ for $-1 \leq x \leq 1$, and $f(x) = 0$ otherwise.

(a) Evaluate the following expression to find the mean of $f(x)$:

$$E(X) = \int_{-1}^{1} \frac{x}{2}\, dx =$$

(b) Evaluate the following expression to find the variance of $f(x)$:

$$V(X) = \int_{-1}^{1} \frac{x^2}{2}\, dx =$$

(c) Find the value for the standard deviation of $f(x)$: SD(X) =

(d) A sample of nine values from the process characterized by $f(x)$ is

{0.3, -0.9, 0.2, 0.8, 0.4, 0.6, -0.9, -0.9, -0.5}.

Find the sample average:

Find the sample median:

Find the sample range:

Find the sample standard deviation:

(e) A second sample of nine values from the process characterized by $f(x)$ is

{0.0, -0.6, -0.2, 0.1, -0.2, -0.9, -0.3, 0.4, 0.6}.

Find the sample average:

Find the sample median:

Find the sample range:

Find the sample standard deviation:

The parameters for $f(x)$ are constant. Statistics may vary from sample to sample. This distinction between statistics and parameters has not always been made clear in statistical textbooks, and is yet another source of confusion for the student.

1.3 Using Statistics to Estimate Parameters

Given a sequence of measurements arising out of a process that displays statistical control, and given that these measurements are arranged in k rational subgroups of size n, one may use the statistics calculated from the data to estimate the parameters of the distributions of the different random variables.

First, with subgrouped data, there are three distinct distributions of interest: these are the Distribution of Individual Values, the Distribution of Subgroup Averages, and the Distribution of Subgroup Ranges. The theoretical Probability Distribution Functions for these three variables, for subgroups of size n = 5, are shown in Figure 1.2.

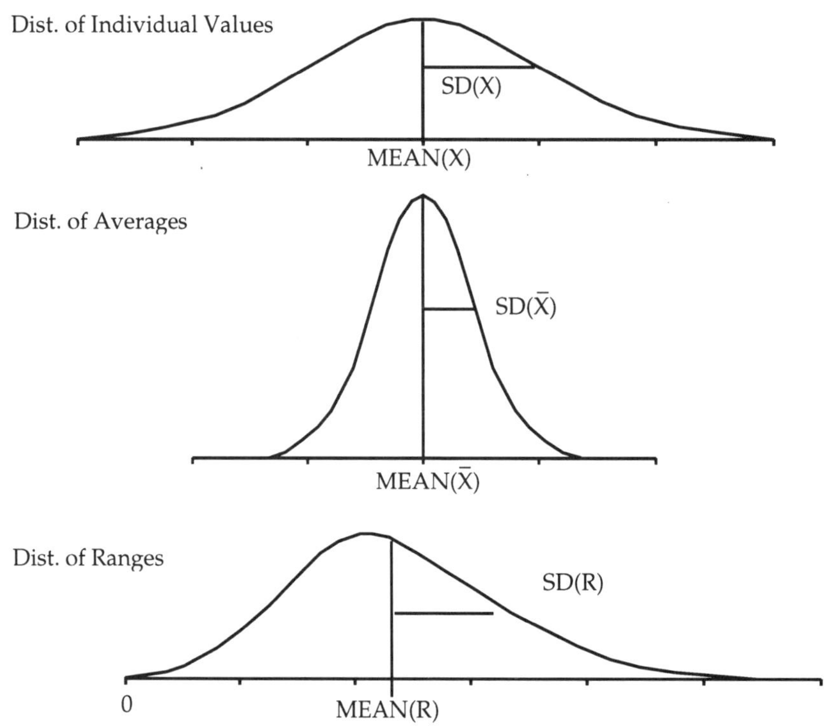

Figure 1.2: Probability Distribution Functions Associated With Subgrouped Data

Each of the Probability Distribution Functions in Figure 1.2 has its own mean and its own standard deviation. Any appreciation of the control chart formulas must begin with an understanding of how the various statistics are used to estimate the six parameters above.

The statistics which are commonly used with Control Charts are the Grand Average and the Average Range. When the data display statistical control these statistics will estimate the following parameters:

> The Grand Average, $\bar{\bar{X}}$, is an Estimate for MEAN($\bar{X}$).
>
> The Average Range, $\bar{R}$, is an Estimate for MEAN(R).

Thus, these two statistics directly estimate two of the six parameters listed above. They also may be used to obtain estimates of the other four parameters: This is done by utilizing some special relationships between these six parameters. These relationships and the estimators for the other parameters are given in the following theorems.

THEOREM ONE:

The Mean of the Distribution of X
is identical to
the Mean of the Distribution of Subgroup Averages:

MEAN(X) = MEAN($\bar{X}$).

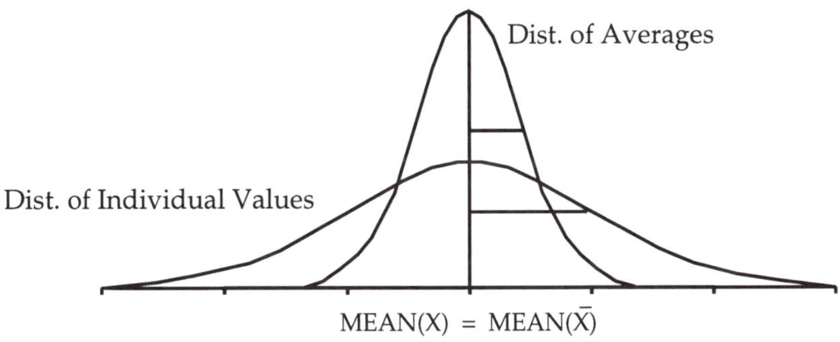

Figure 1.3: Theorem One: A Relationship between Means

CONSEQUENCE OF THEOREM ONE:

Since the Grand Average estimates MEAN($\bar{X}$)
it may also be used to estimate MEAN(X):

> An Estimate of MEAN(X) is $\bar{\bar{X}}$.

THEOREM TWO:

When the Individual Values are normally distributed,
The Mean of the Distribution of the Subgroup Ranges
is equal to
the product of
the Standard Deviation of the Distribution of X
and the constant d_2:

$$\text{MEAN(R)} = d_2 \text{ SD(X)}.$$

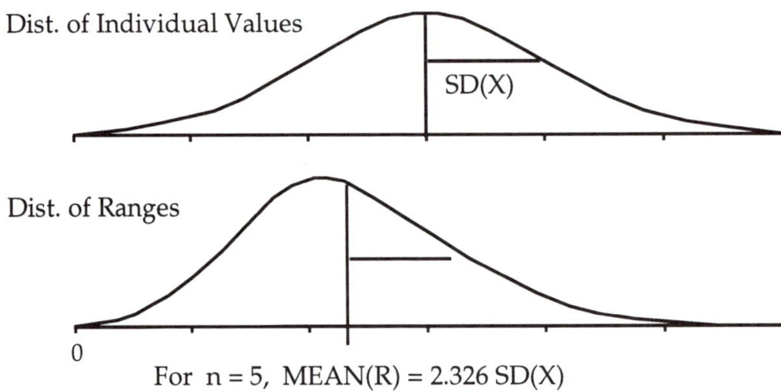

Figure 1.4: Theorem Two: The Relationship between the Mean of the Ranges and SD(X)

CONSEQUENCE OF THEOREM TWO:

Since the Average Range estimates MEAN(R),
the Standard Deviation of the Distribution of X
may be estimated by
the Average Range Divided by the constant d_2:

An Estimate of SD(X) is $\bar{R}/d_2$.

THEOREM THREE:

The Standard Deviation of the Distribution of the Subgroup Averages
is equal to
the Standard Deviation of X
divided by
the square root of the subgroup size:

$$SD(\bar{X}) = \frac{SD(X)}{\sqrt{n}}.$$

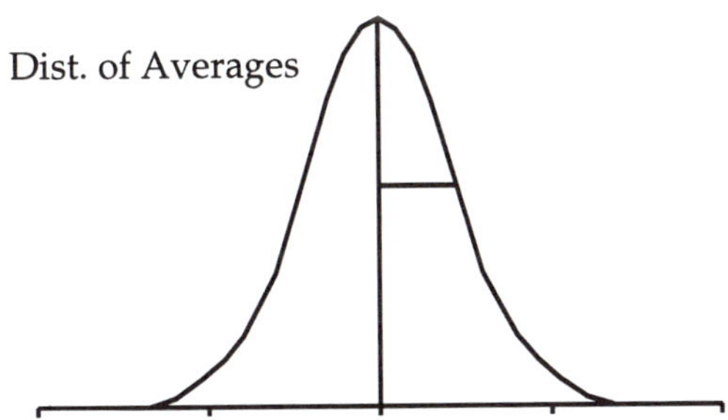

Figure 1.5: Theorem Three: A Relationship between Standard Deviations

CONSEQUENCE OF THEOREM THREE:

An estimate of
the Standard Deviation of the Distribution of the Subgroup Averages
may be obtained from
an Estimate of SD(X)
divided by $\sqrt{n}$:

An Estimate of $SD(\bar{X})$ is $\dfrac{\bar{R}}{d_2 \sqrt{n}}$.

THEOREM FOUR:

When the Individual Values are normally distributed,
the Standard Deviation of the Distribution of Subgroup Ranges
is equal to
the product of
the Standard Deviation of X
and the constant d_3:

$$SD(R) = d_3 \, SD(X).$$

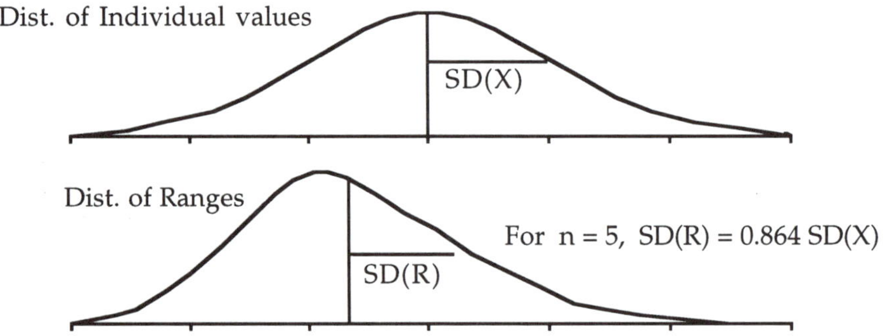

Figure 1.6: Theorem Four: Another Relationship between Standard Deviations

CONSEQUENCE OF THEOREM FOUR:

An estimate of
the Standard Deviation of the Distribution of Subgroup Ranges
may be obtained from
the Estimate of SD(X)
multiplied by the constant d_3:

An Estimate of SD(R) is $\dfrac{d_3 \bar{R}}{d_2}$.

Therefore, the Grand Average is commonly used to estimate two parameters, while the Average Range is commonly used to estimate four different parameters. This practice of using a single statistic to estimate several different parameters is a major source of confusion for the beginning student of statistics. Without making the distinction between statistics and parameters there is absolutely no way out of this confusion. This author is convinced that much of the frustration with statistics experienced by nonmathematicians can be traced to this one source.

1.4 Some Estimates of Dispersion Parameters

The preceding section should serve to illustrate the importance of distinguishing between statistics and parameters. It also should serve to illustrate the complexity that is associated with even the simplest of situations. However, there is yet another distinction that users of statistics will need to know about: the difference between biased and unbiased estimators.

An estimator is said to be ***unbiased*** when it is, on the average, neither too large nor too small. For example, given data collected from a process which displays statistical control, the *average of the data* is an unbiased estimator for the *mean of the distribution* of X.

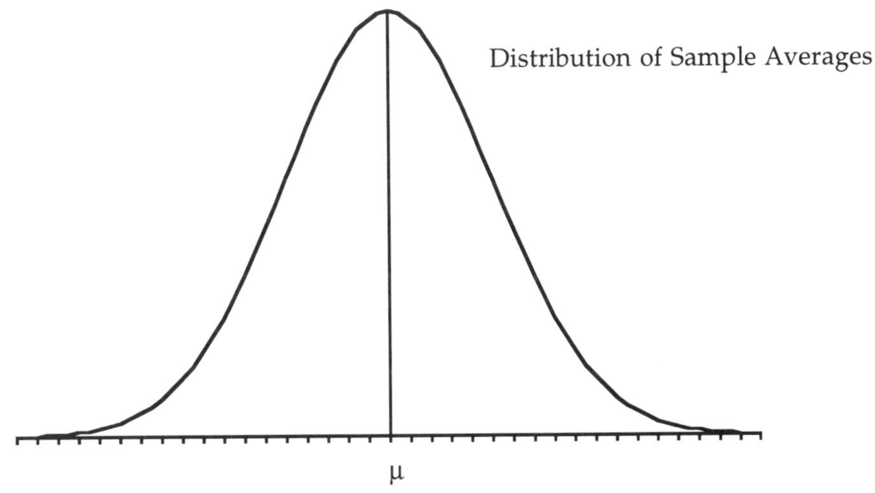

Figure 1.7: $\bar{X}$ is an unbiased estimator for μ

This simply says that, if we constructed a histogram using the $\bar{X}$ values of repeated samples of size n, this histogram would, in the limit, have a balance point equal to the mean of the distribution of X. (This is simply a restatement of Theorem One.)

Thus, in general, a sample statistic is said to be an unbiased estimator for a parameter if the mean of the distribution of the sample statistic is equal to that parameter. If the sample statistic is denoted by the symbol T, and the corresponding parameter for $f(x)$ is θ, then T is an unbiased estimator for θ if and only if θ is the mean of $f(T)$, i.e.,

$$\text{MEAN(T)} = \theta.$$

Any estimator that is *not unbiased* will be said to be ***biased.*** That is, if

$$\text{MEAN(T)} \neq \theta,$$

then T is said to be a biased estimator for θ.

Given data which come from a physical process which displays statistical control, the sample standard deviation, s, may be used to estimate the process standard deviation, SD(X). The distribution which characterizes the behavior of the statistic, s, is shown in the following figure (based on the assumption of normally distributed measurements). Since the mean value for this distribution is not equal to σ, the sample standard deviation is said to be a biased estimator for the process standard deviation.

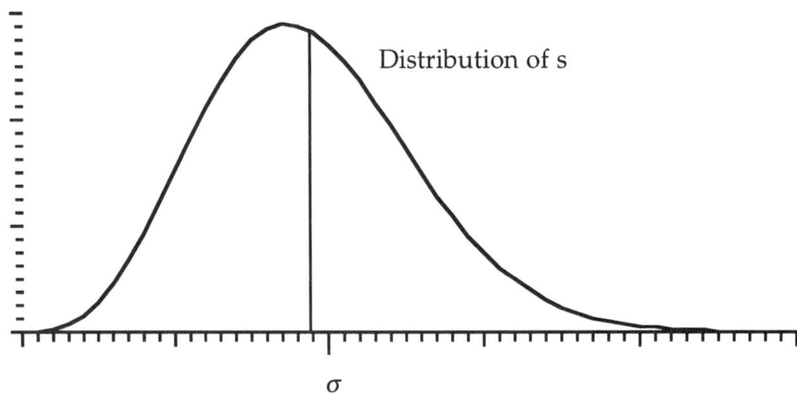

Figure 1.8: The sample standard deviation, *s*, is a biased estimator for SD(X).

Given data which come from a physical process which displays statistical control, the sample range, R, divided by d_2, may be used to estimate the process standard deviation, SD(X).

The distribution which characterizes the behavior of R/d_2 is shown below.

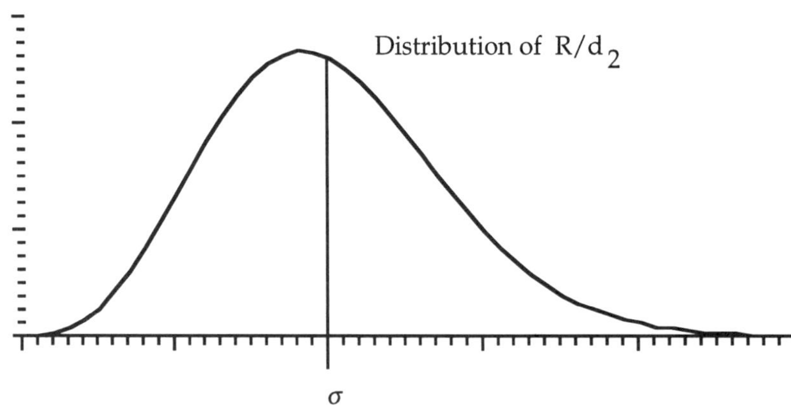

Figure 1.9: R/d_2 is an unbiased estimator of SD(X).

Since the mean value for this distribution is equal to the parameter σ, the statistic (R/d_2) is said to be an unbiased estimator for the process standard deviation, SD(X).

Given the different dispersion statistics [s, s_n, and R], the two different dispersion parameters [SD(X) and V(X)], and this distinction between biased and unbiased estimators, there are many different estimators which could be defined. Ten of the more commonly encountered dispersion estimators are listed in Table 1.1.

The constants c_2, c_4, d_2, and d_2^* are **Bias Correction Factors** which are tabled in the Appendix (Tables A and B, pp.301–303).

Table 1.1: Ten Dispersion Estimators

Dispersion	Estimators for SD(X)		Estimators for V(X)	
Statistic	Biased	Unbiased	Biased	Unbiased
s	s	$\frac{s}{c_4}$	----	s^2
s_n	s_n	$\frac{s_n}{c_2}$	s_n^2	----
R	$\frac{R}{d_2^*}$	$\frac{R}{d_2}$	$\left(\frac{R}{d_2}\right)^2$	$\left(\frac{R}{d_2^*}\right)^2$

Thus, when it comes to estimating dispersion, we are faced with several choices:

(a) which dispersion statistic to use, s, s_n, or R,

(b) which parameter to estimate, SD(X) or V(X), and, occasionally,

(c) whether to use a biased or unbiased estimator.

Each statistical procedure will usually specify which dispersion estimator to use. If, for some reason, a particular estimator cannot be used, approximate results may be obtained by substituting another estimator from the same column of Table 1.1. (Some of these substitutions will be more approximate than others, but that is beyond the scope of this chapter.)

Since the property of being unbiased is preserved only by linear operations, the square or the square root of an unbiased estimator will be a biased estimator. Additionally, those unbiased estimators for V(X) or SD(X) which depend upon Bias Correction Factors are unbiased only if the distribution of X is approximated by a Normal (Gaussian) distribution function. (These estimators may still be used with nonnormal data, but they are not likely to be unbiased estimators.)

Note that the property of being biased or unbiased is a property of the *estimator* (a random variable). When an *estimate* is obtained we have an observed value for the *estimator*. Such observed values can hardly be said to be biased or unbiased. They are what they are, and each value will either be smaller or larger than the unknown value of the parameter. However, it is convenient to denote certain estimates as being "biased" or "unbiased" as a means of designating which formula was used to obtain the value. When these terms are used in this manner they will be set out in quotation marks.

While Table 1.1 may seem rather comprehensive, it is incomplete. Other estimators exist. However, since the purpose of this is to acquaint the reader with the basic distinctions rather than to outline all the possibilities, some estimators were not included here. A more complete listing is given in Table H in the Appendix.

Perhaps the easiest way to compare the 10 estimators in the previous table is to use one data set and calculate the different estimates. The data are ten values obtained at random from a normal distribution having a mean of $\mu = 15$ and a standard deviation of $\sigma = 2.0$. The values are { 14.4, 12.9, 14.1, 18.2, 16.0, 15.2, 17.7, 14.8, 11.8, 13.2 }. The average for these ten values is 14.83. The values for the dispersion statistics are given in Table 1.2.

The bias correction factors for $n = 10$ observations are $c_2 = 0.9227$, $c_4 = 0.9727$, $d_2 = 3.078$, and for one group of 10 values, $d_2^* = 3.178$. The values for the ten estimators in Table 1.1 are listed in Table 1.2 and shown in Figure 1.10. The U and B symbols in the figure refer to "unbiased" and "biased" estimates respectively.

TABLE 1.2: Values of Ten Dispersion Estimates

Dispersion Statistic	Estimates of SD(X) "Biased"	Estimates of SD(X) "Unbiased"	Estimates of V(X) "Biased"	Estimates of V(X) "Unbiased"
$s = 2.0380$	$s = 2.038$	$\frac{s}{c_4} = 2.095$	----	$s^2 = 4.153$
$s_n = 1.9334$	$s_n = 1.933$	$\frac{s_n}{c_2} = 2.095$	$s_n^2 = 3.738$	----
$R = 6.4$	$\frac{R}{d_2^*} = 2.014$	$\frac{R}{d_2} = 2.079$	$\left(\frac{R}{d_2}\right)^2 = 4.322$	$\left(\frac{R}{d_2^*}\right)^2 = 4.056$

For these particular data, the three estimates of SD(X) which come closest to the actual parameter value of 2.0 were obtained using biased estimators. On the other hand, the estimates of V(X) which come closest to the actual value of 4.0 were obtained using unbiased estimators. **The knowledge that an estimate was obtained using an unbiased estimator does not convey any information about the distance between the estimate and the parameter value.** All of the estimates in the table above are reasonably close to the values of the parameters.

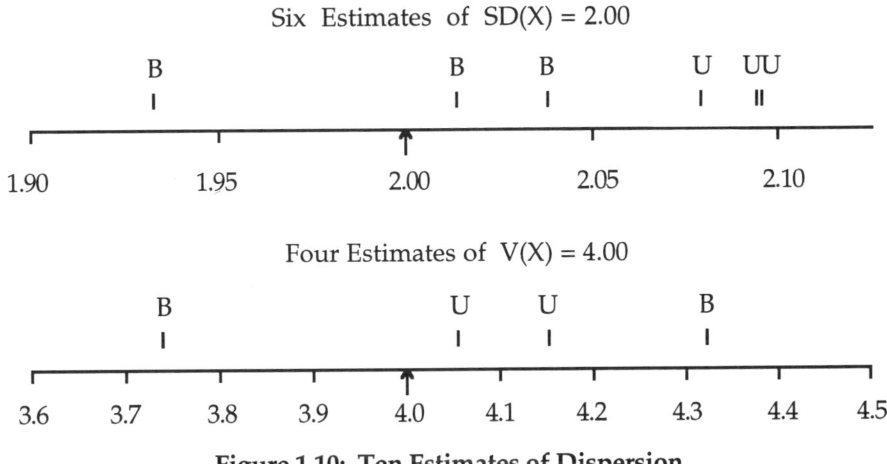

Figure 1.10: Ten Estimates of Dispersion

This proliferation of estimators for the dispersion parameters of the distribution of X is another source of confusion for the student. Given a single sample of n observations obtained from a physical process which displays statistical control, any of the estimators in Table 1.1 may be used to obtain a reasonable estimate of the dispersion of the process.

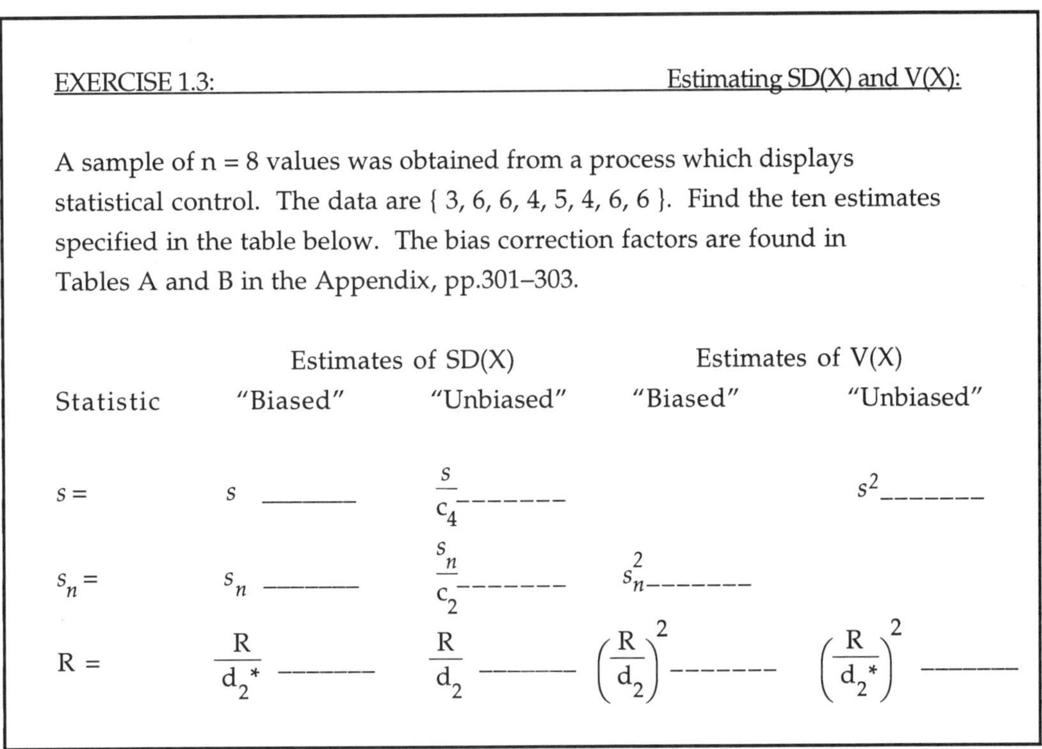

EXERCISE 1.3: Estimating SD(X) and V(X):

A sample of n = 8 values was obtained from a process which displays statistical control. The data are { 3, 6, 6, 4, 5, 4, 6, 6 }. Find the ten estimates specified in the table below. The bias correction factors are found in Tables A and B in the Appendix, pp.301–303.

	Estimates of SD(X)		Estimates of V(X)	
Statistic	"Biased"	"Unbiased"	"Biased"	"Unbiased"
$s =$	s ______	$\frac{s}{c_4}$ ______		s^2 ______
$s_n =$	s_n ______	$\frac{s_n}{c_2}$ ______	s_n^2 ______	
$R =$	$\frac{R}{d_2^*}$ ______	$\frac{R}{d_2}$ ______	$\left(\frac{R}{d_2}\right)^2$ ______	$\left(\frac{R}{d_2^*}\right)^2$ ______

In Exercise 1.3, as in real life, the values for the parameters remain unknown. (In fact, if the underlying process does not display a reasonable degree of statistical control the process parameters simply do not exist.) Assume for the purpose of this illustration that the parameters do exist. Then the estimates above suggest that SD(X) has a magnitude of approximately 1.0 to 1.2 units. Using an unbiased estimator will not guarantee that the error of estimation will be any less in any one case than it would be with a biased estimator. Using one statistic instead of another (e.g. using s instead of R) will not guarantee a smaller error of estimation. In essence, any of the six estimates of SD(X) in Exercise 1.2 could have the smallest error of estimation. Any prejudice in favor of one estimator, or against another one, will usually be based on theoretical considerations which have little to do with whether or not a particular numerical estimate is very good for some specific data set.

When an specific dispersion estimator is used in a formula it will always be best to use that particular estimator. However, when the specific estimator cannot be computed for some reason, we may obtain approximate results by substituting an equivalent estimator for the same parameter. With subgrouped data the catch comes with the word "equivalent."

1.5 Three Ways To Estimate SD(X)

The discussion in the previous section focused on using *one set of n observations* to estimate either SD(X) or V(X). In this section the problem of using *k subgroups of size n* will be considered.

The presence of k subgroups of size n complicates the problem of estimating the standard deviation of the distribution of X beyond that already considered. In addition to having to choose a statistic and an estimator based upon that statistic, the user is now confronted with three different methods for calculating both the basic statistic and the estimate. These three methods are present only when the data consist of subgroups, and for this reason they are seldom presented in basic courses in statistics. Nevertheless, these three methods are extremely important relative to the techniques discussed in this book.

The first of these three methods is essentially the same as the method used in the preceding section. All of the data from the k subgroups of size n are collected into one large group and treated as a single sample.

The second of these three methods will be familiar to those accustomed to using control charts. It consists of treating each subgroup as a separate sample and calculating a dispersion statistic for each subgroup. These separate statistics are then averaged and the average dispersion statistic is then used to form an estimate for one of the dispersion parameters of the distribution of X.

The third of these three methods will probably seem rather strange. It is certainly indirect. Instead of working with the individual values as the first two methods do, the third method works with the subgroup averages. These averages are used to obtain a dispersion statistic, and this dispersion statistic is then used to estimate a dispersion parameter of the distribution of X.

These three methods are outlined in greater detail below, and in each case an example and an exercise are provided to illustrate the method. The reader will probably find it helpful to work each exercise before proceeding to the next method. These exercises are structured to consolidate the reader's understanding of each method. Since these three methods underlie many different techniques, an understanding of these three methods is basic for any student of statistics.

METHOD ONE: (Total or Overall Variation): Given k subgroups of size n, collect all nk observations into one group and calculate one dispersion statistic for this group. This dispersion statistic can then be used as the basis for an estimate of the standard deviation of X.

EXAMPLE 1.1: Method One with Data Set One: (Using s):

Data Set One consists of k = 3 subgroups of size n = 8:

Subgroup	*Measurements*							
1	4	5	5	4	8	4	3	7
2	2	4	3	7	5	4	2	5
3	3	6	6	4	5	4	6	6

Method One effectively collects all 24 observations into one histogram prior to obtaining a dispersion statistic. This one statistic is then used to obtain an estimate of SD(X).

```
                X
                X
                X  X
                X  X  X
             X  X  X  X
          X  X  X  X  X  X
          X  X  X  X  X  X  X              s = 1.551
 ______________________________________
 0   1   2  3  4  5  6  7  8  9  10
```

The sample standard deviation for these 24 values is s = 1.551.
The c_4 bias correction factor for 24 values is c_4 = 0.9892.
Dividing 1.551 by 0.9892 gives an "unbiased" estimate of SD(X) of ***1.568.***

EXERCISE 1.4: Method One with Data Set One: (Using R):

(a) Find the range of the 24 values in Data Set One; R =

(b) Find the value of d_2 for "subgroups" of size 24; d_2 =

(c) An "unbiased" estimate of SD(X) is $\frac{R}{d_2}$ =

When an estimate of the standard deviation of X is obtained by collecting all the data into one group and calculating one dispersion statistic for the group as a whole, the estimate is said to be based on the **Total** or **Overall Variation**. This method of estimation essentially ignores the subgrouping of the data.

METHOD TWO: (Within-Subgroup Variation): Given k subgroups of size n, calculate a separate dispersion statistic for each of the k subgroups, then average these k dispersion statistics. This average dispersion statistic is then used as the basis for an estimate of the standard deviation of X.

EXAMPLE 1.2: Method Two with Data Set One: (Using s):

Data Set One consists of k = 3 subgroups of size n = 8:

Subgroup	*Measurements*	*s*	*R*
1	4 5 5 4 8 4 3 7	1.690	5
2	2 4 3 7 5 4 2 5	1.690	5
3	3 6 6 4 5 4 6 6	1.195	3

Subgroup One:

```
            X
            X  X
         X  X  X     X  X
   ---------------------------
   1  2  3  4  5  6  7  8  9
```

$s_1 = 1.690$

Subgroup Two:

```
      X     X  X
      X  X  X  X     X
   ---------------------------
   1  2  3  4  5  6  7  8  9
```

$s_2 = 1.690$

Subgroup Three:

```
                  X
                  X
            X     X
         X  X  X  X
   ---------------------------
   1  2  3  4  5  6  7  8  9
```

$s_3 = 1.195$

Method Two uses the average of k dispersion statistics to estimate SD(X).

The average sample standard deviation for these 3 subgroups is $\bar{s} = 1.525$.
The c_4 *bias correction factor for 8 values is* $c_4 = 0.9650$.
Dividing 1.525 by 0.9650 gives an "unbiased" estimate of SD(X) of ***1.580.***

EXERCISE 1.5: Method Two with Data Set One: (Using R):

(a) Find the average of the ranges of the three subgroups

in Data Set One; $\bar{R}$ =

(b) Find the value of d_2 for subgroups of size 8; d_2 =

(c) An "unbiased" estimate of SD(X) is $\frac{\bar{R}}{d_2}$ =

When the data are subgrouped, with a separate dispersion statistic calculated for each subgroup, and these statistics are averaged to form the basis for an estimate of SD(X), the estimate is said to be based upon the ***Within-Subgroup Variation***.

***METHOD THREE:* (Between Subgroup Variation): Given k subgroups of size n, calculate one dispersion statistic using the k subgroup averages, use this statistic to obtain an estimate of the standard deviation of the distribution of the subgroup averages, and then scale this estimate up by multiplying by the square root of the subgroup size. The result will be an estimate of the standard deviation of X.**

Method Three starts by finding an estimate of SD($\bar{X}$), and then converts it into an estimate of SD(X). This rather indirect approach may seem to be a lot of trouble, but it provides a useful value for comparison with the other estimates. In essence, Method Three starts off just like Method One, except that it is applied to the *average* values from each subgroup instead of using all of the individual values. The estimate obtained from this first portion is then scaled up by the square root of the subgroup size to obtain an estimate of SD(X). This last operation is based upon the relationship given in Theorem Three, p.21. Inverting the relationship in this theorem we have

$$\sqrt{n}\ \text{SD}(\bar{X}) = \text{SD}(X)$$

and therefore we may compute:

$$\text{Est. SD}(X) = \sqrt{n}\ \{\text{Est. SD}(\bar{X})\}.$$

EXAMPLE 1.3: Method Three with Data Set One: (Using s):

Data Set One consists of k = 3 subgroups of size n = 8:

Subgroup	*Measurements*								$\bar{X}$
1	4	5	5	4	8	4	3	7	5
2	2	4	3	7	5	4	2	5	4
3	3	6	6	4	5	4	6	6	5

Method Three first estimates the standard deviation of $\bar{X}$ using a dispersion statistic computed from the k subgroup averages, and then converts this into an estimate of SD(X) by multiplying by $\sqrt{n}$.

```
             X̄
          X̄  X̄                  s_x̄ = 0.5774.
  ________________________
  1  2  3  4  5  6  7  8  9
```

The sample standard deviation for these 3 subgroup averages is $s_{\bar{X}}$ = 0.5774. The c_4 bias correction factor for 3 values is c_4 = 0.8862.

Dividing 0.5774 by 0.8862 gives an "unbiased" estimate of SD($\bar{X}$) of 0.6515.

Multiplying by $\sqrt{8}$ gives an "unbiased" estimate of SD(X) of ***1.843.***

EXERCISE 1.6: Method Three with Data Set One: (Using R):

(a) Find the range of the three subgroup averages in Data Set One; $R_{\bar{x}}$ =

(b) Find the value of d_2 for subgroups of size 3; d_2 =

(c) An "unbiased" estimate of $SD(\bar{X})$ is $\frac{R_{\bar{x}}}{d_2}$ =

(d) Multiply by $\sqrt{8}$ to get an "unbiased" estimate of SD(X) =

When the data are subgrouped, and the subgroup averages are used to find an indirect estimate of the standard deviation of X, the estimate is said to be based upon the **Between-Subgroup Variation.**

The control chart for Data Set One is shown in Figure 1.11. The formulas and constants for computing control chart limits are given in Table C in the Appendix.

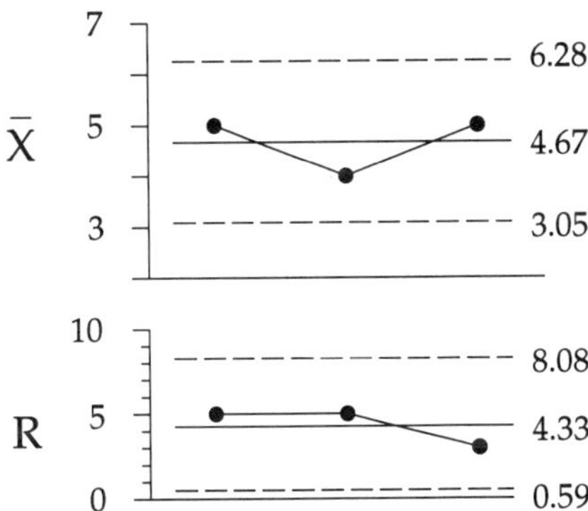

Figure 1.11: Control Chart for Data Set One

The control chart for Data Set One does not reveal any detectable lack of control. Under such conditions, the three methods will generally yield similar estimates of SD(X). The Method One, Method Two, and Method Three estimates based upon the range statistic were, respectively, 1.540, 1.522, and 1.671. The three estimates based upon the sample standard deviation statistic were, respectively, 1.568, 1.580, and 1.843.

But what happens if the k subgroups of size n come from a process that is not in statistical control? To obtain an answer to this question we shall modify the values in Data Set One so that the control chart will display a definite lack of control. This can be done by decreasing each value in subgroup 2 by 2.0 units, and increasing each value in subgroup 3 by 4.0 units. Call these modified data *Data Set Two*. The control chart for Data Set Two is shown in Figure 1.12.

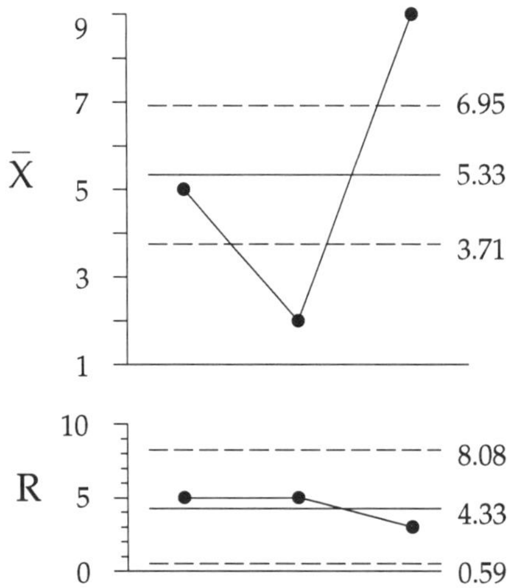

Figure 1.12: Control Chart for Data Set Two

The following Examples and Exercises will consider the results of each of the three methods when they are applied to Data Set Two.

EXAMPLE 1.4: *Method One with Data Set Two: (Using s):*

Data Set Two consists of $k = 3$ subgroups of size $n = 8$:

Subgroup	*Measurements*							
1	4	5	5	4	8	4	3	7
2	0	2	1	5	3	2	0	3
3	7	10	10	8	9	8	10	10

Method One effectively collects all 24 observations into one histogram prior to obtaining a dispersion statistic. This one statistic is then used to obtain an estimate of SD(X).

```
                                        X
             X  X  X           X        X
X        X   X  X  X        X  X        X
X   X    X   X  X  X        X  X   X    X     s = 3.279
----------------------------------------------
0   1    2   3  4  5   6    7  8   9   10
```

The sample standard deviation for these 24 values is $s = 3.279$.
The c_4 bias correction factor for 24 values is $c_4 = 0.9892$.
Dividing 3.279 by 0.9892 gives an "unbiased" estimate of SD(X) of ***3.315.***

EXERCISE 1.7: Method One with Data Set Two: (Using R):

(a) Find the range of the 24 values in Data Set Two; R =

(b) Find the value of d_2 for "subgroups" of size 24; d_2 =

(c) An "unbiased" estimate of SD(X) is $\frac{R}{d_2}$ =

The Total or Overall (Method One) Estimates obtained for Data Set Two are roughly twice the size of those obtained for Data Set One.

Thus, the "signal" introduced by shifting the subgroup averages has affected the Method One estimates.

EXAMPLE 1.5: Method Two with Data Set Two: (Using s):

Data Set Two consists of $k = 3$ subgroups of size $n = 8$:

Subgroup	*Measurements*								*R*	*s*
1	4	5	5	4	8	4	3	7	5	1.690
2	0	2	1	5	3	2	0	3	5	1.690
3	7	10	10	8	9	8	10	10	3	1.195

Each of the subgroup standard deviations, s, and each of the subgroup ranges, R, is computed using $n = 8$ values.

Subgroup One:

```
          X
          X  X
       X  X  X     X  X          s1 = 1.690
 --------------------------
 1  2  3  4  5  6  7  8  9
```

Subgroup Two:

```
X     X  X
X  X  X  X     X                 s2 = 1.690
-----------------------------
0  1  2  3  4  5  6  7  8  9
```

Subgroup Three:

```
                            X
                            X
                      X     X
                   X  X  X  X    s3 = 1.195
 ------------------------------
 1  2  3  4  5  6  7  8  9  10
```

Method Two uses the average of k dispersion statistics to estimate SD(X).

The average sample standard deviation for these 3 subgroups is $\bar{s} = 1.525$.
The c_4 bias correction factor for 8 values is $c_4 = 0.9650$.
Dividing 1.525 by 0.9650 gives an "unbiased" estimate of SD(X) of ***1.580.***

EXERCISE 1.8: Method Two with Data Set Two: (Using R):

(a) Find the average of the ranges of the three subgroups in Data Set Two; $\bar{R} =$

(b) Find the value of d_2 for subgroups of size 8; $d_2 =$

(c) An "unbiased" estimate of SD(X) is $\frac{\bar{R}}{d_2} =$

The Within-Subgroup (Method Two) Estimates for Data Set Two are exactly the same as those obtained for Data Set One. See p.31. Thus, the Method Two Estimates are not affected by the "signal" introduced when the subgroup averages were shifted.

EXAMPLE 1.6: Method Three with Data Set Two: (Using s):

Data Set Two consists of k = 3 subgroups of size n = 8:

Subgroup	*Measurements*								$\bar{X}$
1	4	5	5	4	8	4	3	7	5
2	0	2	1	5	3	2	0	3	2
3	7	10	10	8	9	8	10	10	9

Method Three first estimates the standard deviation of $\bar{X}$ using a dispersion statistic computed from the k subgroup averages, and then converts this into an estimate of SD(X) by multiplying by $\sqrt{n}$.

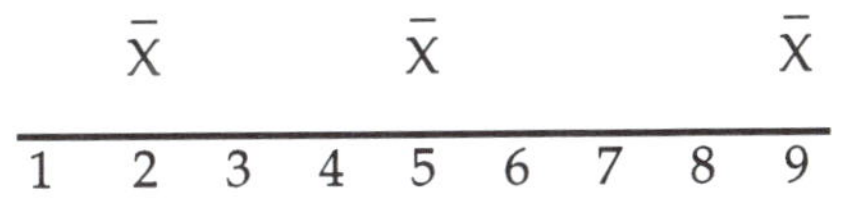

$s_{\bar{X}}$ = 3.512.

The sample standard deviation for these 3 subgroup averages is $s_{\bar{X}}$ = 3.512. The c_4 bias correction factor for 3 values is c_4 = 0.8862.

Dividing 3.512 by 0.8862 gives an "unbiased" estimate of SD($\bar{X}$) of 3.963.

Multiplying by $\sqrt{8}$ gives an "unbiased" estimate of SD(X) of ***11.209.***

EXERCISE 1.9: Method Three with Data Set Two: (Using R):

(a) Find the range of the three subgroup averages in Data Set Two; $R_{\bar{x}}$ =

(b) Find the value of d_2 for subgroups of size 3; d_2 =

(c) An "unbiased" estimate of SD($\bar{X}$) is $\frac{R_{\bar{x}}}{d_2}$ =

(d) Multiply by $\sqrt{8}$ to get an "unbiased" estimate of SD(X) =

Comparing the Between-Subgroup (Method Three) Estimates for Data Set One, pp.32-33, with these for Data Set Two we find that the latter are 6 to 7 times larger than the first. Thus, the "signal" introduced by the shift in the subgroup averages has a severe and dramatic effect upon the Method Three Estimates.

Comparing the estimates from the three methods for Data Set Two we find that, regardless of the statistic used (*s* or *R*) the three estimates vary by over 760 percent! **Thus, for data that do not display statistical control, *the three methods are not equivalent.***

- **Descriptive:** Estimates based upon the **Overall Variation** (Method One) are subject to being inflated if the data do not come from a process that displays statistical control.

- **Comparative:** Estimates based upon the **Between-Subgroup Variation** (Method Three) are subject to being severely inflated if the data do not come from a process that displays statistical control.

- **Analytic:** Estimates based upon the **Within-Subgroup Variation** (Method Two) are unaffected by changes in the subgroup averages. Thus, for subgrouped data, Method Two is the preferred method of estimating the background variation present in the data. ***The use of Method Two is basic to Shewhart's Control Charts, the Analysis of Means, and the Analysis of Variance.***

Thus, when the data are subgrouped, one must specify the **Method of Estimation** that is being used. That Methods One, Two, and Three are not always equivalent is another source of confusion for those working with variation. The difference between these three methods has not been widely appreciated, nor widely taught. As a result, many of the programs for control charts in use today are completely and totally incorrect, since they base control limits on either Method One or Method Three estimates.

Arguing over which dispersion statistic to use, or which estimator to use while not attending to the effect of the different Methods of Estimation is, indeed, like straining out gnats and swallowing a camel. Given subgrouped data, some Method Two Estimate of Dispersion will always be the yardstick against which all other estimates of dispersion will be compared.

1.6 Within-Subgroup Estimates of Dispersion

Since Method Two Estimates play an important role in many analysis techniques, some of the more common Within-Subgroup Estimators are listed here. Each of these estimators uses the *average variation within the subgroups* as the basis for the estimator, and they will, in general, yield similar results.

Estimates based upon the statistic $\bar{R}$ are called **Average Range Estimates.** Four Average Range Estimators are shown in Table 1.3. These include both biased and unbiased estimators for both SD(X) and V(X).

Three dispersion estimators are based upon the **Average Standard Deviation,** $\bar{s}$. The formulas for these three estimators are shown in Table 1.3.

Finally, estimates based upon the **average of the subgroup variances,** $\overline{s^2}$, are called **Pooled Variance Estimates**. The formulas for two such estimators are shown in Table 1.3. (The average of the subgroup variances is also called the "**Mean Square Within**" and is designated by the symbol "MSW.")

Table 1.3: Some Within-Subgroup Estimators for Dispersion Parameters

Name of Estimator	Estimators for SD(X) Biased	Estimators for SD(X) Unbiased	Estimators for V(X) Biased	Estimators for V(X) Unbiased
Pooled Variance	$\sqrt{\overline{s^2}}$	----	----	$\overline{s^2}$
Average Std. Dev.	$\bar{s}$	$\dfrac{\bar{s}}{c_4}$	$(\bar{s})^2$	----
Average Range	$\dfrac{\bar{R}}{d_2^*}$	$\dfrac{\bar{R}}{d_2}$	$\left(\dfrac{\bar{R}}{d_2}\right)^2$	$\left(\dfrac{\bar{R}}{d_2^*}\right)^2$

Other Within-Subgroup (Method Two) Estimators exist, but the nine estimators in Table 1.3 are the ones which the reader is most likely to encounter.

Some Within-Subgroup Estimators for SD(X) are illustrated with two data sets below.

Exercise 1.10 will provide the reader with practice in computing the nine most common Within-Subgroup Dispersion Estimates. The formulas are given in Table 1.3.

EXERCISE 1.10: Estimates of Dispersion for the Coating Weight Data:

Five treatments are compared using subgroups of size n = 4.
The response variable is the coating weight.

Subgroup	1	2	3	4	5
Coating Weights in grams	250	310	250	340	250
	260	330	230	270	240
	230	280	220	300	270
	270	360	260	320	290
Averages	252.5	320	240	307.5	262.5
Ranges	40	80	40	70	50
Std. Dev.	17.078	33.665	18.257	29.861	22.174
s^2	291.667	1133.333	333.333	891.667	491.667

Find the values for the nine estimators of dispersion listed in Table 1.3.

Statistic	Name of Estimator	Estimates of SD(X) "biased"	Estimates of SD(X) "unbiased"	Estimates of V(X) "biased"	Estimates of V(X) "unbiased"
$\overline{s^2}$ =	Pooled Variance	______			______
$\bar{s}$ =	Aver. Std. Dev.	______	______	______	
$\bar{R}$ =	Average Range	______	______	______	______

Examples 1.7 and 1.8 show the effect of an "out-of-control" Range Chart upon three Method Two Estimators.

EXAMPLE 1.7: Different Estimates of SD(X) for the Gauge Study Data:

In a study of a measurement device, two operators measured each of nine dimensions on each of five parts. The data below are one of these nine dimensions. Since one objective of this study was to determine the gauge repeatability, each operator measured each part twice, and these repeat measurements were grouped to form the subgroups. Thus, the treatments consist of the combinations of operators and parts, and there are k = 10 subgroups of size n = 2.

Subgroup	1	2	3	4	5	6	7	8	9	10
Operator	A	A	A	A	A	B	B	B	B	B
Part	1	2	3	4	5	1	2	3	4	5
Values	20	20	25	50	45	20	15	15	45	35
	15	25	25	50	40	20	10	10	20	40
Averages	17.5	22.5	25.0	50.0	42.5	20.0	12.5	12.5	32.5	37.5
Ranges	5	5	0	0	5	0	5	5	25	5
Std. Dev.	3.54	3.54	0	0	3.54	0	3.54	3.54	17.68	3.54
s^2	12.5	12.5	0	0	12.5	0	12.5	12.5	312.5	12.5

The Grand Average is $\bar{\bar{X}}$ = *27.25, and the Average Range is* $\bar{R}$ = *5.5.*
The bias correction factor for n = 2 is 1.128.
An AVERAGE RANGE ESTIMATE OF SD(X) is:

$$\bar{R}/d_2 = 5.5 / 1.128 = \mathbf{4.876.}$$

The average sample standard deviation is $\bar{s}$ = *3.892.*
The bias correction factor for n = 2 is 0.7979.
So an AVERAGE STANDARD DEVIATION ESTIMATE OF SD(X) is:

$$\bar{s}/c_4 = 3.892/0.7979 = \mathbf{4.878.}$$

The average of the subgroup variances is $\overline{s^2}$ = *38.75.*
Thus, the POOLED VARIANCE ESTIMATE OF SD(X) is:

$$\sqrt{\overline{s^2}} = \sqrt{38.75} = \mathbf{6.225.}$$

The Range Chart for these data, on page 62, shows that the variation in Subgroup Nine is excessive. This excessive variation within Subgroup Nine inflates all three estimates in Example 1.7. However, they are not all inflated by the same amount. One way to determine the extent to which these estimates are inflated is to compare them with estimates which do not use the data from Subgroup Nine.

EXAMPLE 1.8: Revised Estimates of SD(X) for the Gauge Study Data:

Subgroup	1	2	3	4	5	6	7	8	9	10
Ranges	5	5	0	0	5	0	5	5		5
Std. Dev.	3.54	3.54	0	0	3.54	0	3.54	3.54		3.54
s^2	12.5	12.5	0	0	12.5	0	12.5	12.5		12.5

Deleting the values of Subgroup 9 from the computations, the three Within-Subgroup Estimates of SD(X) are recalculated below:

The Average Range for k = 9 subgroups of size n = 2 is $\bar{R} = 3.333$.
The bias correction factor for subgroups of size n = 2 is 1.128.
Thus, our AVERAGE RANGE ESTIMATE OF SD(X) becomes:

$$\bar{R}/d_2 = 3.333/1.128 = \mathbf{2.955.}$$

The earlier estimate of 4.876 was **65%** *larger than this value.*

The Average Sample Standard Deviation is now $\bar{s} = 2.360$.
The bias correction factor is 0.7979.
So our AVERAGE STANDARD DEVIATION ESTIMATE OF SD(X) is:

$$\bar{s}/c_4 = 2.360/0.7979 = \mathbf{2.958.}$$

The earlier estimate of 4.878 was **65%** *larger than this value.*

The Average of the Subgroup Variances is $\overline{s^2} = 8.333$.
The POOLED VARIANCE ESTIMATE OF SD(X) is:

$$\sqrt{\overline{s^2}} = \sqrt{8.333} = \mathbf{2.887.}$$

The earlier estimate of 6.225 was **116%** *larger than this value.*
Thus the excessive subgroup variance of Subgroup Nine inflated the Pooled Variance Estimate by 116%, while it inflated the Average Range Estimate and the Average Standard Deviation Estimate by only 65%.

Both the Average Range Estimators for SD(X) and the Average Standard Deviation Estimators for SD(X) will be less sensitive to extreme values of subgroup variation than the Pooled Variance Estimator for SD(X).

When the Range Chart is in control, the various Within-Subgroup Estimators for SD(X) will generally be quite similar.

When the Range Chart is out-of-control the Pooled Variance Estimators for both SD(X) and V(X) will tend to be more severely inflated than the other estimators. So even though the Pooled Variance Estimator for SD(X) is common in many applications, the robustness of the other two estimators makes them the preferred estimators for use with ANOM charts and Control Charts (Shewhart, 1931, p.302). This is the reason that Control Charts and ANOM charts are less sensitive to nonhomogeneous variation than is the Analysis of Variance.

Even though Method Two (Within-Subgroup) Estimators may be inflated by inconsistent within-subgroup variation, they are still the most robust, and most reliable, estimates of dispersion available for use with subgrouped data.

1.7 Degrees of Freedom

Any discussion of dispersion statistics and estimators must eventually get around to something called ***degrees of freedom***. Degrees of freedom are important for the following reason: in order to use certain tables (Tables D, E and F in the Appendix) the reader will have to have a value for the degrees of freedom for different estimates of dispersion. Therefore, since one must deal with degrees of freedom, it is appropriate to consider just what this strange term encompasses. It is the purpose of this section to briefly outline what degrees of freedom are, what they do, and how to find the appropriate value for the degrees of freedom for a given dispersion statistic.

In a narrow sense, degrees of freedom simply describe the amount of data used to compute certain dispersion statistics. Therefore, one way of interpreting degrees of freedom is that they are a function of n and k. This interpretation will provide simple ways of obtaining values for degrees of freedom for different dispersion statistics.

On another level, degrees of freedom are a measure of the dispersion of a dispersion statistic. The degrees of freedom for s is inversely proportional to V(s), the variance of the distribution of s. As the degrees of freedom go up, the variance of s goes down.

Therefore degrees of freedom provide a useful way of describing the relative efficiency of the different estimators. For example, with a single sample of size 15, (k = 1 and n = 15), one could estimate SD(X) using either the sample standard deviation, s, or the sample range, R. Table 1.4 indicates that the unbiased estimator based upon the sample standard deviation will have 14 degrees of freedom. Table 1.4 and Table B in the Appendix indicate that the unbiased estimator based upon the range will have 10.8 degrees of freedom. Corresponding to these different degrees of freedom one can see a difference between the distributions which describe the behavior of these two unbiased estimators for SD(X) in Figure 1.13.

Since both estimators are unbiased estimators, these two distributions have the same mean value (shown by the vertical line). However, these two distributions do display different amounts of dispersion about the mean. The distribution with 10.8 degrees of freedom has a greater amount of variation about the mean than the distribution with 14 degrees of freedom. In other words, the estimates will vary more when there are fewer degrees of freedom.

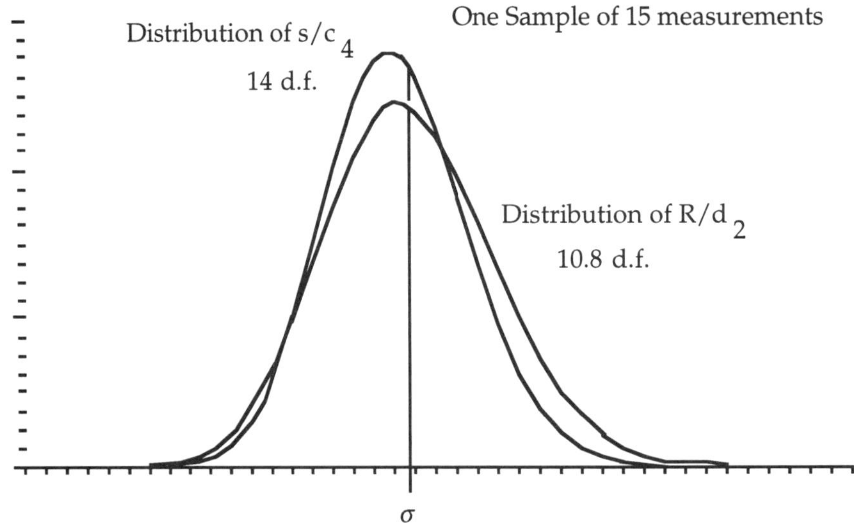

Figure 1.13: Comparison of Distributions of Two Unbiased Method One Estimators for SD(X)

With a sample of size n = 15 one can begin to see the lower relative efficiency of the range statistic compared to the sample standard deviation. In the long run, when n is large, the range estimates will have a larger average error of estimation than will the standard deviation estimates. However, as may be seen from the figure above, it will be hard to detect the difference between these two estimators in practice.

Strictly speaking, only s, s^2, s_n and $s_n{}^2$ have distributions which are naturally indexed by degrees of freedom. Therefore, the degrees of freedom for these statistics are well defined. Any single value of s, s^2, s_n, *or* $s_n{}^2$ will be said to have degrees of freedom equal to (N-1), where N is the number of values used to compute the statistic. In this application, degrees of freedom will always be integral-valued.

The sample range, R, does not have a distribution which is naturally indexed by degrees of freedom. However, it is possible to closely approximate the distribution of the range by transforming the distribution of some sample standard deviation (Patnaik, 1950). When this is done, the degrees of freedom for the approximating distribution will be said to be the "effective degrees of freedom" for the range statistic. Thus, in this manner, the notion of degrees of freedom can be extended to non-s-type dispersion statistics and the estimators based on these statistics. In this context, degrees of freedom do not have to be integral-valued.

Thus, in a general sense, the degrees of freedom for an estimator for a dispersion parameter may be said to characterize (1) the number of observations in the data and (2) the efficiency with which the estimator uses these observations.

The traditional symbol for degrees of freedom is the lower case Greek letter "nu" = ν.

The degrees of freedom for the various estimators for SD(X) and V(X) are indicated in the following tables.

These tables are defined in the context of subgrouped data. Therefore, let the symbol k denote the number of subgroups, let the symbol n denote the subgroup size, and let the symbol N denote the product of n and k:

$$N = nk.$$

The symbol $\nu_{B(k,n)}$ denotes the "effective degrees of freedom" for the Average Range Estimators. These values are found by using Table B in the Appendix.

Table 1.4: Some Total or Overall (Method One) Estimators for Subgrouped Data

Given k subgroups of size n, with a total of N = nk observations, and dispersion statistics computed using all N values:

Dispersion Statistic	Estimators for SD(X) Biased	Estimators for SD(X) Unbiased	Estimators for V(X) Biased	Estimators for V(X) Unbiased	Degrees of Freedom
s	s	$\frac{s}{c_4}$	----	s^2	(N-1)
R	$\frac{R}{d_2^*}$	$\frac{R}{d_2}$	$\left(\frac{R}{d_2}\right)^2$	$\left(\frac{R}{d_2^*}\right)^2$	$\nu_{B(1,N)}$

Use Bias Correction Factors for one subgroup of size N.

$\nu_{B(1,N)}$ = degrees of freedom from Table B for 1 subgroup of size N

The reader is reminded that **all Method One Estimators are subject to being inflated whenever the data display a lack of statistical control**. This method of estimation is therefore discouraged except in circumstances where the user knows, beyond a reasonable doubt, that the data have come from a physical process which displays statistical control. Failing such certainty, it will be much better to use some Method Two Estimator.

Table 1.5: Some Within-Subgroup (Method Two) Estimators for Subgrouped Data

Given k subgroups of size n, and k dispersion statistics:

Name of Estimator	Estimators for SD(X) Biased	Estimators for SD(X) Unbiased	Estimators for V(X) Biased	Estimators for V(X) Unbiased	Degrees of Freedom
Pooled Variance	$\sqrt{\overline{s^2}}$	----	----	$\overline{s^2}$	k(n-1)
Average Std. Dev.	$\bar{s}$	$\frac{\bar{s}}{c_4}$	$(\bar{s})^2$	----	ν_a
Average Range	$\frac{\bar{R}}{d_2^*}$	$\frac{\bar{R}}{d_2}$	$\left(\frac{\bar{R}}{d_2}\right)^2$	$\left(\frac{\bar{R}}{d_2^*}\right)^2$	$\nu_{B(k,n)}$

These estimators are all based upon an average of k dispersion statistics
Use Bias Correction Factors for k subgroups of size n
ν_a = degrees of freedom from formula in text below
$\nu_{B(k,n)}$ = degrees of freedom from Table B for k subgroups of size n

The symbol ν_a will denote the "equivalent degrees of freedom" for the Average Standard Deviation Estimators [Palm and Wheeler (1990)]. These degrees of freedom may be found according to the following formula:

$$\text{If } n = 2 \text{ or } n = 3, \quad \text{then } \nu_a = \nu_{B(k,n)}$$
$$\text{If } n \geq 4, \quad \text{then } \nu_a = [k(n-1) - 0.2(k-1)].$$

These degrees of freedom may be nonintegral, but in use they will be rounded to the nearest integer.

The symbol $\nu_{B(k,n)}$ denotes the "effective degrees of freedom" for the Average Range Estimators. These values are found by using Table B in the Appendix. Except for those cases where k(n-1) is small, ν_B will be approximately equal to 0.9[k(n-1)]. These degrees of freedom may be nonintegral, but in use they will be rounded to the nearest integer.

Within-Subgroup (Method Two) Estimators will be unaffected by changes in process location. The Average Standard Deviation Estimators and Average Range Estimators are more robust to nonhomogeneous variation within the subgroups than are the Pooled Variance Estimators (see Examples 1.7 and 1.8).

Table 1.6: Some Between-Subgroup (Method Three) Estimators for Subgrouped Data

Given k subgroups of size n:

Dispersion Statistic	Estimators for SD(X) Biased	Estimators for SD(X) Unbiased	Estimators for V(X) Biased	Estimators for V(X) Unbiased	Degrees of Freedom
$s_{\bar{x}}$	$s_{\bar{x}} \sqrt{n}$	$\frac{s_{\bar{x}}}{c_4} \sqrt{n}$	----	$n s_{\bar{x}}^2$	$(k-1)$
$R_{\bar{x}}$	$\frac{R_{\bar{x}}}{d_2^*} \sqrt{n}$	$\frac{R_{\bar{x}}}{d_2} \sqrt{n}$	$n\left(\frac{R_{\bar{x}}}{d_2}\right)^2$	$n\left(\frac{R_{\bar{x}}}{d_2^*}\right)^2$	$\nu_{B(1,k)}$

These dispersion statistics are computed using the k subgroup averages

Use Bias Correction Factors for one subgroup of size k

$\nu_{B(1,k)}$ = degrees of freedom from Table B for 1 subgroup of size k

The reader is reminded that **all Between-Subgroup (Method Three) Estimators are subject to severe inflation whenever the data fail to display statistical control** (with respect to the subgroup averages). For this reason Method Three Estimates are not usually computed except for purposes of comparison with Method Two Estimates.

The reader should note that when the data are properly considered to be one subgroup of size n, (i.e. k = 1), Method Two and Method Three disappear. We can use only Method One, and the formulas in Table 1.4 are appropriate. (In this case, N = n.)

By the use of Tables 1.4, 1.5, and 1.6, the reader can find the appropriate degrees of freedom for the more common estimates of dispersion. Moreover, these tables define those estimators which can be thought of as being equivalent in practice.

EXERCISE 1.11: Method One Estimates for Data Set Three:

Three subgroups of size four are:

Subgroup	Values			
I	1	1	4	2
II	8	7	8	9
III	4	3	5	8

Method One:

```
                                  X
X           X                     X
X   X   X   X   X           X     X     X
-------------------------------------------
1   2   3   4   5     6     7     8     9
```

(a) Obtain estimates of SD(X) using the Unbiased Method One Estimators:

(i) $\dfrac{R}{d_2} =$

(ii) $\dfrac{s}{c_4} =$

(b) Obtain estimates of V(X) using the Unbiased Method One Estimators:

(i) $\left(\dfrac{R}{d_2^*}\right)^2 =$

(ii) $s^2 =$

(c) Find the degrees of freedom for each of the estimates above:

EXERCISE 1.12: Method Two Estimates for Data Set Three:

Three subgroups of size four are:

Subgroup	Values				$\bar{X}$	R	s
I	1	1	4	2	2	3	1.4142
II	8	7	8	9	8	2	0.8165
III	4	3	5	8	5	5	2.1602

Method Two:

	1	2	3	4	5	6	7	8	9
I	X X	X		X					
II							X	X X	X
III			X	X	X			X	

(a) Obtain estimates of SD(X) using the following Method Two Estimators:

(i) $\frac{\bar{R}}{d_2} =$

(ii) $\frac{\bar{s}}{c_4} =$

(iii) $\sqrt{\overline{s^2}} =$

(b) Obtain estimates of V(X) using the Unbiased Method Two Estimators:

(i) $\left(\frac{\bar{R}}{d_2^*}\right)^2 =$

(ii) $\overline{s^2} =$

(c) Find the degrees of freedom for each of the estimates above:

EXERCISE 1.13: Method Three Estimates for Data Set Three:

Three subgroups of size four are:

Subgroup	Values				$\bar{X}$
I	1	1	4	2	2
II	8	7	8	9	8
III	4	3	5	8	5

Method Three: $\bar{X}$ at 2, $\bar{X}$ at 5, $\bar{X}$ at 8 on a scale 1 2 3 4 5 6 7 8 9

(a) Obtain estimates of SD(X) using the Unbiased Method Three Estimators:

(i) $\frac{R_{\bar{x}}}{d_2}\sqrt{n} =$

(ii) $\frac{s_{\bar{x}}}{c_4}\sqrt{n} =$

(b) Obtain estimates of V(X) using the Unbiased Method Three Estimators:

(i) $n\left(\frac{R_{\bar{x}}}{d_2^*}\right)^2 =$

(ii) $n\,s_{\bar{x}}^2 =$

(c) Find the degrees of freedom for each of the estimates above:

1.8 Time Series Data

When data consist of a series of k single values spread out over time, it is possible to think of the data as a sequence of k subgroups of size n = 1. With this structure in the data, it is not possible to use Method Two to estimate SD(X), and the distinction between Method Three and Method One disappears. However, because of the danger of inflation that exists with Method One, it is still not advisable to use Method One with data of this type unless *one knows for certain that the data are in statistical control.* Since such knowledge will be very rare, and since time series data are, by definition, collected at different times, it is best to use a technique for estimating SD(X) that is insensitive to a lack of control.

While the *technique* of Method Two cannot be used directly with k subgroups of size n = 1, it is possible to utilize the *philosophy* of Method Two by using a two-point moving range (sometimes called the method of successive differences). By using the successive differences between the observations as ranges from pseudo-subgroups of size n = 2, one can obtain an estimate of SD(X) that is much more insensitive to a lack of control than any of the Method One estimates. It is this insensitivity to a lack of control that makes the two-point moving range suitable for calculating control limits to check for a lack of control. If the time series happens to be out-of-control, the limits will be less inflated when based upon a two-point moving range than they will when they are based upon any Method One estimate.

The two-point moving range works better than the Method One estimates of SD(X) simply because of the structure of the formula of the estimate. Rather than being based upon *one dispersion statistic*, the estimate is based upon the *average of several dispersion statistics.* Moreover, with Method One, an extreme value will make a contribution to the dispersion statistic that is *proportional to the square of its deviation* from the average, while with the two-point moving range an extreme value makes a contribution that is only *proportional to its deviation* from the average. It is these two characteristics of the two-point moving range that make it the preferred method of estimating SD(X) for time-series data.

Finally, the two-point moving range is 100% efficient. There is no advantage to computing a two point moving standard deviation, since, for subgroups of size n = 2,

$$s = \frac{R}{\sqrt{2}} .$$

The two-point moving range is fully equivalent to a two-point moving standard deviation, only easier to use.

1.9 Summary

While many details are presented in this chapter, there are two major points.

First of all, the choice of statistic for estimating dispersion is relatively unimportant. In most circumstances the estimates based on the different statistics will be quite similar.

Secondly, the choice of computational method for estimating dispersion is all important. The only method of computation which is robust to a lack of statistical control is Method Two, the Within-Subgroup Method. This is why Control Charts, the Analysis of Means, and the Analysis of Variance are all built upon the use of Within-Subgroup Estimators.

Of course, the Within-Subgroup approach does require that the subgroups be rational, or that the data can be placed in a rational time order, but this requirement is usually quite easy to meet in practice. (Some authors have neatly ignored this point when advocating Method One over Method Two: they have shown that with irrational subgroups one can get larger estimates from Method Two than from Method One! Such anomalies disappear when the subgrouping is made rational.) In a control chart context, rational subgroups are very important. In fact, obtaining rational subgroups is the key to effective control charts.

With experimental data, the experiment itself will almost naturally result in rational subgroups. (Each treatment combination defines a subgroup.) Thus, in performing an experiment, the major planning question concerns which subgroups will be obtained in the experiment, rather than how to organize the data into subgroups.

CHAPTER TWO

THE ANALYSIS OF MEANS

How can one compare the effect of k different treatments (such as k different combinations of operating conditions, or k different formulations) upon some response variable? A traditional approach is to apply each one of the treatments in turn, collecting n independent observations of the response variable for each treatment. Of course, it is logical to arrange such data into k subgroups of size n so that the subgroup averages will reflect the effects of the different treatments upon the response variable.

Unfortunately, comparing the treatments is slightly more complex than merely comparing the subgroup averages. Background variation (noise) alone will guarantee that the subgroup averages will differ to some extent. The question of interest is not whether the subgroup averages are different, but whether the different treatments **caused** some of the differences between the subgroup averages. Therefore, *any analysis of the subgroup averages must seek to separate those differences which may represent signals from those differences which* *merely represent noise.* This is traditionally done by filtering out the noise and then interpreting those differences which remain as *potential* signals.

This is why any analysis of experimental data which does not seek to filter out the noise will inevitably yield confusing and even contradictory results.

The remainder of this book is concerned with different techniques for filtering out the noise in a set of observations collected under differing conditions. Some fairly simple techniques for doing this are given in this chapter.

2.1 Comparing k Treatments Using Control Charts

Shewhart's Control Chart for Averages makes an allowance for natural variation by setting control limits on either side of the Grand Average. Any subgroup average that deviates from the Grand Average by more than three standard deviations is said to be detectably different from the Grand Average. Thus, when used with experimental data, the control chart will identify those treatments which are likely to have detectable effects upon the response. If the subgroup average is above the upper control limit, the treatment can be said to have increased the response above the Grand Average by a detectable amount, and if the subgroup average is below the lower control limit, the treatment can be said to have lowered the response below the Grand Average by a detectable amount.

Moreover, because the Control Chart is a fairly conservative procedure, it will yield relatively few False Alarms. Those subgroup averages which fall outside the control limits are very likely to represent signals. Of course, the only proof that they do represent signals will come from the traditional confirmatory tool of the scientific method: the nontrivial replication of results. If the same treatment produces the same effect in subsequent experiments or in practice (i.e. production), then it is a real effect. If the same treatment does not produce the same effect in subsequent work, then it was merely noise instead of signal. Thus, all that one can ever accomplish with a single experiment is to identify *potential* signals. Real signals will persist over time, and false signals will not persist. This is why, even though no single experiment can be confirmatory, a sequence of experiments can provide an overwhelming body of evidence that a given relationship exists.

EXAMPLE 2.1: *Control Charts for the Coating Weight Data:*

Five treatments are compared using subgroups of size n = 4.
The response variable is the coating weight in grams.

Subgroup	1	2	3	4	5
Weights	250	310	250	340	250
	260	330	230	270	240
	230	280	220	300	270
	270	360	260	320	290
Averages	252.5	320	240	307.5	262.5
Ranges	40	80	40	70	50

The Grand Average is $\bar{\bar{X}} = 276.5$*, and the Average Range is* $\bar{R} = 56$.

Using Table C, p. 304, the control chart constants for subgroups of size n = 4 are:

$$A_2 = 0.729,$$
$$D_3 = 0.0,$$
$$D_4 = 2.282,$$

and the formulas in Table C give:

$$UCL_{\bar{X}} = [\,276.5 + 0.729\,(56)\,] = 317.32,$$
$$LCL_{\bar{X}} = [\,276.5 - 0.729\,(56)\,] = 235.68,$$
$$UCL_R = [\,2.282\,(56)\,] = 127.8$$

The average for subgroup two is above the upper control limit.

Thus it would appear that Treatment 2 results in coatings that are detectably greater than the Grand Average of 276.5 grams.

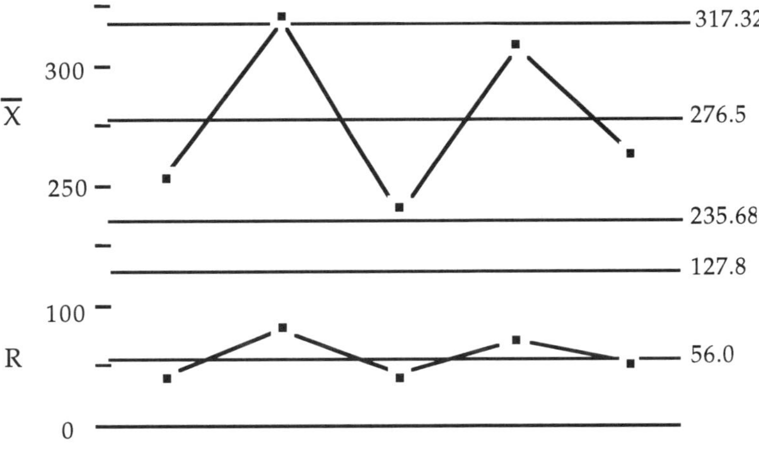

Figure 2.1: Control Chart for the Coating Weight Data

Thus Shewhart's control charts provide a simple way of analyzing experimental data when there are n observations per treatment. Before this use of control charts can be characterized more fully it will be necessary to make some additional distinctions.

2.2 Experimental Data vs. Production Data

While control charts may be used with experimental data, they are commonly used with production data. This distinction between production data and experimental data will be helpful in understanding the difference between Control Charts and the Analysis of Means.

Production data are characterized by routine. They therefore will tend to be relatively inexpensive and relatively abundant. In many cases it will be easy to obtain additional data whenever such data are needed to verify or confirm some phenomenon. Moreover, with production data, the process is assumed to be stable until we have evidence to the contrary. We shall require strong evidence to the effect that the process is out-of-control before we will be willing to act. In other words, with production data, we are looking for something that we do not really want to find. For this reason Control Charts are set up to be fairly conservative.

On the other hand, **experimental data** are characterized by uniqueness. They are obtained in the presence of special conditions, special materials, and special techniques. This, along with the involvement of technical personnel, tends to limit the amount of experimental data available while it makes that data relatively more expensive than production data. If additional experimental data should be needed, an additional experiment will be necessary. Moreover, since the different treatments represent actual changes in the process, there is the hope on the part of the experimenter that some of these changes will produce some corresponding changes in the response variable. Therefore, in analyzing experimental data, one is shifting through the data looking for changes which are thought to exist, and an exploratory approach is justified.

This is why Control Charts have a potential shortcoming as a tool for analyzing experimental data. Industrial experiments will generally involve the exploratory analysis of a limited amount of data that is, *a priori*, thought to contain real differences. Control Charts are set up for the analysis of ongoing streams of data that, hopefully, contain no real differences. So, when a Control Chart is used to analyze experimental data those differences identified as potential signals by the Control Chart are likely to represent real effects, but some real differences may be missed.

Therefore, the questions asked when interpreting experimental data differ from those asked when interpreting production data. This will justify a slightly different approach to the analysis of experimental data than that which is appropriate for production data.

The essence of the analysis of experimental data is the decision regarding the relationship between a factor and the response variable. For example, if changes in the level of Factor A result in changes in the level of the Response Variable Y, then Factor A is said to have an effect upon the Response Variable. If changes in Factor A do not result in changes in the level of the Response Variable Y, then Factor A does not have an effect upon the Response Variable. The experimenter must decide which is the case. Thus one has the decision matrix shown in Figure 2.2.

DECISION MADE	POSSIBLE STATES OF NATURE: Factor A actually affects Response Y	Factor A has no actual effect upon Response Y
1. Factor A affects Y	Correct Decision	False Alarm
2. Factor A does not affect Y	Missed Signal	Correct Decision

Figure 2.2: Consequences of Different Decisions

A study will be said to be **Conservative** when the analysis is performed in such a way that there should be very few False Alarms. Given a finite amount of data, the only way to minimize False Alarms will be to minimize the number of times that the first decision is made. By avoiding the first decision, one will automatically minimize the opportunities for a False Alarm. (The Control Chart, when used with finite amounts of data, tends to be a fairly conservative procedure.) A Conservative analysis is appropriate when confirmation from subsequent experiments is not readily available. This is typical in biomedical research and agricultural research.

A study will be said to be **Exploratory** when the analysis is performed in such a way that there should be few Missed Signals. Given a finite amount of data, this is traditionally done by using decision rules which encourage the first decision: "Factor A affects Y." By making the first decision more often, the second decision will be made less often, and there will be fewer opportunities for a missed signal.

Of course, with an ongoing stream of production data there is a third option: postpone the decision until more data are available. However, in an experimental situation, with a finite amount a data available, one will essentially opt for one of the two decisions in Figure 2.2, and then will take action according to this decision.

If one had perfect foreknowledge, then one could always make the right decision. One would adapt one's decision rule to each specific situation, and thereby avoid both False Alarms and Missed Signals. Since such knowledge is generally unavailable to mere mortals, we must devise some way of obtaining decision rules which is both reasonable and reasonably efficient. Thus we come to the probability approach.

In trying to mathematically model the options listed in Figure 2.2, it has been traditional to denote the risk of a False Alarm by the symbol α. In a similar vein, the risk of a Missed Signal is denoted by the symbol β. Under the assumptions that:

(1) the measurements do not display any discreteness,

(2) the measurements come from a process that displays *exact* stability,

(3) the measurements are normally and independently distributed,

it is possible to come up with a probability model for the decision problem. In this theoretical world, it is possible to obtain decision rules that will yield specific values for α and β. Conservative decision rules would have a small value for α (say $\alpha = 0.01$), and Exploratory decision rules would have larger values for α (say $\alpha \geq 0.10$). (Larger values of α correspond to decision rules that favor the first decision in Figure 2.2, and therefore correspond to a smaller value for β.)

While such probability models, with their nice, exact values for α and β, may be true in their own world, they are not exactly true in the real world. This is due to the shortcomings of the assumptions. Actual data are always discrete, they are never normally distributed, and processes do not display *exact* stability.

However, given the necessity of filtering out at least some of the noise, and lacking any other guidelines on how to obtain a decision rule for separating potential signals from background noise, we will use decision rules based upon the probabilistic models. The probabilistic models can be thought of as providing a theoretical approximation which can be used as a guide for practice. The decision rules obtained in this manner are probably not the optimum decision rules for any given real world situation, but they at least have the property that they are logical and appropriate in the theoretical world, and they are reasonably efficient in practice.

Thus, in describing the decision rules based upon probability models, one will have to specify a value for α. While this value is the theoretical probability of a False Alarm in the mathematical world, it cannot be safely interpreted as such in the real world. In practice, the α-level for a procedure simply tells the reader how the experimenter defined the decision rule for interpreting his data. Conservative analyses will use a small α-level, and Exploratory analyses will use a large α-level. Therefore, while the α-level does not actually give the probability of a False Alarm, or translate into a confidence level for a set of results, it does rank the sensitivity of the analysis. For a given data set, larger α-levels will yield a more sensitive analysis and smaller α-levels will yield a less sensitive analysis.

In the industrial context, where today's experimental results can often be verified by tomorrow's production, most experimental studies should use the exploratory approach. With the exploratory approach one is less likely to miss any effects that might be of interest, and any False Alarms will be quickly identified as such as subsequent data accumulate.

This is the justification of the Analysis of Means (ANOM). The Control Chart already provides a fairly conservative procedure for analyzing experimental data. The Analysis of Means will be used when a more exploratory approach is desired.

The Analysis of Means is a simple generalization of the control chart for subgroup averages that allows the user to adjust the width of the limits to fit each particular application. Therefore, unlike the Control Chart limits, the ANOM decision limits are set up for a particular number of subgroups. Any change in the number of subgroups will require a change in the decision limits. This is why the Analysis of Means cannot be used with ongoing streams of production data.

Moreover, since ANOM decision limits are defined for a particular number of subgroups, the procedure requires the specification of an α-level. As noted above, the α-level will merely indicate the approach being taken by the experimenter and will specify the way in which the ANOM decision limits are computed.

Thus, ANOM is different from the Control Chart for Subgroup Averages in two physical aspects: (a) ANOM is limited to a finite number of subgroups, and (b) ANOM requires the specification of an overall α-level for the procedure. The first of these differences prevents one from using the ANOM technique with production data. The second of these differences lets the user adjust the sensitivity of the ANOM procedure.

2.3 The Decision Limits for ANOM

Given *any number of subgroups* of size n, the control limits for the Average Chart are

$$(\text{Grand Average}) \pm 3\ (\text{Est. SD}(\bar{X})).$$

Given *exactly k subgroups* of size n, the ANOM decision limits are

$$(\text{Grand Average}) \pm H\ (\text{Est. SD}(\bar{X})).$$

Thus, instead of using three standard deviation limits, ANOM decision limits are set at a distance on either side of the Grand Average which is determined by the value for H. In order to obtain a value for H we must first specify three quantities:

(i) ν = the degrees of freedom for the estimate of SD($\bar{X}$)
(ii) α = the overall α-level for the ANOM procedure, and
(iii) k = the number of subgroup averages being compared.

The values for the degrees of freedom will be found either by a formula or from a table (see Section 1.7, p.43). The overall α-level will be selected by the analyst, depending upon whether the analysis is exploratory or conservative in nature. The value of k will remain fixed for any one analysis.

The procedure for finding the ANOM decision limits for k subgroups of size n will be as follows:

1. **Estimate the standard deviation of X:** Use some Within-Subgroup Estimator for SD(X), and find the degrees of freedom for this estimator, ν. (See Table 1.5, p.46).

2. **Estimate the standard deviation of the subgroup averages:** Find Est. SD($\bar{X}$) by dividing [Est. SD(X)] by $\sqrt{n}$. The degrees of freedom for this estimate will be exactly the same as the degrees of freedom for Est. SD(X).

3. **Find value for H:** Choose α, then use α, k, and v to find value for H (see Table D, p. 305–308).

4. **Find ANOM decision limits:** Compute $\bar{\bar{X}} \pm H\ (\text{Est. SD}(\bar{X}))$.

These limits are plotted on a running record of the subgroup averages. Averages that fall outside the decision limits are said to be detectably different from the Grand Average. They represent potential signals.

EXAMPLE 2.2: ANOM Chart for the Coating Weight Data:

Subgroup	*1*	*2*	*3*	*4*	*5*
Coating Weights in grams	*250*	*310*	*250*	*340*	*250*
	260	*330*	*230*	*270*	*240*
	230	*280*	*220*	*300*	*270*
	270	*360*	*260*	*320*	*290*
Averages	*252.5*	*320*	*240*	*307.5*	*262.5*
Ranges	*40*	*80*	*40*	*70*	*50*

The Grand Average is $\bar{\bar{X}} = 276.5$, *and the Average Range is* $\bar{R} = 56$.

1. *Using Table B, p.302, Est. SD(X) =* $\bar{R}/d_2^* = 56 / 2.096 = 26.72$ *grams, and this estimate has* $v = 13.9$ *degrees of freedom.*

2. *Est.* $SD(\bar{X}) = \dfrac{Est.SD(X)}{\sqrt{n}} = \dfrac{26.72}{\sqrt{4}} = 13.36$ *grams,*
 with $v = 14$ *d.f. (to nearest integer).*

3. *With* $\alpha = 0.10$, $k = 5$, $v = 14$ *, simple linear interpolation in Table D, p.306, gives the value of H to be approximately 2.26.* — interpolated

4. *So Decision Limits are :* $276.5 \pm 2.26\ (13.36) = 246.3$ *to 306.7*

The ANOM chart is shown in Figure 2.3.
Subgroups 2 and 4 have averages that are detectably greater than the Grand Average of 276.5, and Subgroup 3 has an average that is detectably less than the Grand Average of 276.5.

The critical values in Table D (the H values) were derived for use with the Pooled Variance Estimator for SD(X):

$$\text{Est. SD(X)} = \sqrt{\overline{s^2}}$$

As shown in the previous example, other Method Two Estimators may be used in lieu of the Pooled Variance Estimator. (See Table 1.5, p.46.)

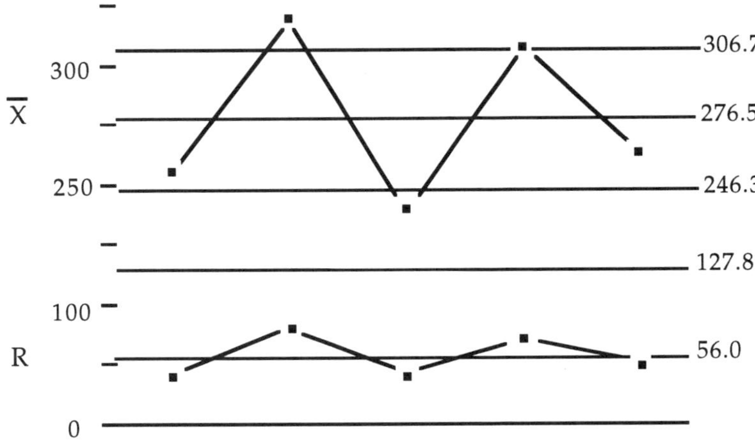

Figure 2.3: ANOM Chart for Coating Weight Data

In theory, the Biased Average Range Estimator for SD(X), ($\bar{R}/d_2^*$), will be a slightly better substitute for the Pooled Variance Estimator for SD(X) than the unbiased estimators, ($\bar{R}/d_2$) and ($\bar{s}/c_4$).

In practice, these or other Method Two Estimators for SD(X) may be used in lieu of the Pooled Variance Estimator with no serious consequences as long as the degrees of freedom are also adjusted.

The estimate of SD($\bar{X}$) obtained in step two of the ANOM calculations will have the same degrees of freedom as the estimate of SD(X) since division by $\sqrt{n}$ does not alter the amount of information contained in an estimate, .

The overall α-level for the ANOM procedure is chosen to reflect the approach the experimenter wishes to use. This value, along with the number of subgroups, k, and the degrees of freedom, v, will then be used in Table D of the Appendix (pp.305–308) to find the appropriate value of H. It will usually suffice to use simple linear interpolation in Table D to find values for H.

Finally, comparing the Control Chart approach with the ANOM approach, the ANOM approach allows the analysis to be suited to the situation. With The Coating Weight Data, the Control Chart identified one treatment as different from the Grand Average, while ANOM categorized the five treatments into three groups. While the Control Chart used a conservative approach, the ANOM used an exploratory approach. Thus, the ANOM procedure is more likely to have identified all real treatment effects. It also has an increased chance of getting some False Alarms.

EXAMPLE 2.3: ANOM Chart for the Gauge Study Data:

The Gauge Study Data are described in Example 1.7, p. 41.

Subgroup	1	2	3	4	5	6	7	8	9	10
Operator	A	A	A	A	A	B	B	B	B	B
Part	1	2	3	4	5	1	2	3	4	5
Values	20	20	25	50	45	20	15	15	45	35
	15	25	25	50	40	20	10	10	20	40
Averages	17.5	22.5	25.0	50.0	42.5	20.0	12.5	12.5	32.5	37.5
Ranges	5	5	0	0	5	0	5	5	25	5

These k = 10 subgroups of size n = 2 have $\bar{\bar{X}} = 27.25$ *and* $\bar{R} = 5.5$.

1. *An Estimate of SD(X) is* $= \bar{R}/d_2^* = 5.5 / 1.159 = 4.746,\ \nu = 9.0$

2. *Thus,* $\text{Est. SD}(\bar{X}) = \dfrac{4.746}{\sqrt{2}} = 3.356,$ *with* $\nu = 9$.

3. *With an* α*-level of 0.10, k = 10, and* $\nu = 9$*; H is 2.91.*

4. *The ANOM decision limits are* $27.25 \pm 2.91\ (3.356) = 17.48$ *to* 37.02.
   The upper control limit for the Range Chart is

$$UCL_R = D_4\bar{R} = 3.268\ (5.5) = 17.97.$$

Subgroups 4, 5, and 10 have averages above the UDL. Subgroups 1, 7 and 8 have averages below the LDL. And Subgroup 9 has an excessively large range. Interpreting the overall graph, there are detectable differences between the parts and detectable differences between the operators.

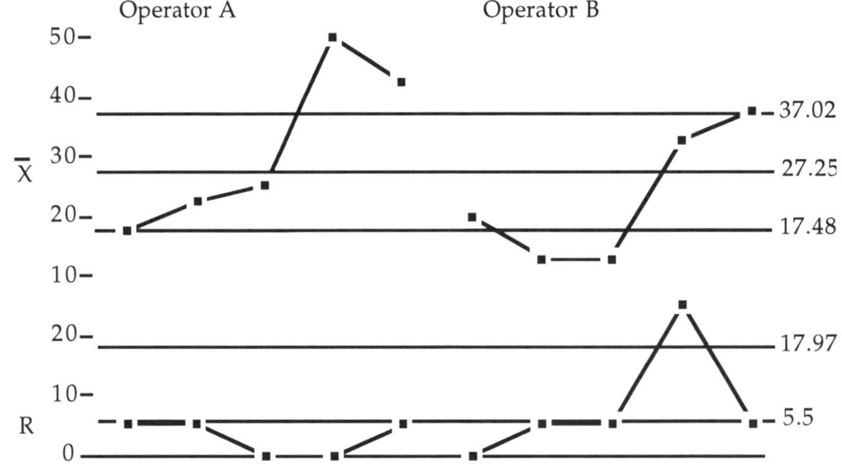

Figure 2.4: ANOM Chart for Gauge Study Data

2.4 The Role of the Range Chart with ANOM

The ordinary Control Chart for Ranges is a useful companion to the ANOM chart. With it we can check for consistency between the subgroup ranges. Since the ANOM decision limits depend upon the estimate of SD(X), it is important that the experimenter be aware of any inconsistency between the subgroup ranges. Since subgroups will tend to be small, it is difficult for the Range Chart to detect decreased variation, but not too hard to detect excessive variation.

Whenever one or more ranges fall above the upper control limit on the Range Chart, the estimate of SD(X) will be inflated regardless of the statistic used for the estimate. When this happens, it is usually advisable to revise the estimate of SD(X) by eliminating the out-of-control ranges. This revised estimate will be an appropriate value for use in finding the ANOM decision limits. Of course, such decision limits should be applied only to those averages that correspond to the ranges that are within the control limits.

The reason for this restriction is twofold. First, when a subgroup has an out-of-control range, it is definitely different from the other subgroups. Since increased variation is generally bad, these subgroups are not likely to be of much interest, and they may be excluded from further analysis. Secondly, if the variation is known to be different for certain subgroups, and these subgroups are excluded from the estimation of SD(X), then the ANOM decision limits will not be appropriate for the averages of these subgroups. (If there are two or more such subgroups, separate decision limits could be calculated for these subgroup averages.)

EXAMPLE 2.4: Revised ANOM limits for the Gauge Study Data:

Figure 2.4 shows the range for Subgroup 9 to be out-of-control. Since this out-of-control range will inflate the Average Range and the decision limits, we might wish to delete Subgroup 9 and revise the limits. With $k = 9$ and $n = 2$, we find a Grand Average of 26.667 and an Average Range of 3.333.

The revised estimate of SD(X) is $\bar{R}/d_2^* = 3.333 / 1.163 = 2.866, \quad \nu = 8.1$

Thus, $$\text{Est. } SD(\bar{X}) = \frac{2.866}{\sqrt{2}} = 2.027, \text{ with } \nu = 8.1.$$

With an α-level of 0.10, $k = 9$, and $\nu = 8$, H is 2.90.
So the revised ANOM decision limits are $26.667 \pm 2.90\ (2.027) = 20.79$ to 32.54.
These ANOM limits do not apply to Subgroup 9.
The revised upper control limit for the Range Chart is 3.268(3.333) = 10.89.
This control limit is plotted against the range for Subgroup 9 in order to emphasize the difference between the range of Subgroup 9 and the other ranges.

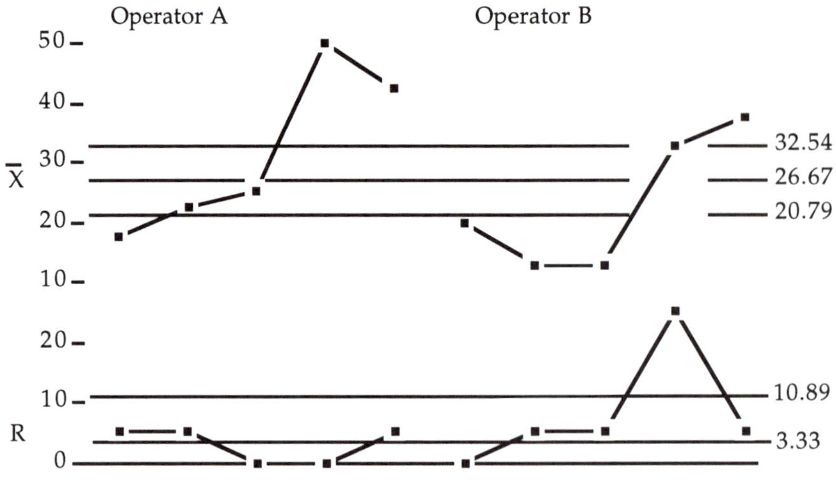

Figure 2.5: Revised ANOM Chart for Gauge Study Data

While the inflated ANOM limits found in the original analysis identified several subgroup averages as different from the Grand Average, the revised limits are even more revealing. The original analysis was simply more conservative than it was thought to be.

Thus, while a useful analysis may be obtained without using the Range Chart, the Range Chart can identify inconsistencies in variation. Ignoring such inconsistencies in variation will result in an ANOM chart that is less sensitive than it should be. Using the Range Chart to check for consistent variation within the subgroups provides an opportunity to refine the ANOM limits when appropriate.

The exercises on the following pages should help the student master the ANOM procedure. The Reflectivity Data will also be used in later examples and exercises. The Steel Parts Data are the data used by Ellis Ott when he introduced the ANOM technique in 1967.

EXERCISE 2.1: ANOM Chart for the Reflectivity Data:

Ten treatments are used to evaluate the reflectivity of a certain coating. These treatments consisted of various combinations of Concentration and Process Temperature. Two levels of Concentration for one particular ingredient were used (5 grams per liter and 10 grams per liter), while five levels of Temperature were used (75, 100, 125, 150, and 175 degrees C). These two factors were fully crossed, with each combination corresponding to exactly one subgroup. There were three pieces produced and measured for each subgroup.

Subgroup	1	2	3	4	5	6	7	8	9	10
Concentration	5	5	5	5	5	10	10	10	10	10
Temperature	75	100	125	150	175	75	100	125	150	175
Measured	35	31	30	28	19	38	36	39	35	30
Reflectivity	39	37	31	20	18	46	44	32	47	38
	36	36	33	23	22	41	39	38	40	31
Averages	36.67	34.67	31.33	23.67	19.67	41.67	39.67	36.33	40.67	33.0
Ranges	4	6	3	8	4	8	8	7	12	8

The Grand Average is 33.7333. The Average Range is 6.80.

(a) Plot the running record for the Subgroup Averages and the Subgroup Ranges on the graph paper provided on page 67.

(b) Compute the appropriate limits using the worksheet on page 66, and plot these limits on page 67.

(c) Interpret the ANOM Plot.

Analysis of Means Worksheet

A. BASIC INFORMATION:

Data Set Name ______________________________ $k =$ ____ $n =$ ____

Grand Average $= \bar{\bar{X}} =$ [ ] Average Range $= \bar{R} =$ ____________

B. FIND THE CONTROL LIMITS FOR THE RANGE CHART: (Table C)

$LCL_R = D_3\bar{R} =$ $UCL_R = D_4\bar{R} =$

Are any subgroup ranges outside these limits?

C. ESTIMATE THE STANDARD DEVIATION OF X: (Using k = _____ subgroups of size n = ____)

Use Table B to find: $\nu =$ _____ $d_2^* =$ ________

$$\text{EST. SD(X)} = \frac{\bar{R}}{d_2^*} = \frac{\qquad}{\qquad} = \underline{\qquad\qquad}$$

D. ESTIMATE THE STANDARD DEVIATION OF THE SUBGROUP AVERAGES:

$$\text{EST. SD}(\bar{X}) = \frac{\text{EST. SD(X)}}{\sqrt{n}} = \frac{\qquad}{\qquad} = \boxed{\qquad}$$

The degrees of freedom for this estimate is the same as in Step C above.

E. FIND THE VALUE FOR H:

Choose $\alpha =$ _____ . From Step C above, $\nu =$ _____.
The number of subgroup averages being compared is $k =$ _____.

From Table D, the value for H is = [ ]

F. FIND THE DECISION LIMITS FOR THE ANOM CHART:

The decision limits for the ANOM chart are $\bar{\bar{X}} \pm H\left[\text{EST. SD}(\bar{X})\right]$:

$UDL_{\bar{X}} =$

$LDL_{\bar{X}} =$

Exercise 2.1:
One Way ANOM for Reflectivity Data:

Exercise 2.3
Main-Effect ANOMs for Reflectivity Data

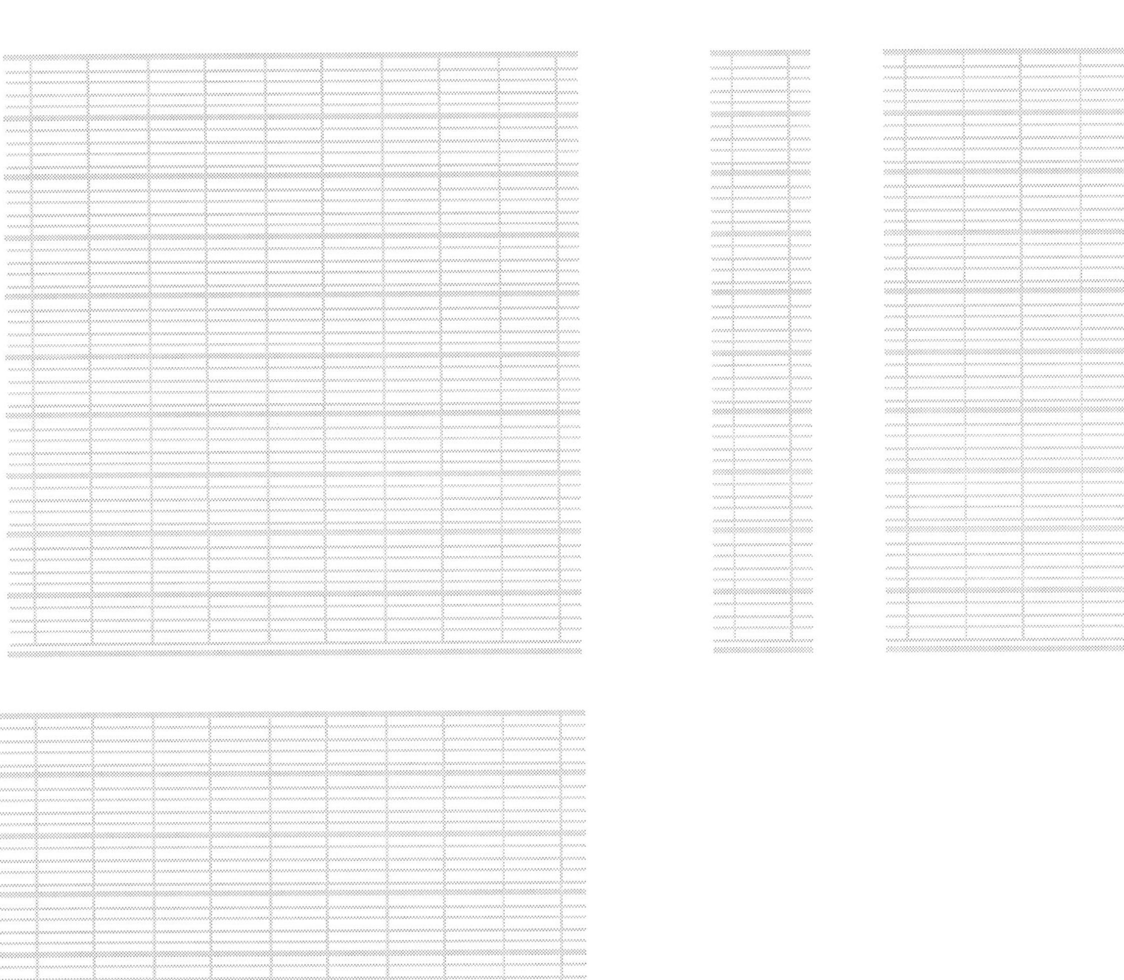

ANOM Plots for Reflectivity Data

EXERCISE 2.2: ANOM Chart for the Steel Parts Data:

Twenty-four treatments were studied for their effect upon the length of some steel parts. Each treatment consisted of different combinations of three factors.

1. The type of Heat Treatment had two levels in this study (W and L).
2. The parts were processed on four Machines (A, B, C, and D).
3. The parts were processed at three Times (8 am, 11 am, and 3 pm).

Four parts were produced under each treatment combination, giving a total of 96 parts.

Subgroup	1	2	3	4	5	6	7	8	9	10	11	12
Heat Treat	W	W	W	W	W	W	W	W	W	W	W	W
Machine	A	B	C	D	A	B	C	D	A	B	C	D
Time	8	8	8	8	11	11	11	11	3	3	3	3
Lengths:	6	7	2	6	3	7	2	9	6	6	2	8
	9	9	1	6	6	8	3	7	4	11	-1	5
	1	5	0	7	1	4	1	11	9	10	6	4
	3	5	4	3	-1	8	0	6	5	4	1	10
Averages	4.75	6.5	1.75	5.5	2.25	6.75	1.5	8.25	6.0	7.75	2.0	6.75
Ranges	8	4	4	4	7	4	3	5	5	7	7	6
s^2 values	12.25	3.67	2.92	3.00	8.92	3.58	1.67	4.92	4.67	10.92	8.67	7.58

Subgroup	13	14	15	16	17	18	19	20	21	22	23	24
Heat Treat	L	L	L	L	L	L	L	L	L	L	L	L
Machine	A	B	C	D	A	B	C	D	A	B	C	D
Time	8	8	8	8	11	11	11	11	3	3	3	3
Lengths:	1	4	0	4	1	4	0	6	3	7	0	3
	6	5	-1	5	3	6	2	4	0	8	-2	4
	0	3	0	5	1	1	-1	9	6	10	4	7
	4	6	1	4	-2	3	1	3	7	0	-4	0
Averages	2.75	4.5	0.0	4.5	0.75	3.5	0.5	5.5	4.0	6.25	-0.5	3.5
Ranges	6	3	2	1	5	5	3	6	7	10	8	7
s^2 values	7.58	1.67	0.67	0.33	4.25	4.33	1.67	7.00	10.00	18.92	11.67	8.33

The grand average is 3.958.
The average range is 5.29.
The average s^2 value is 6.2153.

Compute the ANOM decision limits for these data. (Use $\alpha = 0.10$.)
(A computation worksheet is given on the next page.)

Analysis of Means Worksheet

A. BASIC INFORMATION:

Data Set Name ________________________________ $k =$ ____ $n =$ ____

Grand Average = $\bar{\bar{X}}$ = [ ] Average Range = $\bar{R}$ = ____________

B. FIND THE CONTROL LIMITS FOR THE RANGE CHART: (Table C)

$LCL_R = D_3 \bar{R} =$ $UCL_R = D_4 \bar{R} =$

Are any subgroup ranges outside these limits?

C. ESTIMATE THE STANDARD DEVIATION OF X: (Using k = _____ subgroups of size n = _____)

Use Table B to find: $\nu =$ _____ $d_2^* =$ _______

$$\text{EST. SD(X)} = \frac{\bar{R}}{d_2^*} = \frac{\qquad}{} = \_\_\_\_\_\_$$

D. ESTIMATE THE STANDARD DEVIATION OF THE SUBGROUP AVERAGES:

$$\text{EST. SD}(\bar{X}) = \frac{\text{EST. SD(X)}}{\sqrt{n}} = \frac{\qquad}{} = \boxed{\qquad}$$

The degrees of freedom for this estimate is the same as in Step C above.

E. FIND THE VALUE FOR H:

Choose $\alpha =$ _____. From Step C above, $\nu =$ _____.

The number of subgroup averages being compared is $k =$ _____.

From Table D, the value for H is = [ ]

F. FIND THE DECISION LIMITS FOR THE ANOM CHART:

The decision limits for the ANOM chart are $\bar{\bar{X}} \pm H\left[\text{EST. SD}(\bar{X})\right]$:

$UDL_{\bar{X}} =$

$LDL_{\bar{X}} =$

2.5 ANOM With Multifactor Studies

Some studies will involve more than one factor. When this happens the treatments become treatment combinations, that is, each treatment will consist of some combination of levels of two or more factors. In fact this happened with the Gauge Study Data, the Reflectivity Data, and the Steel Parts Data.

In the Steel Parts Data, the three factors involved were (a) the method of heat treatment (2 methods), (b) the machine used (4 machines), and (c) the time of day the parts were processed (3 times of day). If every level of each factor is to occur with every level of each other factor, there will have to be (2 x 4 x 3) = 24 treatments. Such a study is said to be "fully crossed." Each of the 24 treatments will represent exactly one combination of the levels of the three factors.

EXAMPLE 2.5: One Way ANOM for the Steel Parts Data:

The data and the context were given in Exercise 2.2, p.68. The ANOM chart shows four subgroups to be detectably different from the Grand Average. Subgroups 8 and 10 are above the upper decision limit, and subgroups 15 and 23 are below the lower decision limit. However, the multiplicity of factors makes it difficult to interpret just what is happening without some additional analysis.

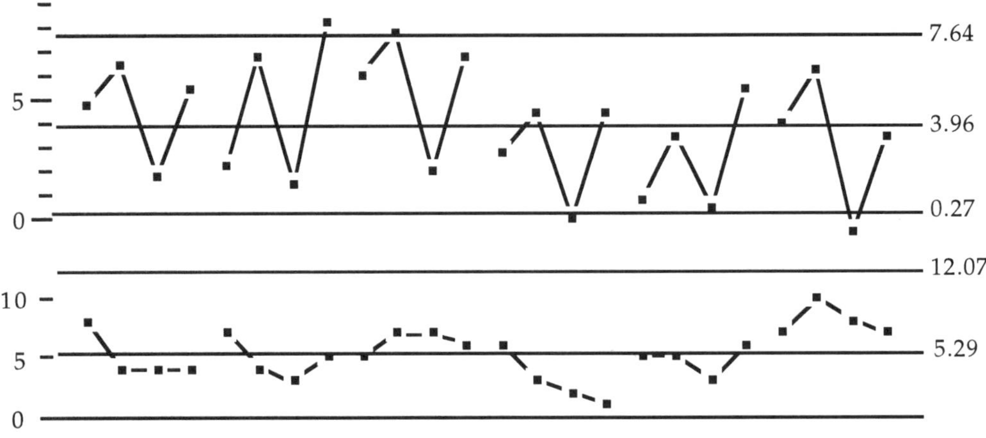

Figure 2.6: One-Way ANOM Chart for the Steel Parts Data

With a fully crossed study it is possible to obtain separate ANOM charts for each of the factors. These separate charts will explicitly consider the effect of each factor by itself.

The one twist to obtaining these "Main-Effect ANOM charts" is the estimation of SD(X). Regardless of the factor being studied, the estimate of SD(X) remains the same as for the "One-Way ANOM chart."

For each Main-Effect ANOM chart, the estimate of SD($\bar{X}$) will be calculated from the initial estimate of SD(X), and it will have the same number of degrees of freedom as the initial estimate of SD(X). (A Main Effect ANOM simply repeats steps D., E., and F. on the ANOM worksheet for each main effect in the study.)

EXAMPLE 2.6: Main Effect ANOMs for the Steel Parts Data:

A. Heat Treatment ANOM:

Each level of the Heat Treatment factor occurs with 12 subgroups, therefore there are 48 observations for each level of Heat Treatment. The average for each set of 48 values gives the Heat Treatment averages:

Heat Treatment	*W*	*L*
averages	*4.979*	*2.938*

The Est. SD(X) from the One-Way ANOM was 2.559, with 66 d.f.

The Est. $SD(\bar{X}) = \dfrac{2.559}{\sqrt{48}} = 0.369.$

With $\alpha = 0.10$, $k = 2$, *and* $v = 66$, *the value for H is 1.18.*

ANOM limits are

$3.958 \pm 1.18\,(0.369) = 3.522$ *to* 4.394.

Both values exceed the limits, thus there is a detectable difference between the two heat treatments.

B. Machine ANOM:

The Machine averages are obtained using the 24 values that correspond to each of the four machines.

Machine	*A*	*B*	*C*	*D*
Average	*3.417*	*5.875*	*0.875*	*5.667*

The Est. $SD(\bar{X}) = \dfrac{2.559}{\sqrt{24}} = 0.5224.$

With $\alpha = 0.10$, $k = 4$, *and* $v = 66$; *H is 1.94.*

ANOM limits are

$3.958 \pm 1.94\,(.5224) = 2.945$ *to* 4.971.

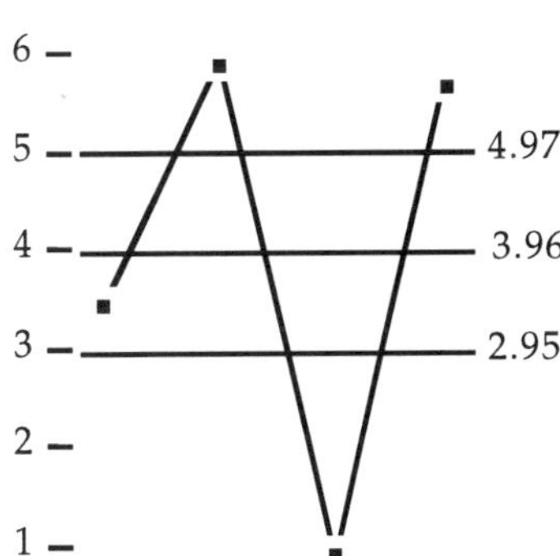

Machines B and D exceed the upper decision limit, and machine C is below the lower decision limit. Thus, the apparent differences between machines may be interpreted as being real.

C. Time of Day ANOM:
The Time of Day averages are based on the 32 observations obtained respectively, at 8am, 11am, and 3pm.

Time of Day	*8*	*11*	*3*
Averages	*3.781*	*3.625*	*4.469*

The Est. $SD(\bar{X}) = \frac{2.559}{\sqrt{32}} = 0.452.$

With $\alpha = 0.10,\ k = 3,$ *and* $v = 66;$ *H is 1.71.*
ANOM limits are

$$3.958 \pm 1.71\,(0.454) = 3.184 \text{ to } 4.732$$

None of the averages fall outside the limits, so there is no detectable effect due to the time of day.

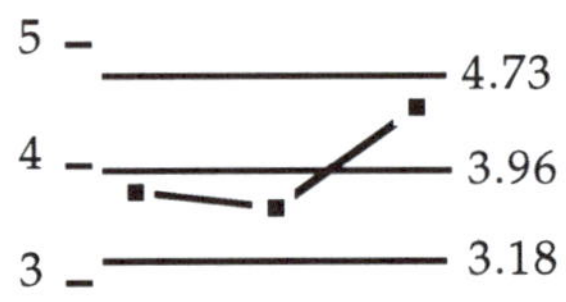

The One-Way ANOM described in section 2.3 compares each subgroup with the Grand Average. The Main-Effect ANOM described in this section compares the average response for each level of each factor with the Grand Average. Thus, these two ANOM techniques answer slightly different questions. The One-Way ANOM will remain the basic procedure, while the Main Effect ANOM (and the ANOMR procedure given in Section 2.8) will be used as follow-up analyses.

> EXERCISE 2.3 Main Effect ANOMs for the Reflectivity Data:
> Obtain the Main-Effect ANOMs for the Reflectivity Data (p.65).
> (Space for these plots is provided on p.67.)

2.6 Interaction Effects and ANOM

Two factors will be said to interact when the effect of one factor differs at different levels of the other factor. That is, the effect of Factor A in the presence of one level of Factor B is *not the same* as the effect of Factor A in the presence of another level of Factor B. This difference in the effects would be described as an AB interaction effect.

Clearly, interaction effects will have an impact upon how any experimental study should be interpreted. If a factor is involved in an interaction, then that factor is important even though it might not show up as important in a Main-Effect ANOM plot. Therefore, a way of detecting and interpreting interaction effects is needed.

If there are interaction effects in the data, they will show up on the One-Way ANOM plot as nonparallelness between corresponding line segments. For example, the line showing the effect of Factor A moving from level A1 to level A2 while Factor B is at level B1 should be parallel to the line showing the effect of Factor A going from level A1 to level A2 while Factor B is at level B2. If these line segments are dramatically nonparallel, then there may well be an AB interaction effect.

Therefore, the first place to look for the possibility of interaction effects is in the One-Way ANOM plot. If corresponding line segments are dramatically nonparallel, then expect to find interactions. The problem with this rather straightforward approach is that random variation (background noise) will generally create a certain amount of nonparallelness between corresponding line segments. This random nonparallelness must be filtered out before a given pair of nonparallel line segments can be said to represent a potential interaction effect.

While methods for doing this with ANOM exist, this author prefers to use the Analysis of Variance to identify significant interactions, and then to use the ANOM plot to interpret these interactions.

However, before getting lost in computations for possible interaction effects, it is best to use the ANOM plots to see if there are likely to be any interactions. In addition to looking for nonparallel line segments, we may also compare the Main-Effect ANOM plots with the One-Way ANOM plot.

Consider only those Main-Effect ANOM Plots which have one or more points outside the decision limits. If the patterns of variation seen on these plots combine to give a pattern of variation that is quite similar to the one found on the One-Way ANOM Plot, then it is unlikely that there are any interaction effects present in the data. On the other hand, if the sum of the Main-Effect patterns of variation does not fully explain the variation seen on the One-Way ANOM Plot, then the unexplained variation is likely to be due to interaction effects.

Consider the Steel Parts Data: Detectable differences in Length were found for both Heat Treatment and Machine. The patterns of variation from these Main-Effect ANOM Plots, drawn to the same scale, are:

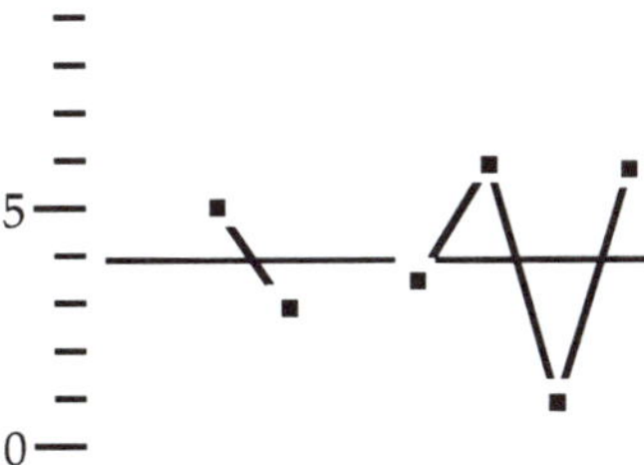

Figure 2.8: Patterns of Variation for Main-Effects in the Steel Parts Data

When these two patterns of variation are combined and extended across all 24 subgroups in the Steel Parts Data we get the following pattern:

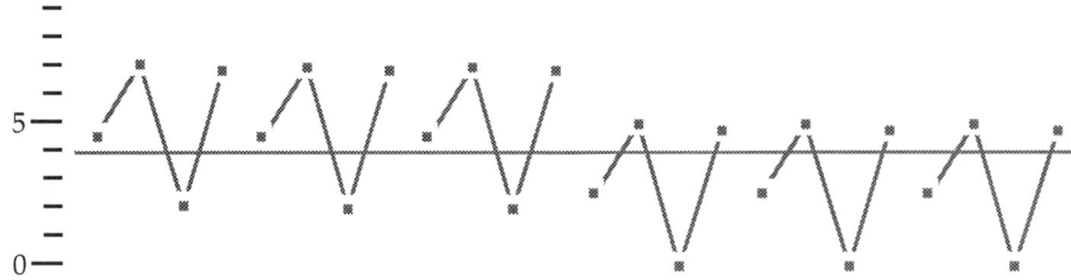

Figure 2.9: Expected Pattern for One-Way ANOM for Steel Parts Data

Superimposing the actual pattern of the One-Way ANOM over the expected pattern in Figure 2.9 we get Figure 2.10. The strong similarity between the expected and the actual pattern does not leave much to be explained by interaction effects. Thus, it is unlikely that there are any large interaction effects present in these data.

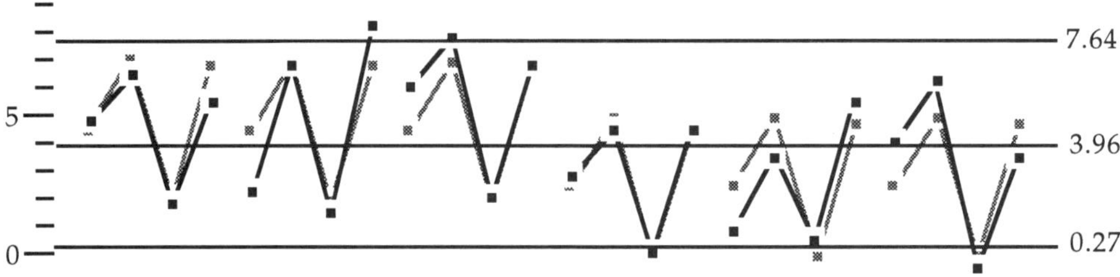

Figure 2.10: Expected and Actual One-Way ANOM Patterns of Variation for Steel Parts Data

This same approach for the Reflectivity Data (page 65) yields Figure 2.11. The expected pattern (in gray) does not completely explain the variation shown in the actual pattern (in black). In particular, the last two subgroup averages appear to be displaced upward. Thus, it is likely that there is an interaction effect in these data.

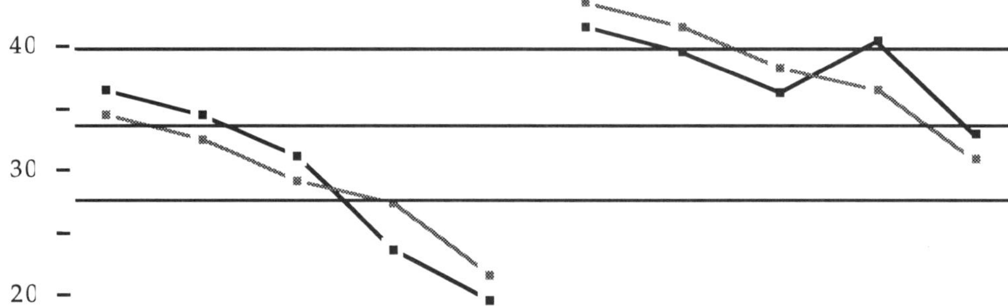

Figure 2.11: Expected and Actual Patterns of Variation for the Reflectivity Data

This simple comparison between the Main-Effect ANOM Plots and the One-Way ANOM Plot is one of the easiest ways to detect gross interaction effects. If the Analysis of Variance detects interaction effects, the interactions will usually be made intelligible by the ANOM plots.

2.7 ANOM With Unequal Subgroup Sizes

If the subgroups do not all have the same amount of data let n_j denote the number of values in the jth subgroup. With unequal subgroup sizes the One-Way ANOM procedure given above must be modified in two ways: the estimate of SD(X) will have to be obtained as the weighted average of the estimates from each separate subgroup, and there will have to be a different set of Decision Limits for each distinct value of n_j.

Since the averaging of the subgroup ranges is predicated upon equal subgroup sizes, it is no longer possible to average all k subgroup ranges and then estimate SD(X) from this one value. With unequal subgroup sizes one will have to estimate SD(X) for each subgroup. Since a weighted average will be required, it is easiest to use the s^2 values for each subgroup to estimate V(X) and then to find the square root.

From Table 1.5, p.46, the estimate of V(X) for the jth subgroup will be:

$$\text{Est. V(X)} = s_j^2 \text{ with } (n_j - 1) \text{ d.f.}$$

Next, each of these estimates is multiplied by its own degrees of freedom, and the sum of these products is divided by the sum of the individual degrees of freedom:

$$\text{Combined Est. V(X)} = \frac{\sum_{j=1}^{k} \left[(n_j - 1)\, s_j^2 \right]}{\sum_{j=1}^{k} (n_j - 1)}$$

The degrees of freedom for this estimate will be given by the sum in the denominator. The combined estimate for SD(X) will be the square root of this estimate of V(X). This estimate may be divided by the square root of the different values for n_j to obtain the appropriate estimates for SD($\bar{X}_j$).

The different values for Est. SD($\bar{X}_j$) will result in different sets of ANOM Decision Limits, but the H value will be the same for all these sets of limits.

The following example uses a modified form of the Coating Weight Data to illustrate how one would obtain the One-Way ANOM when the subgroup sizes are not equal.

EXAMPLE 2.7: ANOM for Modified Coating Weight Data:

Subgroup	1	2	3	4	5
Coating Weights in grams	250	310	250	340	250
	260	330	230	270	240
		280		300	270
		360			290
Averages	255	320	240	303.333	262.5
s_j^2	50	1133.333	200	1233.333	491.667
$d.f. = n_j - 1$	1	3	1	2	3

The Average of all 15 values is 282.0.
*(Note that this is **not** the average of the subgroup averages.)*

1. $\text{Est. } SD(X) = \sqrt{\dfrac{50+3(1133.333)+200+2(1233.333)+3(491.667)}{1+3+1+2+3}}$

$$= \sqrt{\frac{7591.667}{10}}$$

$$= \sqrt{759.1667} = 27.55 \text{ with 10 d.f.}$$

2. For $n = 2$, Est. $SD(\bar{X}) = \dfrac{27.55}{\sqrt{2}} = 19.48$ grams, with 10 d.f.

For $n = 3$, Est. $SD(\bar{X}) = \dfrac{27.55}{\sqrt{3}} = 15.91$ grams, with 10 d.f.

and for $n = 4$, Est. $SD(\bar{X}) = \dfrac{27.55}{\sqrt{4}} = 13.78$ grams, with 10 d.f.

3. $\alpha = 0.10$, $k = 5$, $v = 10$, give $H = 2.36$ (from Table D).
4. For $n = 2$, ANOM Decision Limits are $282 \pm 2.36(19.48) = 236.03$ to 327.97;
   for $n = 3$, ANOM Decision Limits are $282 \pm 2.36(15.91) = 244.45$ to 319.55;
   and for $n = 4$, ANOM Decision Limits are $282 \pm 2.36(13.78) = 249.48$ to 314.52.

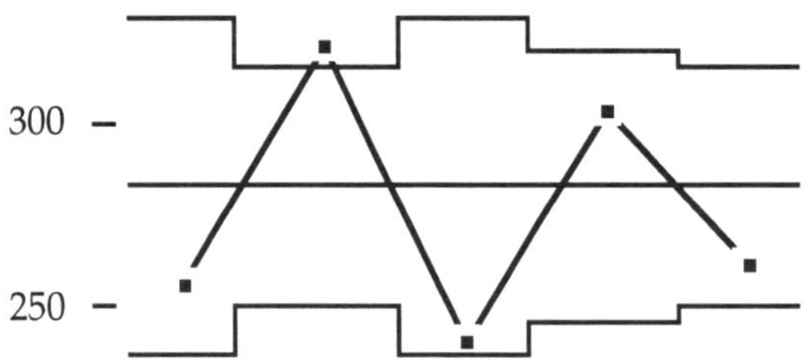

Figure 2.12: ANOM for Unequal Subgroup Sizes

2.8 The Analysis of Mean Ranges

The ANOM technique may be extended to the Analysis of Mean Ranges (ANOMR). Just as the Main-Effect ANOM compared averages for different levels of a factor in the experiment, the ANOMR compares average ranges for different levels of a factor. Thus, the ANOMR procedure is similar to the Main-Effect ANOM: it is intended for balanced, fully-crossed multifactor studies.

Given data consisting of k subgroups of size n, with two or more fully-crossed factors, and given that one of these factors (Factor A) has L levels, there will be exactly m = k/L subgroups for each level of Factor A. These m subgroups will each have a subgroup range, and these m ranges may be averaged to get an Average Range for each level of Factor A.

Approximate ANOMR Decision Limits may be found using:

$$\bar{\bar{R}} \pm H\,[\,\text{Est. SD}(\bar{R})\,]$$

where the value of H depends upon the α-level, the degrees of freedom for the estimate of the standard deviation of the Average Ranges, and the number of Average Ranges being compared, and where the double-bar notation is used to denote the Overall Average Range (the average of the ranges of all k subgroups of size n).

Just as with the Main-Effect ANOM, the ANOMR will use the estimate of SD(X) obtained in the One-Way ANOM as the foundation for the Decision Limits. Building on this foundation, we note that from Theorem Four, p.22:

$$\text{Est. SD}(R) = \frac{d_3\,\bar{\bar{R}}}{d_2}$$

where the d_2 and d_3 values are found based on the original subgroup size, n.

This estimated standard deviation is converted into an estimated standard deviation for the Average Ranges by dividing by the square root of m (m = the number of subgroups for each level of Factor A = the number of ranges averaged to get each Average Range).

For each Main Effect in the experiment, the steps for the ANOMR procedure are:

1. **Estimate the standard deviation for subgroup ranges:** Based upon n = subgroup size for the One-Way ANOM, find values for d_2 and d_3, and compute

$$\text{Est. SD}(R) = \frac{d_3\,\bar{\bar{R}}}{d_2}.$$

This estimate has degrees of freedom equal to $\nu_{B(k,n)}$. (See Table 1.5, page 46.)

2. **Estimate the standard deviation for Main Effect Average Ranges:** If the Main Effect has L levels, then there should be m = k/L subgroups for each level of the Main Effect. Obtain the Average Range for each level of the Main Effect by averaging the m ranges for each

level, and estimate the standard deviation of these Main Effect Average Ranges by

$$\text{Est. SD}(\bar{R}) = \frac{\text{Est. SD}(R)}{\sqrt{m}}.$$

This estimate will have the same degrees of freedom as the estimate in step 1.

3. **Obtain the critical value from Table D:** Choose the α-level, then use α, ν, and L to find a value for H from Table D, pp.305–308. (The symbol k in Table D represents the number of averages being compared, which in this case is equal to the number of levels, L, for the Main Effect being considered.)

4. **Compute the ANOMR Decision Limits:**

$$\bar{\bar{R}} \pm H\,[\,\text{Est. SD}(\bar{R})]$$

Any Main Effect Average Range which falls outside these limits may be said to be detectably different from the Overall Average Range. Thus, such Average Ranges will identify levels of the Main Effect which have either greater variation or lesser variation than the other levels.

Example 2.8 illustrates this procedure using the Steel Parts Data. This approximate ANOMR procedure may be used with Table D to obtain ANOMR Decision Limits for α-levels of $\alpha = 0.25$, 0.10, 0.05 and 0.01.

A theoretically more precise version of the ANOMR procedure is given by Neil Ullman (1989) with an acknowledgment to R. Hogg. In Ullman's ANOMR procedure the decision limits for a particular Main Effect are calculated using the formulas:

$$UDL_{\bar{R}} = H_U\,\bar{\bar{R}}$$

$$LDL_{\bar{R}} = H_L\,\bar{\bar{R}}$$

where the H_U and H_L factors depend upon four quantities:

(1) α = the α-level for the procedure,

(2) L = the number of levels for the Main Effect,

(3) m = the number of subgroup ranges averaged to find each of the Average Ranges (the number of subgroups per level of the Main Effect), and

(4) n = the original subgroup size in the One-Way ANOM.

The values for H_U and H_L are given in Table D2 for α values of $\alpha = 0.01$ and $\alpha = 0.05$.

Example 2.9 illustrates that there is little practical difference between the two ANOMR procedures. The difference between the computed limits is usually quite small, and it disappears as m increases.

Since Ullman's Tables include α-levels of 0.01 and 0.05 only, the reader will have to use the approximate procedure when an exploratory analysis is desired. In the following examples the decision limits are based upon $\alpha = 0.05$ to permit comparisons between the two ANOMR procedures.

Example 2.8: *Analysis of Mean Ranges for Steel Parts Data:*

A. Heat Treatment ANOMR

Each level of Heat Treatment is used with 12 subgroups, (m = 24/2 = 12). The averages of the 12 ranges for each Heat treatment are:

Heat Treatment	*W*	*L*
Average Ranges	*5.333*	*5.25*

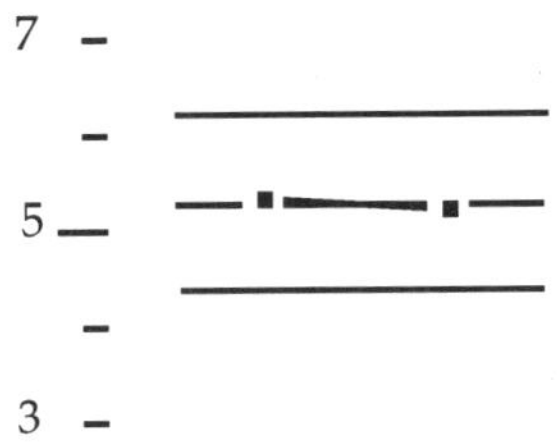

$$\text{Est. } SD(R) = \frac{0.8798\,(5.29167)}{2.059} = 2.261$$

with v = 65.9 degrees of freedom.

$$\text{So, Est } SD(\bar{R}) = \frac{2.261}{\sqrt{12}} = 0.653$$

With $\alpha = 0.05$, $v = 66$, and $L = 2$, $H = 1.41$

So the ANOMR Decision Limits are: $5.29 \pm 1.41\ (0.653) = 4.37$ to 6.21, and there is no evidence that the Heat Treatments result in different amounts of variation in the length of the steel parts.

B. Machine ANOMR

Each Machine is used with 6 subgroups, (m = 24/4 = 6). The averages for the 6 ranges for each of these Machines are:

Machine	*A*	*B*	*C*	*D*
Average Ranges	*6.333*	*5.5*	*4.5*	*4.833*

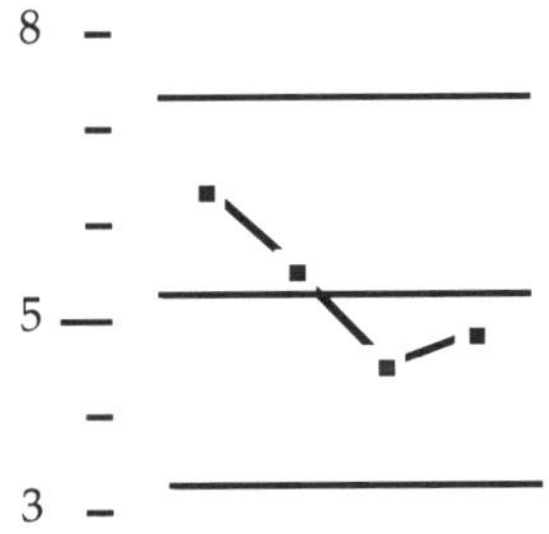

$$\text{Est. } SD(R) = \frac{0.8798\,(5.29167)}{2.059} = 2.261$$

with v = 65.9 degrees of freedom.

$$\text{So, Est } SD(\bar{R}) = \frac{2.261}{\sqrt{6}} = 0.923$$

With $\alpha = 0.05$, $v = 66$, and $L = 4$, $H = 2.20$

So the ANOMR Decision Limits are: $5.29 \pm 2.20\ (0.923) = 3.26$ to 7.32, and there is no evidence that the different Machines result in different amounts of variation in the length of the steel parts.

C. Time of Day ANOMR

Each Time of Day occurs with 8 subgroups, (m = 24/3 = 8). The averages for the 8 ranges for each Time of Day are:

Time of Day	*8*	*11*	*3*
Average Ranges	*4.0*	*4.75*	*7.125*

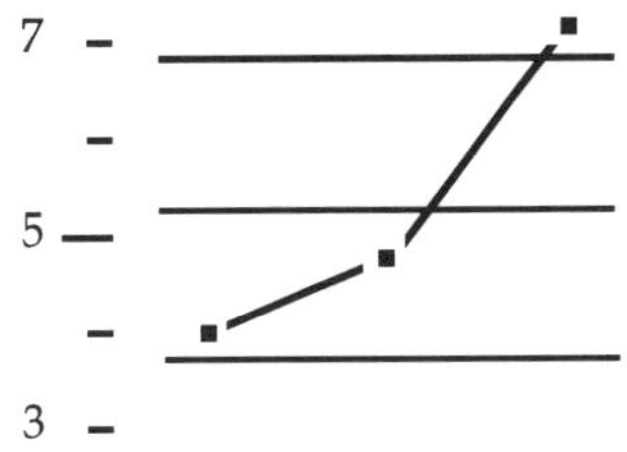

$$\text{Est. } SD(R) = \frac{0.8798\,(5.29167)}{2.059} = 2.261$$

with v = 65.9 degrees of freedom.

$$\text{So, Est } SD(\bar{R}) = \frac{2.261}{\sqrt{8}} = 0.799$$

With $\alpha = 0.05$, $v = 66$, and $L = 3$, $H = 1.96$

So the ANOMR Decision Limits are: $5.29 \pm 1.71\ (0.799) = 3.73$ to 6.86, and the

Average Range for 3 p.m. is above the upper decision limit. This is evidence that the variation in the lengths of the steel parts was detectably greater at 3 p.m. than it was earlier. Thus, the variation in length appears to have changed during the day. Whether this is a reproducible (real) effect, a False Alarm or an indication of a lack of control in the manufacturing process is not clear from the data.

Example 2.9: *(Alternative) Analysis of Mean Ranges for Steel Parts Data:*

A. Heat Treatment ANOMR *(Ullman)*

With n = 4, m = 12, L = 2 and α = 0.05, Ullman's formulas yield

$$UDL_{\bar{R}} = H_U\, \bar{\bar{R}} = 1.170\,(5.2917) = 6.19$$

$$LDL_{\bar{R}} = H_L\, \bar{\bar{R}} = 0.830\,(5.2917) = 4.39$$

Compare these limits with the earlier values of 4.37 to 6.21 (shown by the shaded lines).

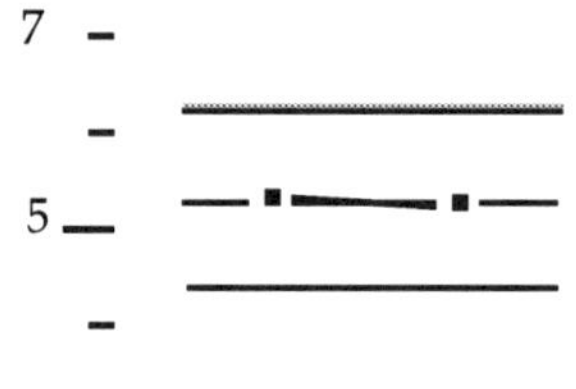

B. Machine ANOMR *(Ullman)*

With n = 4, m = 6, L = 4 and α = 0.05, Ullman's formulas yield

$$UDL_{\bar{R}} = H_U\, \bar{\bar{R}} = 1.400\,(5.2917) = 7.41$$

$$LDL_{\bar{R}} = H_L\, \bar{\bar{R}} = 0.653\,(5.2917) = 3.46$$

Compare these limits with the earlier values of 3.26 to 7.32 (shown by the shaded lines).

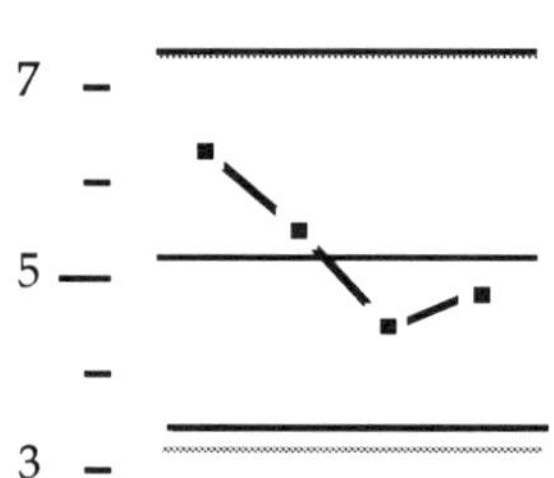

C. Time of Day ANOMR *(Ullman)*

With n = 4, m = 8, L = 3 and α = 0.05, Ullman's formulas yield

$$UDL_{\bar{R}} = H_U\, \bar{\bar{R}} = 1.305\,(5.2917) = 6.91$$

$$LDL_{\bar{R}} = H_L\, \bar{\bar{R}} = 0.719\,(5.2917) = 3.80$$

Compare these limits with the earlier values of 3.73 to 6.86 (shown by the shaded lines).

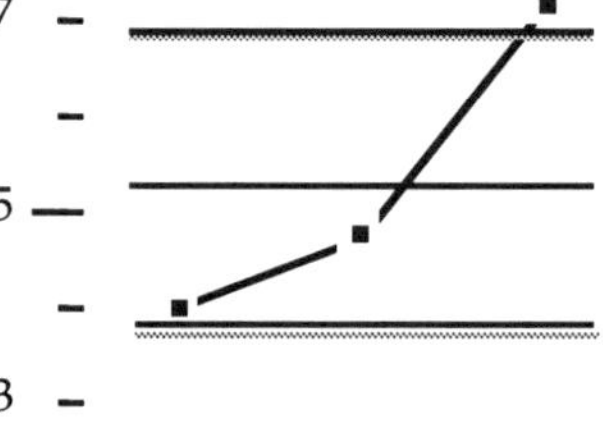

While either ANOMR procedure may be used to compare the Average Ranges for the different levels of each of the Main Effects in a multifactor ANOM, it is not always interesting to do so. Radical changes in variation will generally be visible on the ordinary Range Chart obtained with the One-Way ANOM. Therefore, the use of ANOMR is entirely a discretionary matter.

Exercise 2.4 ANOMR for Reflectivity Data:

Perform an ANOMR for Concentration and Temperature Main Effects in the Reflectivity Data. Use either method, and let $\alpha = 0.05$.

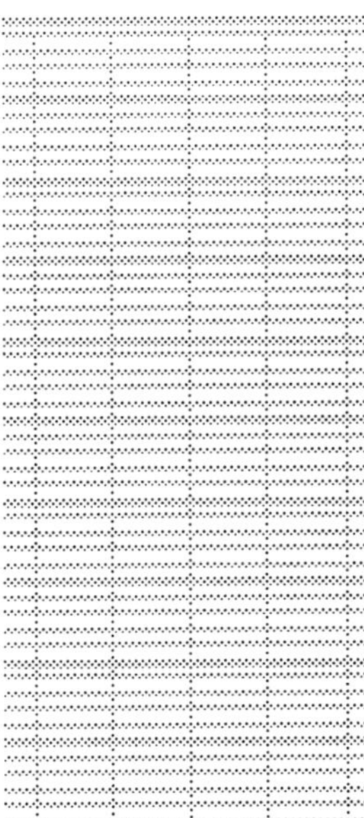

2.9 Summary of Uses of ANOM

You may have spent weeks collecting the data. You may have spent days analyzing the data. It is now time to communicate the results of the analysis and you have 30 seconds to do so before you loose the attention of your audience. How can you beat the 30-second rule? The only way you can consistently do so is to use graphs. And ANOM is primarily a graph. In fact, it is a graph that will be very easy to understand by those who are accustomed to control charts. This ability to quickly and effectively communicate the results of an experiment is perhaps the most important characteristic of ANOM.

ANOM is appropriate wherever means are being compared. The One-Way ANOM compares each average with the Grand Average. While equal subgroup sizes are desirable, they are not absolutely necessary for a One-Way ANOM.

The One-Way ANOM may be used with experiments where the factors are not fully crossed. When some treatment combinations are missing the ANOM plot will, in effect, have gaps in the running record. These gaps will make the interpretation of the ANOM plot difficult. Any time an ANOM plot has such gaps, it should be considered to be incomplete, and interpreted accordingly. Such incomplete ANOM plots must be used with caution.

The Main Effect ANOM compares the average responses at each level of a single factor with the Grand Average. Thus, a Main Effect ANOM will require that the factors be fully crossed and that the same number of observations occur in each subgroup. This special ANOM procedure is therefore more restricted in application than the One-Way ANOM. When it can be used, it may be combined with the One-Way ANOM to obtain some idea of the likelihood of interaction effects in the data.

ANOMR compares the Average Ranges for each level of a single factor with the Overall Average Range. Once again the data need to be balanced and fully-crossed.

Moreover, because of the graphic nature of ANOM, it also provides a supplement to the traditional Analysis of Variance approach.

Finally, the ANOM techniques given in this chapter will often provide a useful follow-up analysis and visualization device for screening designs and other fractional factorial experiments.

CHAPTER THREE

THE ANALYSIS OF VARIANCE

As outlined in Chapter Two, both the Control Chart and the ANOM chart provide simple, powerful, and graphic ways of comparing the averages of k subgroups of size n. This chapter outlines a purely numerical technique for making this same comparison. Since this approach is based upon estimates of V(X), it is called the Analysis of Variance. Why should we bother with another technique? The graphic techniques are powerful and easy to interpret, but they are difficult to apply to certain complex situations. The Analysis of Variance does not suffer this same limitation.

3.1 The Concept of Analysis of Variance

The basic idea of the Analysis of Variance (ANOVA) is to estimate the variance of X using the three different methods cited in Chapter One. If the estimates are all quite similar, then there is little likelihood that there are any detectable differences between the subgroup averages. However, if there is a substantial difference between the estimates, then it is very likely that some of the subgroup averages are detectably different.

Just as both Shewhart's Control Charts and ANOM use the Within-Subgroup Variation to set up their respective limits, ANOVA will use the Within-Subgroup Variation (Method Two) as the yardstick as to what should be if there is no detectable difference between the subgroup averages. In this sense, all three procedures are built upon the same foundation. However, while both Control Charts and ANOM Charts set limits on the variation for subgroup averages, ANOVA compares the **Between-Subgroup Estimate of V(X)** with the **Within-Subgroup Estimate of V(X)**. These two estimates are used because the Between-Subgroup Variation is the most sensitive to differences between the subgroup averages, while the Within-Subgroup Variation is completely insensitive to differences between the subgroup averages. In obtaining these estimates it is traditional to use the unbiased estimators for V(X) which are based upon s^2 (See the upper right-hand estimates in Tables 1.4, 1.5, and 1.6, pp.45–47).

Data Set One will be used both to illustrate the different methods of estimating variation which are used in the Analysis of Variance procedure and to introduce the One-Way ANOVA Table.

EXAMPLE 3.1: Three Estimates of V(X) for Data Set One:

Three subgroups of size n = 8 have the following values:

Subgroup	*Measurements*	$\bar{X}$	s^2
1	4 5 5 4 8 4 3 7	5	2.856
2	2 4 3 7 5 4 2 5	4	2.856
3	3 6 6 4 5 4 6 6	5	1.429

Method One *(the Total Variation Estimate) uses the sample variance based on all nk = 24 values:*

```
        X
        X
        X X
        X X X
      X X X X
    X X X X X X
    X X X X X X X
  ---------------
0 1 2 3 4 5 6 7 8
```

therefore,

$$\text{Est. } V(X) = s^2 = \frac{55.3333}{23} = \mathbf{2.406.}$$

Method Two *(the Within-Subgroup Estimate) begins with a sample variance for each subgroup:*

```
        X
        X X
      X X X   X X
-------------------
0 1 2 3 4 5 6 7 8 9
```

Subgroup 1 has $s^2 = \frac{20}{7} = 2.856,$

```
    X   X X
    X X X X   X
-------------------
0 1 2 3 4 5 6 7 8 9
```

Subgroup 2 has $s^2 = \frac{20}{7} = 2.856,$

```
                X
                X
          X     X
       X  X  X  X
_______________________________
 0  1  2  3  4  5  6  7  8  9
```

Subgroup 3 has $s^2 = \frac{10}{7} = 1.429$

And the Within Subgroup Estimate is the average of these variances:

$$\text{Est. } V(X) = \overline{s^2} = \frac{1}{3}\left(\frac{20}{7} + \frac{20}{7} + \frac{10}{7}\right) = \frac{50}{21} = \mathbf{2.381.}$$

Method Three *(the Between-Subgroup Estimate) begins with the sample variance of the subgroup averages:*

```
             X̄

          X̄  X̄
_______________________________
 0  1  2  3  4  5  6  7  8  9
```

For the three subgroup averages,

$$s_{\bar{x}}^2 = \frac{0.66667}{2},$$

So the Between Subgroup Estimate of V(X) is:

$$\text{Est. } V(X) = n\, s_{\bar{x}}^2 = 8\left(\frac{0.66667}{2}\right) = \frac{5.3333}{2} = \mathbf{2.667.}$$

So for Data Set One, the Between-Subgroup Estimate of V(X), 2.667, is quite similar to the Within-Subgroup Estimate of V(X), 2.381. This similarity suggests that the variation between the subgroup averages may be accounted for by the background variation. (See the Control Chart for Data Set One on p.33.)

Note that each estimate involves a ratio, and these ratios have a special property: the numerators and denominators sum up.

Method Three:	Est. V(X) = 5.3333 / 2 = 2.667	Between Subgroups
Method Two:	Est. V(X) = 50 / 21 = 2.381	Within Subgroups
Method One:	Est. V(X) = 55.3333 / 23 = 2.406	Total Variation

The numerator of the Method Three Estimate adds to the numerator of the Method Two Estimate to yield the numerator of the Method One Estimate. Likewise, the denominator of the Method Three Estimate adds to the Denominator of the Method Two Estimate to yield the denominator of the Method One Estimate. The actual estimates themselves, however, do not add up.

The Analysis of Variance makes this special relationship between these numerators and denominators visible by placing the quantities above in an ANOVA Table. The numerators are generically referred to as "Sums of Squares," the denominators are called "Degrees of Freedom," and the actual ratios (the estimates of V(X)) are called "Mean Squares." Each Method receives a separate row in the ANOVA Table, so that the result looks like:

ONE-WAY ANOVA FOR DATA SET ONE:

Source	Sum of Squares	D.F.	Mean Squares	F-Ratio
Between	5.3333	2	2.667	2.667/2.381 =1.12
Within	50.0000	21	2.381	
Total	55.3333	23	2.406	

3.2 The ANOVA Table

The One-Way ANOVA Table shown above consists of five columns and three rows. Each row presents the results of one of the three methods of estimation outlined earlier. The first entry (Source) in each row identifies the method of estimation summarized in that row. The second entry (Sum of Squares or SS) in each row is the numerator of the estimate of V(X). The third entry (D.F. or Degrees of Freedom) in each row is the denominator for the estimate of V(X). The fourth entry (Mean Squares or MS) in each row is the estimate of V(X) itself. Finally, the Between-Subgroup Estimate is compared with the Within-Subgroup Estimate by the F-Ratio in the last column.

The reason that ANOVA is built upon the s^2 type estimators is the special way that the numerators and denominators add up. For this reason, it is unusual to use any other estimators from the tables in Chapter One in the Analysis of Variance. However, before the existence of computers and electronic calculators, the unbiased estimators for V(X) based upon the Range were created to provide an approximate ANOVA. (Patnik, 1950)

Originally, the Sums of Squares and the Degrees of Freedom were included in the ANOVA table because the additivity of these terms could be exploited to save a considerable amount of hand computation. Now they are traditional, and provide a quick check on the correctness of the computations.

Unlike the Sums of Squares and the Degrees of Freedom, the Mean Squares do NOT add up. This is why the Method One Estimate of V(X) is traditionally left off the table. It is not explicitly used, and it is not the sum of the other mean squares, so it serves no purpose in the ANOVA table.

Finally, the F-ratio is the value used to evaluate the similarity of the estimates. It consists of the Mean Square Between divided by the Mean Square Within.

$$\text{F-Ratio} = \frac{\text{MSB}}{\text{MSW}}$$

When the F-ratio is close to 1.0, the estimates will be said to be similar. When the F-ratio is large, the estimates are dissimilar, and this dissimilarity is taken to be an indication of potentially real differences between the subgroup averages. The definition of "large" is an F-ratio that exceeds the upper critical value of an F-distribution. This upper critical value of an F-distribution will depend upon

(a) the choice of an overall α-level for the procedure, and

(b) the degrees of freedom for the two estimates used.

The F-distribution and the interpretation of the F-ratio will be covered later in this chapter. Once again the reader is reminded that the α-level merely specifies the manner in which the experimenter defines the decision rule for interpreting the data. It does not actually define the risk of a False Alarm, the confidence level, or the practical importance of the results.

Thus, the ANOVA table is a formalized way of presenting these numbers so that certain relationships can be seen and exploited. It came into use when the calculations were done by hand and the exploitation of relationships was crucial to reducing the amount of work involved in performing an ANOVA. While this need is not as great today, the formal structure of the ANOVA table is helpful in organizing the results for presentation. As can be seen from the description above, the ANOVA table essentially decomposes each estimate of V(X) into a ratio, and presents both the estimates and their decomposition in a traditional format.

Data Set Two will be used to provide an additional illustration of the use of the three methods for estimating V(X).

<u>*EXAMPLE 3.2:*</u> <u>*Three Estimates of V(X) for Data Set Two:*</u>

Three subgroups of size n = 8 have the following values:

Subgroup	*Measurements*									$\bar{X}$	s^2
1	4	5	5	4	8	4	3	7	5	2.856	
2	0	2	1	5	3	2	0	3	2	2.856	
3	7	10	10	8	9	8	10	10	9	1.429	

Method One *(the Total Variation Estimate) is based on all nk = 24 values:*

```
                                        X
             X   X   X         X        X
X        X   X   X   X     X   X        X
X   X    X   X   X   X     X   X   X    X
0   1    2   3   4   5  6  7   8   9   10
```

$$\textit{therefore, Est. } V(X) = s^2 = \frac{247.3333}{23} = \mathbf{10.754.}$$

Method Two *(the Within-Subgroup Estimate) begins with a variance for each subgroup:*

```
            X
            X  X
         X  X  X     X  X
------------------------------
0  1  2  3  4  5  6  7  8  9
```

Subgroup 1 has $s^2 = \frac{20}{7} = 2.856,$

```
X     X  X
X  X  X  X     X
------------------------------
0  1  2  3  4  5  6  7  8  9
```

Subgroup 2 has $s^2 = \frac{20}{7} = 2.856,$

```
                              X
                              X
                        X     X
                     X  X  X  X
---------------------------------
0  1  2  3  4  5  6  7  8  9  10
```

Subgroup 3 has $s^2 = \frac{10}{7} = 1.429$

And the Within Subgroup Estimate is the average of these variances:

$$\text{Est. } V(X) = \overline{s^2} = \frac{1}{3}\left(\frac{20}{7} + \frac{20}{7} + \frac{10}{7}\right) = \frac{50}{21} = \mathbf{2.381.}$$

Method Three *(the Between-Subgroup Estimate) begins with the variance of the subgroup averages:*

```
      _        _              _
      X        X              X
---------------------------------
0  1  2  3  4  5  6  7  8  9  10
```

For the three subgroup averages,

$$s_{\bar{x}}^2 = \frac{24.66667}{2}$$

So the Between Subgroup Estimate of V(X) is:

$$\text{Est. } V(X) = n\, s_{\bar{x}}^2 = 8\left(\frac{24.66667}{2}\right) = \frac{197.3333}{2} = \mathbf{98.667.}$$

So for Data Set Two, the Between-Subgroup Estimate of V(X), 98.667, is substantially larger than the Within-subgroup Estimate of V(X), 2.381. This suggests that the difference between the subgroup averages is greater than that which could have occurred by chance. These data support a decision that there is a definite difference between the subgroup averages.

EXERCISE 3.1: ANOVA for Data Set Two:

Based on the results shown in Example 3.2, write out the One-Way ANOVA Table for Data Set Two.

SOURCE	SS	DF	MS	F
Between				
Within				
Total				

EXERCISE 3.2 ANOVA for Data Set Three:

The data shown in Exercises 1.11, 1.12, and 1.13, pp.48–50, were used to obtain various estimates of SD(X) and V(X).
In each exercise, the last estimate of V(X) is the estimate used in ANOVA.
(a) List the three estimates of V(X) in the Mean Square Column below:
(b) List the Degrees of Freedom in the DF column below:
(c) Multiply DF by MS to obtain the Sums of Squares:
(d) Find the F-ratio for these data.

Source	SS	DF	MS	F
Between				
Within				
Total				

3.3 The Computation of ANOVA Values

Exercise 3.1 above, in conjunction with Example 3.2, p.87, should suggest a way of computing the values for the ANOVA Table.

The values for the Total row of the ANOVA table can be found by computing the s^2 value using all nk observations. Since these values will not explicitly be used in computing the F-ratio, their computation is optional. If they are computed, they will provide a check on the correctness of the other computations, but otherwise they are not needed. Given the ease with which a keystroke error can occur, the computation of the values in the Total row is generally advisable so the other computations can be verified.

The values for the Within-Subgroup row of the ANOVA Table can be found by calculating an s^2 value for each subgroup individually. Since those calculators that compute s^2 will also compute the average at the same time, one should be careful to extract and record these subgroup averages along with the s^2 values. Once these k dispersion statistics have been found, they are averaged to get the Mean Square Within. Since each s^2 term has exactly (n-1) degrees of freedom, the Mean Square Within has k(n-1) degrees of freedom. Multiplying the Mean Square Within by its degrees of freedom yields the Within-Subgroup Sum of Squares.

The values for the Between-Subgroup row of the ANOVA table are found by computing the s^2 value using the k subgroup averages, and then multiplying it by the subgroup size, n. The resulting value is the Mean Square Between. Since k values are used to find this s^2 value, the estimate has (k-1) degrees of freedom. Multiplying the Mean Square by the degrees of freedom will yield the Between-Subgroup Sum of Squares.

These computations are outlined on the ANOVA Worksheet given below. The quantities in boxes are the values that are transferred to the ANOVA Table. This ANOVA Worksheet is intended for use with balanced data (subgrouped data are said to be balanced when all the subgroups contain the same number of observations). An extra copy of this worksheet is given in the Appendix.

If the subgroup sizes should happen to be unequal, the computations described above must be modified in three ways:

(1) The Mean Square Within must be found by forming the ratio of two sums. The numerator of this ratio will be the sum of the numerators of the k separate subgroup variances, and the denominator will be the sum of the denominators of the k separate subgroup variances. These sums are, respectively, the Within-Subgroup Sum of Squares and the Within-Subgroup Degrees of Freedom. (See Section 2.7, p.75, for an illustration of this.)

(2) The Total Degrees of Freedom is equal to N-1 where N is the total number of observations in all the subgroups, and the Total Sum of Squares is still equal to [s^2] x [Total DF].

(3) The Between Subgroup entries may be obtained from these others by appropriate subtractions.

For those who appreciate the preciseness of algebraic notation, the formulas for the different values in the ANOVA table are shown below. (Those who are not algebra-friendly may quickly turn to the next page.)

First consider the case of equal subgroup sizes:

Let X_{ij} = The ith observation in the jth subgroup, $i = 1$ to n, and $j = 1$ to k.

$\bar{X}_j$ = The average of the jth subgroup, $j = 1$ to k.

$\bar{\bar{X}}$ = The grand average of all nk observations.

ONE-WAY ANOVA FOR k SUBGROUPS OF SIZE n

SOURCE	SS	DF	MS	F-Ratio
Between	$SSB = n\sum_{j=1}^{k} (\bar{X}_j - \bar{\bar{X}})^2$	(k-1)	$\frac{SSB}{k-1}$	$\frac{MSB}{MSW}$
Within	$SSW = \sum_{j=1}^{k}\sum_{i=1}^{n} (X_{ij} - \bar{X}_j)^2$	k(n-1)	$\frac{SSW}{k(n-1)}$	
Total	$SST = \sum_{j=1}^{k}\sum_{i=1}^{n} (X_{ij} - \bar{\bar{X}})^2$	(nk-1)		

For unequal subgroup sizes, n becomes a function of the value of j, and the formula for the Between Sum of Squares is modified to become

$$\sum_{j=1}^{k} \left[n_j \, (\bar{X}_j - \bar{\bar{X}})^2 \right]$$

while the Within-Subgroup Degrees of Freedom becomes

$$\sum_{j=1}^{k} (n_j - 1)$$

and the Total Degrees of Freedom becomes

$$\left[\sum_{j=1}^{k} n_j \right] - 1.$$

Computational formulas have been derived to minimize the work when doing ANOVA by hand, but with the advent of inexpensive electronic calculators, these formulas are not really necessary today. In most instances it is easier to remember how to use the calculator than it is to remember the computational formulas. In addition, many programs for performing the ANOVA calculations are available for any type of computer system. Thus, the computational forms are not given here.

Analysis of Variance Worksheet

Data Set Name ______________________________ k=____ n=____

The values that are enclosed in boxes will be inserted into the ANOVA table.

I. TOTAL or OVERALL VARIATION

A. Find s^2 using all nk observations = s^2 = ____________

B. Check Value = Grand Average = $\bar{\bar{X}}$ = ____________

C. Total Degrees of Freedom = (nk - 1) = [ ]

D. Total Sum of Squares = (nk-1) s^2 = TSS = [ ].

II. WITHIN-SUBGROUP VARIATION

A. Obtain the Subgroup Average and Subgroup Variance for each of the k subgroups:

Subgroup Averages, $\bar{X}$			Subgroup Variances, s^2		
1____	6____	11____	1____	6____	11____
2____	7____	12____	2____	7____	12____
3____	8____	13____	3____	8____	13____
4____	9____	14____	4____	9____	14____
5____	10____	15____	5____	10____	15____

B. Average these s^2 values to get the Mean Square Within = $\overline{s^2}$ =

MSW = [ ]

C. Within-Subgroup Degrees of Freedom =

k (n - 1) = [ ]

D. Within-Subgroup Sum of Squares =

k(n-1) MSW = SSW = [ ].

III. BETWEEN-SUBGROUP VARIATION

A. Calculate s^2 using the k subgroup averages = $s_{\bar{x}}^2$ = ____________

B. Check Value = Grand Average = $\bar{\bar{X}}$ = ____________

(Does this value match the value in I. B. above? If not, you have made a mistake.)

C. Find the Mean Square Between by multiplying the $s_{\bar{x}}^2$ value in III.A. by n =

n $s_{\bar{x}}^2$ = MSB = [ ]

D. Between-Subgroup Degrees of Freedom =

(k - 1) = [ ]

E. Between-Subgroup Sum of Squares =

(k-1) MSB = SSB = [ ]

(Pythagoras claims that (III.E.) + (II.D.) = (I.D.). If they don't, guess who's wrong.)

The use of this worksheet will be illustrated with the Gauge Study Data.

EXAMPLE 3.3: ANOVA for the Gauge Study Data:

The Gauge Study Data are described in Example 1.7, p.41.

Subgroup	1	2	3	4	5	6	7	8	9	10
Operator	A	A	A	A	A	B	B	B	B	B
Part	1	2	3	4	5	1	2	3	4	5
Values	20	20	25	50	45	20	15	15	45	35
	15	25	25	50	40	20	10	10	20	40
Averages	17.5	22.5	25.0	50.0	42.5	20.0	12.5	12.5	32.5	37.5
s^2	12.5	12.5	0	0	12.5	0	12.5	12.5	312.5	12.5

The ANOVA Worksheet would have the following values:

I.A. $s^2 = 177.56579$
I.B. *average* = 27.25
I.C. $nk - 1 = 19$
I.D. *TSS* = 3373.75

II.A. *given above*
II.B. *MSW* = 38.75
II.C. $k(n-1) = 10$
II.D. *SSW* = 387.5

III.A. $s_{\bar{x}}^2 = 165.902778$
III.B. *average* = 27.25
III.C. *MSB* = 331.805556
III.D. $(k-1) = 9$
III.E. *SSB* = 2986.25

So the ANOVA Table for these data would be

SOURCE	SS	DF	MS	F
Between	2986.25	9	331.805556	8.56
Within	387.5	10	38.75	
Total	3373.75	19		

EXERCISE 3.3: ANOVA for the Coating Weight Data:

The Coating Weight Data consist of 20 observations of the weight of a given coating. Four pieces were produced under each of five different treatment combinations. The weights are expressed in grams.

Subgroup	1	2	3	4	5
Coating	250	310	250	340	250
Weights	260	330	230	270	240
in grams	230	280	220	300	270
	270	360	260	320	290

Use the ANOVA Worksheet on the following page, and fill in the ANOVA table for these data.

SOURCE	SS	DF	MS	F
BETWEEN				
WITHIN				
TOTAL				

Analysis of Variance Worksheet

Data Set Name ______________________________ k=____ n=____

The values that are enclosed in boxes will be inserted into the ANOVA table.

I. TOTAL or OVERALL VARIATION

A. Find s^2 using all nk observations = s^2 = ____________

B. Check Value = Grand Average = $\bar{\bar{X}}$. = ____________

C. Total Degrees of Freedom = (nk - 1) = [ ]

D. Total Sum of Squares = (nk-1) s^2 = TSS = [ ].

II. WITHIN-SUBGROUP VARIATION

A. Obtain the Subgroup Average and Subgroup Variance for each of the k subgroups:

Subgroup Averages, $\bar{X}$			Subgroup Variances, s^2		
1____	6____	11____	1____	6____	11____
2____	7____	12____	2____	7____	12____
3____	8____	13____	3____	8____	13____
4____	9____	14____	4____	9____	14____
5____	10____	15____	5____	10____	15____

B. Average these s^2 values to get the Mean Square Within = $\overline{s^2}$ =

MSW = [ ]

C. Within-Subgroup Degrees of Freedom =

k (n - 1) = [ ]

D. Within-Subgroup Sum of Squares =

k(n-1) MSW = SSW = [ ].

III. BETWEEN-SUBGROUP VARIATION

A. Calculate s^2 using the k subgroup averages = $s_{\bar{x}}^2$ = ____________

B. Check Value = Grand Average = $\bar{\bar{X}}$ = ____________

(Does this value match the value in I. B. above? If not, you have made a mistake.)

C. Find the Mean Square Between by multiplying the $s_{\bar{x}}^2$ value in III.A. by n =

n $s_{\bar{x}}^2$ = MSB = [ ]

D. Between-Subgroup Degrees of Freedom =

(k - 1) = [ ]

E. Between-Subgroup Sum of Squares =

(k-1) MSB = SSB = [ ]

(Pythagoras claims that (III.E.) + (II.D.) = (I.D.). If they don't, guess who's wrong.)

3.4 Interpreting The F-Ratio

Given k subgroups of size n where the differences between the subgroup averages are merely due to random variation, the ratio of the Mean Square Between to the Mean Square Within should be in the neighborhood of 1.0. This happened with Data Set One, p.86.

Given k subgroups of size n where there are real differences between the subgroup averages, the ratio of the Mean Square Between to the Mean Square Within should be much greater than 1.0. This happened with Data Set Two, p.89.

Thus, the F-ratio provides a way to compare two different estimates of V(X). If the subgroup averages differ only because of background variation, then both estimates are estimates of V(X), but if there is a real difference between some of the subgroup means, the Between-Subgroups Estimate of V(X) is likely to be inflated.

Moreover, it is the structure of the F-ratio which makes it robust. The statements above are true without regard to any distributional assumptions.

However, when it comes to defining a decision rule for deciding when the F-ratio is close to 1.0 and when it is far from 1.0, we must again use theory as a guide.

Under the assumption of nondiscrete, independent, normally distributed individual values, with common variance, and with the side condition that the subgroup averages differ only because of background variation, it is possible to define a probability model for the behavior of the F-ratio. This probability model is the F-distribution.

This distribution will allow one to obtain a critical value for evaluating the F-ratio. If the F-ratio calculated in the ANOVA table exceeds the F-distribution critical value, then it is reasonable to decide that some of the subgroup averages differ by a detectable amount. On the other hand, if the computed F-ratio does not exceed the F-distribution critical value, then it is reasonable to decide that the differences between the subgroup averages may be considered to be due to background variation alone.

Being defined by the ratio of positive quantities, the F-distribution is defined only for positive values. Like most distributions of this type, the F-distribution is positively skewed. The defining parameters of the F-distribution are the degrees of freedom for each of the quantities in the F-ratio, thus the F-distribution has two degrees of freedom:

ν_1 will be the degrees of freedom for the numerator of the F-ratio, and
ν_2 will be the degrees of freedom for the denominator of the F-ratio.

These two parameters will define the F-distribution. Given these two degrees of freedom, and an α-level for the analysis, one can look up a critical value in the Table F in the Appendix. This critical value will then provide a guideline for interpreting the calculated F-ratio.

In Section 3.1, the ANOVA table for Data Set One had a calculated F-ratio of 1.12. The degrees of freedom for this ratio were 2 and 21. Using Table F in the Appendix, with an α-level of 0.10, the critical value of the F-distribution is, approximately, 2.58. Since the F-ratio does not exceed this value, the data contain no strong evidence that the subgroup means are detectably different. The usual way of expressing this result is to say that the F-ratio is *nonsignificant*. This is a negative finding, and such a finding should never be turned into a positive statement. It simply states that the data analyzed do not provide any strong and clear-cut evidence in favor of the subgroup averages differing by more than can be attributed to background variation. (The very fact that we have performed an experiment is, *a priori*, reason to believe that the subgroup averages actually differ by some amount, and a nonsignificant F-ratio simply says that any such differences are not easily detectable in the data.)

In Exercise 3.1, you were supposed to have found the F-ratio for Data Set Two. This F-ratio is based upon 2 and 21 degrees of freedom and has a value of 41.44. With $\alpha = 0.10$, the critical value will still be approximately 2.58. Thus, there is substantial evidence that the subgroup means are detectably different in Data Set Two. The usual way of expressing this result is to say that the F-ratio is *significant*. Once again, the terminology tends to suggest more than it really means.

In Example 3.3, the ANOVA table for the Gauge Study Data gave an F-ratio of 8.56. This F-ratio is based upon 9 and 10 degrees of freedom, and with $\alpha = 0.10$, the critical value for the F-distribution is 2.35. Thus, there is evidence that the subgroup means are detectably different in the Gauge Study Data.

Classifying F-ratios as either *significant* or *nonsignificant* can be confusing. The common misconception is that a significant F-ratio will correspond to subgroup differences that are of practical importance. This is not the case. When an F-ratio is said to be significant, it simply means that there are *detectable* differences between the subgroup means. A significant F-ratio does not indicate which differences are detectably different. A significant F-ratio does not indicate how large these differences may actually be. A significant F-ratio simply indicates that some differences between the subgroup means are large enough to be detected in spite of the noise in the data. Thus, a significant F-ratio is merely a green light for the investigator to look further.

On the other hand, a nonsignificant F-ratio suggests that any differences which may exist between the subgroups are obscured by the background variation.

Finally, the *size* of the F-ratio does not indicate the absolute size of the subgroup differences. It simply indicates the relative signal strength compared to the noise level. That is, a large F-ratio will signify that the differences between the subgroup averages are large relative to V(X).

The most straightforward way of interpreting an F-ratio is as a Signal-to-Noise ratio. The numerator is inflated by any signals which may be present, and the denominator estimates the background noise level.

The robustness of the F-ratio can be seen in Example 3.3. From Example 2.3, p.62, quite a bit is already known about the Gauge Study Data. In particular, the Range Chart was shown to be out of control. Subgroup 9 shows excessive variation, which is a violation of the assumption that the individual measurements be identically distributed except for possible differences between the subgroup means. From Examples 1.7 and 1.8, pp.41–42, the effect of deleting Subgroup 9 from the data can be seen. This deletion would cause the Mean Square Within to drop from 38.75 to 8.333. Thus, the F-ratio found in Example 3.3 had a denominator that was *inflated* by approximately 365%! In spite of the resulting deflation of the F-ratio, it was found to be "significant."

Examples 1.7 and 1.8 show that ANOVA is more sensitive to such inflation than ANOM or Control Charts. (Recall that while ANOVA uses a pooled variance estimator, ANOM and Control Charts can use average range estimators.) Thus, the ANOVA technique may miss some differences that other techniques will detect, but, in spite of this, the ANOVA technique for detecting differences between subgroup averages is still very robust and useful.

3.5 Making Sense of a One-Way ANOVA

The ANOVA presented above is called a One-Way ANOVA. This name comes from the fact that the Total Sum of Squares was split "one-way" into two components. This One-Way ANOVA provides an F-ratio that can be used to check for the presence of detectable differences between the subgroup averages. A significant F-ratio is thus broadly equivalent to being told that the Average Chart is out-of-control, or being told that some subgroup averages fall outside the ANOM decision limits. In short, it is not specific enough to be of much use in and of itself. Thus, some follow-up analysis is needed with a One-Way ANOVA.

Tukey's Post Hoc Test is one such follow-up procedure.

Tukey's Post Hoc Test provides a way to make all possible pairwise comparisons among k subgroup averages when the subgroup size is constant. The yardstick for these comparisons is Tukey's **Honestly Significant Difference.** [Scheffe' (1953, 1959), Tukey (1953)].

Given k subgroups of size n, the Honestly Significant Difference is

$$HSD = q\left[\text{ Est. SD}(\bar{X})\right] = q\sqrt{\frac{MSW}{n}}$$

where q is a constant that depends on α, k, and ν. (Here, as before, α = the overall α-level for the procedure, k = the number of subgroup averages being compared, and ν = the degrees of freedom for the dispersion estimate used.) The values for q are the critical values for the Studentized Range Distribution. These are given in Table E in the Appendix, p.313–315. Any two subgroup averages that differ by more than the HSD will be said to be detectably different.

EXAMPLE 3.4: Tukey's Post Hoc Test for the Coating Weight Data:

Subgroup	*1*	*2*	*3*	*4*	*5*
Coating Weights in grams	*250*	*310*	*250*	*340*	*250*
	260	*330*	*230*	*270*	*240*
	230	*280*	*220*	*300*	*270*
	270	*360*	*260*	*320*	*290*
Averages	*252.5*	*320*	*240*	*307.5*	*262.5*
Ranges	*40*	*80*	*40*	*70*	*50*

Here k = 5, n = 4, and MSW = 628.333 with 15 degrees of freedom (see Exercise 3.3, p.94). Given α = 0.10, the Honestly Significant Difference is:

$$HSD = 3.83\sqrt{\frac{628.33}{4}} = 3.83\,(12.533) = 48.00$$

Any two treatment averages that differ by 48.0 units or more will be said to be different at the α = 0.10 level.

Treatments	*Difference*	*Detectable?*	*Treatments*	*Difference*	*Detectable?*
1-2	*-67.5*	*YES*	*1-3*	*12.5*	*no*
1-4	*-55*	*YES*	*1-5*	*-10*	*no*
2-3	*80*	*YES*	*2-4*	*12.5*	*no*
2-5	*62.5*	*YES*	*3-4*	*-67.5*	*YES*
3-5	*-22.5*	*no*	*4-5*	*45.0*	*no*

While the listing of all possible pairwise comparisons can be carried out as above, it is usually easier to plot the averages on a line and look for distances that exceed the HSD value.

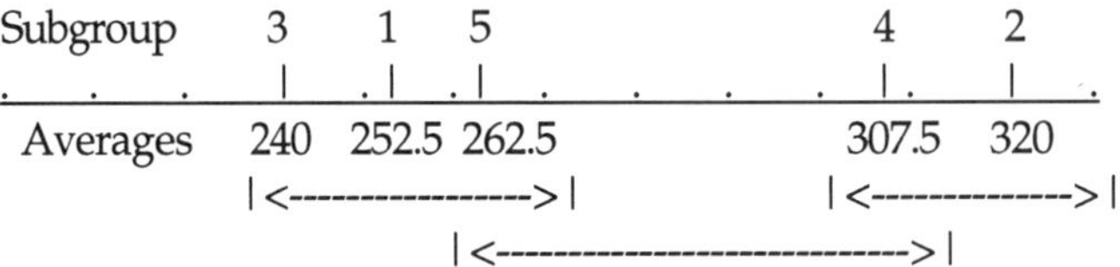

Figure 3.1: Tukey Post Hoc Test for Coating Weight Data

By underlining the groups that are NOT detectably different, the detectable differences can be read right off the graph. *Treatments not underlined by the same line are detectably different.* Treatments 3, 1, and 5 are underlined by one line, therefore they are *not* detectably different from each other. Treatment 3 is not underlined by the lines beneath Treatments 4 and 2, therefore Treatment 3 *is* detectably different from Treatments 4 and 2. Treatment 1 is not underlined by the lines beneath Treatments 4 and 2, therefore Treatment 1 *is* detectably different from Treatments 4 and 2. Since Treatments 4 and 5 are underlined by one and the same line, they are *not* detectably different. But Treatment 5 is not underlined by the line beneath Treatment 2, so Treatment 5 *is* detectably different from Treatment 2. All of these Statements can be summarized as follows. *Treatments 3 and 1 are seen to be detectably different from treatments 4 and 2, while treatment 5 is detectably different from treatment 2.* This approach has two advantages over the list method. The first is the graphic presentation of the results, the second is the greater economy of words needed to present the results.

3.6 The Difference Between ANOM and ANOVA

The Analysis of Means computes a bound for the amount by which a single subgroup average is likely to deviate from the Grand Average due to chance alone. This may be seen in Example 2.2, p.60. There the ANOM Decision Limits were placed at ± 2.26 (13.36) = ± 30.2 units on either side of the Grand Average. Those subgroup averages which differed from the Grand Average by more than this amount were said to be *detectably* different from the Grand Average.

The Analysis of Variance works in a different manner: instead of drawing a line and looking for subgroup averages which cross over the line, ANOVA accumulates the overall deviation from the Grand Average into a single statistic. Again we consider the Coating Weight Data. The subgroup averages are plotted against the Grand Average in Figure 3.2. The vertical lines show the deviation of each subgroup average from the Grand Average. These deviations are, respectively, –24, 43.5, –36.5, 31, and –14.

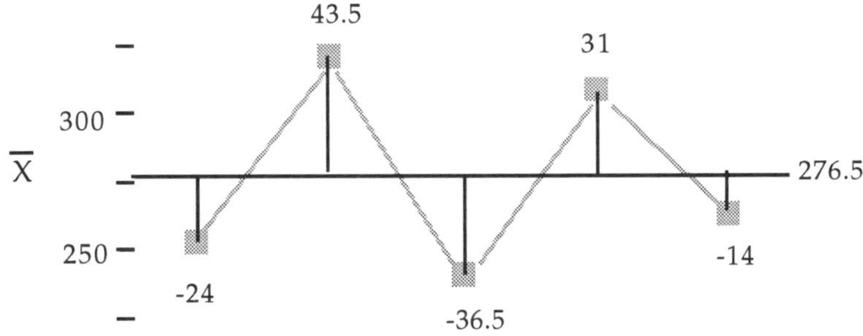

Figure 3.2: Deviations from Grand Average for Coating Weight Data

Squaring these deviations, multiplying by the size of each subgroup, and summing the products we get the Sum of Squares Between Subgroups:

$$\begin{aligned} SSB &= \{ 4(-24)^2 + 4(43.5)^2 + 4(-36.5)^2 + 4(31)^2 + 4(-14)^2 \} \\ &= 4 \{ 576 + 1892.25 + 1332.25 + 961 + 196 \} \\ &= 19{,}830.0 \end{aligned}$$

which is the quantity from the ANOVA Table that is inflated whenever the subgroup averages are detectably different.

So while ANOM focuses on ***each*** subgroup's individual deviation from the Grand Average, ANOVA accumulates and quantifies the total amount by which ***all*** of the subgroup averages deviate from the Grand Average. While ANOM seeks to identify which subgroup averages differ from the Grand Average, ANOVA measures the overall tendency of the subgroup averages to deviate from the Grand Average.

In the following exercise, and occasionally throughout the rest of this text, the reader will see a bar above the final digit in a number. This is a standard notation for a repeating digit. Thus, the first subgroup average in Exercise 3.4 is actually 36.6666666... This excessive precision is useful in computing the quantities in the ANOVA table since some of them can be quite sensitive to round-off errors.

EXERCISE 3.4: ANOVA for the Reflectivity Data:

Obtain the One-Way ANOVA table for these data:

Subgroup	1	2	3	4	5	6	7	8	9	10
Concentration	5	5	5	5	5	10	10	10	10	10
Temperature	75	100	125	150	175	75	100	125	150	175
Measured	35	31	30	28	19	38	36	39	35	30
Reflectivity	39	37	31	20	18	46	44	32	47	38
	36	36	33	23	22	41	39	38	40	31
Averages	$36.\bar{6}$	$34.\bar{6}$	$31.\bar{3}$	$23.\bar{6}$	$19.\bar{6}$	$41.\bar{6}$	$39.\bar{6}$	$36.\bar{3}$	$40.\bar{6}$	$33.\bar{0}$
s^2 values	$4.\bar{3}$	$10.\bar{3}$	$2.\bar{3}$	$16.\bar{3}$	$4.\bar{3}$	$16.\bar{3}$	$16.\bar{3}$	$14.\bar{3}$	$36.\bar{3}$	$19.\bar{0}$

The Grand Average is $33.7\bar{3}$. The average s^2 value is $14.\bar{0}$.

Source	Sums of Squares	degrees of freedom	Mean Squares	F-ratio
Between				
Within				
Total				

EXERCISE 3.5: Tukey's Post Hoc Test for the Reflectivity Data:

Perform Tukey's Post Hoc Test Using the Reflectivity Data. ($\alpha = 0.10$).

Analysis of Variance Worksheet

Data Set Name ______________________________ k=____ n=____

The values that are enclosed in boxes will be inserted into the ANOVA table.

I. TOTAL or OVERALL VARIATION

A. Find s^2 using all nk observations = s^2 = ____________

B. Check Value = Grand Average = $\bar{\bar{X}}$ = ____________

C. Total Degrees of Freedom = (nk - 1) = [ ]

D. Total Sum of Squares = (nk-1) s^2 = TSS = [ ].

II. WITHIN-SUBGROUP VARIATION

A. Obtain the Subgroup Average and Subgroup Variance for each of the k subgroups:

Subgroup Averages, $\bar{X}$			Subgroup Variances, s^2		
1____	6____	11____	1____	6____	11____
2____	7____	12____	2____	7____	12____
3____	8____	13____	3____	8____	13____
4____	9____	14____	4____	9____	14____
5____	10____	15____	5____	10____	15____

B. Average these s^2 values to get the Mean Square Within = $\overline{s^2}$ =

MSW = [ ]

C. Within-Subgroup Degrees of Freedom =

k (n - 1) = [ ]

D. Within-Subgroup Sum of Squares =

k(n-1) MSW = SSW = [ ].

III. BETWEEN-SUBGROUP VARIATION

A. Calculate s^2 using the k subgroup averages = $s_{\bar{X}}^2$ = ____________

B. Check Value = Grand Average = $\bar{\bar{X}}$ = ____________

(Does this value match the value in I. B. above? If not, you have made a mistake.)

C. Find the Mean Square Between by multiplying the $s_{\bar{X}}^2$ value in III.A. by n =

n $s_{\bar{X}}^2$ = MSB = [ ]

D. Between-Subgroup Degrees of Freedom =

(k - 1) = [ ]

E. Between-Subgroup Sum of Squares =

(k-1) MSB = SSB = [ ]

(Pythagoras claims that (III.E.) + (II.D.) = (I.D.). If they don't, guess who's wrong.)

EXERCISE 3.6: ANOVA and Tukey Post Hoc for the Steel Parts Data:

Obtain the One-Way ANOVA and a Tukey Post Hoc Analysis for these data:

Subgroup	1	2	3	4	5	6	7	8	9	10	11	12
Heat Treat	W	W	W	W	W	W	W	W	W	W	W	W
Machine	A	B	C	D	A	B	C	D	A	B	C	D
Time	8	8	8	8	11	11	11	11	3	3	3	3
Lengths:	6	7	1	6	6	8	3	7	5	10	-1	10
	9	9	2	6	3	7	2	9	4	11	2	5
	1	5	0	7	1	4	1	11	9	6	6	4
	3	5	4	3	-1	8	0	6	6	4	1	8
Averages	4.75	6.5	1.75	5.5	2.25	6.75	1.5	8.25	6.0	7.75	2.0	6.75
s^2 values	12.25	$3.\bar{6}$	$2.91\bar{6}$	3.0	$8.91\bar{6}$	$3.58\bar{3}$	$1.\bar{6}$	$4.91\bar{6}$	$4.\bar{6}$	$10.91\bar{6}$	$8.\bar{6}$	$7.58\bar{3}$

Subgroup	13	14	15	16	17	18	19	20	21	22	23	24
Heat Treat	L	L	L	L	L	L	L	L	L	L	L	L
Machine	A	B	C	D	A	B	C	D	A	B	C	D
Time	8	8	8	8	11	11	11	11	3	3	3	3
Lengths:	4	6	-1	4	3	6	2	9	6	8	0	4
	6	5	0	5	1	4	0	4	0	7	-2	3
	0	3	0	5	1	1	-1	6	3	10	4	7
	1	4	1	4	-2	3	1	3	7	0	-4	0
Averages	2.75	4.5	0.0	4.5	0.75	3.5	0.5	5.5	4.0	6.25	-0.5	3.5
s^2 values	$7.58\bar{3}$	$1.\bar{6}$	$0.\bar{6}$	$0.\bar{3}$	4.25	$4.\bar{3}$	$1.\bar{6}$	7.0	10.0	$18.91\bar{6}$	$11.\bar{6}$	$8.\bar{3}$

The grand average is 3.958. The average s^2 value is 6.2153.

Source	Sums of Squares	d.f.	Mean Squares	F-ratios
Between				
Within				
Total				

Analysis of Variance Worksheet

Data Set Name ______________________________ k=____ n=____

The values that are enclosed in boxes will be inserted into the ANOVA table.

I. TOTAL or OVERALL VARIATION

A. Find s^2 using all nk observations = s^2 = ____________

B. Check Value = Grand Average = $\bar{\bar{X}}$ = ____________

C. Total Degrees of Freedom = (nk - 1) = [ ]

D. Total Sum of Squares = (nk-1) s^2 = TSS = [ ].

II. WITHIN-SUBGROUP VARIATION

A. Obtain the Subgroup Average and Subgroup Variance for each of the k subgroups:

Subgroup Averages, $\bar{X}$			Subgroup Variances, s^2		
1____	6____	11____	1____	6____	11____
2____	7____	12____	2____	7____	12____
3____	8____	13____	3____	8____	13____
4____	9____	14____	4____	9____	14____
5____	10____	15____	5____	10____	15____

B. Average these s^2 values to get the Mean Square Within = $\overline{s^2}$ =

MSW = [ ]

C. Within-Subgroup Degrees of Freedom =

k (n - 1) = [ ]

D. Within-Subgroup Sum of Squares =

k(n-1) MSW = SSW = [ ].

III. BETWEEN-SUBGROUP VARIATION

A. Calculate s^2 using the k subgroup averages = $s_{\bar{x}}^2$ = ____________

B. Check Value = Grand Average = $\bar{\bar{X}}$ = ____________

(Does this value match the value in I. B. above? If not, you have made a mistake.)

C. Find the Mean Square Between by multiplying the $s_{\bar{x}}^2$ value in III.A. by n =

n $s_{\bar{x}}^2$ = MSB = [ ]

D. Between-Subgroup Degrees of Freedom =

(k - 1) = [ ]

E. Between-Subgroup Sum of Squares =

(k-1) MSB = SSB = [ ]

(Pythagoras claims that (III.E.) + (II.D.) = (I.D.). If they don't, guess who's wrong.)

EXERCISE 3.7: The Tensile Strength Data:

The effect of Catalyst, Curing Time, and Curing Temperature upon the Tensile Strength of a particular rubber compound was examined by means of the following experiment. The measurements were made to the nearest 50 psi, and expressed in units of 10 psi (that is, 11750 psi is recorded as 1175 below). Five pieces were measured for each of the eight treatment combinations.

Subgroup	1	2	3	4	5	6	7	8
Catalyst	1%	1%	1%	1%	2%	2%	2%	2%
Time	2m	2m	6m	6m	2m	2m	6m	6m
Temp	100	120	100	120	100	120	100	120
Tensile	950	1130	1150	1090	1180	1055	1190	990
Strengths	1065	1175	1165	1150	1175	1100	1185	1015
	970	1160	1125	1185	1180	1110	1185	940
	995	1165	1125	1170	1195	1135	1200	980
	1010	1170	1190	1165	1190	1120	1210	990
Averages	998	1160	1151	1152	1184	1104	1194	983
Ranges	115	45	65	95	20	80	25	75
s^2 values	1932.5	312.5	767.5	1357.5	67.5	917.5	117.5	745.0

Find the One-Way ANOVA table for these data.
Perform a Tukey Post Hoc Test for these data.
Find the One-Way ANOM chart for these data.
Find the Main-Effect ANOM charts for these data.

Source	Sums of Squares	d.f.	Mean Squares	F-ratios
Between				
Within				
Total				

Analysis of Variance Worksheet

Data Set Name ______________________________ k=____ n=____

The values that are enclosed in boxes will be inserted into the ANOVA table.

I. TOTAL or OVERALL VARIATION

A. Find s^2 using all nk observations = s^2 = __________

B. Check Value = Grand Average = $\bar{\bar{X}}$ = __________

C. Total Degrees of Freedom = (nk - 1) = [ ]

D. Total Sum of Squares = (nk-1) s^2 = TSS = [ ].

II. WITHIN-SUBGROUP VARIATION

A. Obtain the Subgroup Average and Subgroup Variance for each of the k subgroups:

Subgroup Averages, $\bar{X}$			Subgroup Variances, s^2		
1____	6____	11____	1____	6____	11____
2____	7____	12____	2____	7____	12____
3____	8____	13____	3____	8____	13____
4____	9____	14____	4____	9____	14____
5____	10____	15____	5____	10____	15____

B. Average these s^2 values to get the Mean Square Within = $\overline{s^2}$ =

MSW = [ ]

C. Within-Subgroup Degrees of Freedom =

k (n - 1) = [ ]

D. Within-Subgroup Sum of Squares =

k(n-1) MSW = SSW = [ ].

III. BETWEEN-SUBGROUP VARIATION

A. Calculate s^2 using the k subgroup averages = $s_{\bar{x}}^2$ = __________

B. Check Value = Grand Average = $\bar{\bar{X}}$ = __________

(Does this value match the value in I. B. above? If not, you have made a mistake.)

C. Find the Mean Square Between by multiplying the $s_{\bar{x}}^2$ value in III.A. by n =

n $s_{\bar{x}}^2$ = MSB = [ ]

D. Between-Subgroup Degrees of Freedom =

(k - 1) = [ ]

E. Between-Subgroup Sum of Squares =

(k-1) MSB = SSB = [ ]

(Pythagoras claims that (III.E.) + (II.D.) = (I.D.). If they don't, guess who's wrong.)

One Way ANOM for Tensile Strength Data: Main Effects ANOMs:

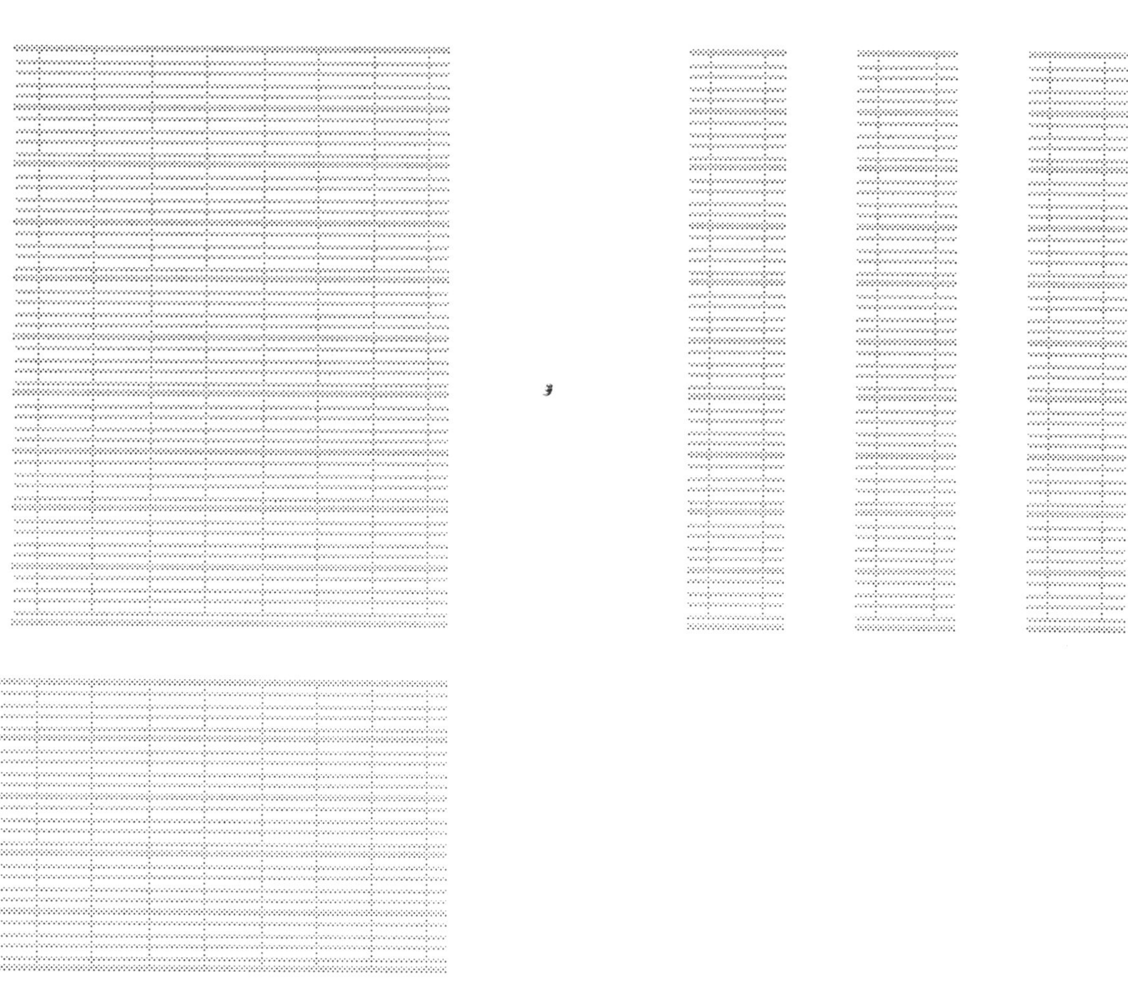

ANOM Plots for Tensile Strength Data

Analysis of Means Worksheet

A. BASIC INFORMATION:

Data Set Name ______________________ k = ____ n = ____

Grand Average = $\bar{\bar{X}}$ = [ ] Average Range = $\bar{R}$ = __________

B. FIND THE CONTROL LIMITS FOR THE RANGE CHART: (Table C)

$LCL_R = D_3\bar{R}$ = $UCL_R = D_4\bar{R}$ =

Are any subgroup ranges outside these limits?

C. ESTIMATE THE STANDARD DEVIATION OF X: (Using k = ____ subgroups of size n = ____)

Use Table B to find: ν = ____ d_2^* = ______

$$\text{EST. SD(X)} = \frac{\bar{R}}{d_2^*} = \text{\_\_\_\_\_\_\_\_} = \text{\_\_\_\_\_\_\_\_\_\_}$$

D. ESTIMATE THE STANDARD DEVIATION OF THE SUBGROUP AVERAGES:

$$\text{EST. SD}(\bar{X}) = \frac{\text{EST. SD(X)}}{\sqrt{n}} = \text{\_\_\_\_\_\_\_\_} = \boxed{}$$

The degrees of freedom for this estimate is the same as in Step C above.

E. FIND THE VALUE FOR H:

Choose α = ____. From Step C above, ν = ____.

The number of subgroup averages being compared is k = ____.

From Table D, the value for H is = [ ]

F. FIND THE DECISION LIMITS FOR THE ANOM CHART:

The decision limits for the ANOM chart are $\bar{\bar{X}} \pm H\left[\text{EST. SD}(\bar{X})\right]$:

$UDL_{\bar{X}}$ =

$LDL_{\bar{X}}$ =

CHAPTER FOUR

CONTRASTS

The One-Way ANOVA does not make specific comparisons between the subgroup averages. It simply compares two estimates of the variance of X. If these two estimates of V(X) are inconsistent, then there are some differences between subgroup averages that are too large to be reasonably attributed to chance variation. Thus, while the One-Way ANOVA does detect excessive variation on the part of the subgroup averages, it does not identify just which subgroup averages differ from the rest.

Both ANOM and Control Charts will identify subgroup averages that are detectably different from the grand average. Tukey's Post Hoc Test allows direct comparisons between two subgroup averages. A technique for making general comparisons between subgroup averages will be outlined in this chapter.

4.1 Comparisons and Contrasts

Whenever two things are compared it is natural to end up with some statement of the form "A is greater than B, A is equal to B, or A is less than B." Thus, a ***comparison*** can be said to create an order relationship between two things or between two groups of things.

A ***contrast*** is just a special way of making a comparison. A contrast compares two things or groups of things by comparing their *difference* with *zero*. Thus, while a comparison places two things in an order relationship, a contrast places the difference of two things in an order relationship with zero. A contrast will express the relation *A is greater than B* as *(A - B) is* *greater than zero*. Likewise, *A is less than B* becomes *(A - B) is less than zero*, while *A equals B* becomes *(A - B) equals zero*. This may seem a bit awkward, but it has one advantage. By fixing one end of the comparison at zero, we reduce the task of making a comparison to that of computing a difference. Contrasts are therefore easier to work with than comparisons, but they are more awkward to express in English. For this reason, we shall speak in terms of comparisons, but work with the corresponding contrasts.

Given a set of k objects to be compared, such as k subgroup averages, a contrast is defined on this set of k objects by any set of k coefficients $\{ c_1, c_2, \dots , c_j, \dots, c_k \}$ which sum to zero. This is the one condition required in order for a set of coefficients to define a contrast. Other conditions will create special types of contrasts, but this one condition is basic.

Letting C denote a contrast, the algebraic expression of a contrast is

$$C = c_1 \mu_1 + c_2 \mu_2 + \dots c_j \mu_j + \dots c_k \mu_k$$

or

$$C = \sum_{j=1}^{k} c_j \mu_j$$

subject to the condition that

$$\sum_{j=1}^{k} c_j = 0.$$

In defining a comparison in terms of a contrast there are two things to keep in mind. First of all, those subgroups that are assigned a coefficient of zero will be excluded from the comparison. Secondly, it is the *signs* of the coefficients that actually define the comparison, rather than the values chosen. The absolute values of the coefficients can be anything as long as the sum of the coefficients is zero.

EXAMPLE 4.1: Contrasts for the Lab Comparison Data:

Four laboratories are compared in their ability to perform a particular test.
Twenty samples are prepared from one master batch.
Five samples are sent at random to each lab for analysis.
The response is the percentage of dissolved potassium.
Labs A and B are internal labs, while labs C and D are external labs.

Lab	*A*	*B*	*C*	*D*
Values	55.9	58.7	60.7	62.7
	56.1	61.4	60.3	64.5
	57.3	60.9	60.9	63.1
	55.2	59.1	61.4	59.2
	58.1	58.2	62.3	60.3
Averages	56.52	59.66	61.12	61.96
s^2	1.352	1.983	0.592	4.668

Source	*SS*	*DF*	*MS*
Labs	85.9255	3	28.64
Within	34.38	16	2.15
Total	120.3055	19	

The F-ratio is 13.33, which exceeds the 0.05 critical value of 3.24.

Among the many possible comparisons that could be defined, three of particular interest are:
Compare internal labs, A vs B,
Compare external labs, C vs D,
Compare internal and external labs, (A, B) vs (C,D)

Coefficients that define these contrasts are given in the table below:

Lab:	A	B	C	D
Contrast	c_1	c_2	c_3	c_4
A vs B	1	-1	0	0
C vs D	0	0	1	-1
A,B vs C,D	1	1	-1	-1

In each case, it is the signs of the coefficients that define the comparison, and the magnitudes of the coefficients are chosen so that the sum of the coefficients will be zero. When a particular laboratory was not involved in the comparison of interest, it was assigned a coefficient of zero.

Contrasts are very versatile. Any comparison that is meaningful for a set of subgroup averages can be expressed as a contrast. The key to effectively using contrasts is the ability to translate comparisons into corresponding sets of coefficients and back again. This ability comes only with practice.

EXAMPLE 4.2: *Contrasts for the Coating Weight Data:*

Assume that the five treatments consisted of combinations of viscosities for the primer and the coating compound, as shown below.

Treatment	1	2	3	4	5
Primer Viscosity	20	20	25	30	30
Coating Viscosity	15	25	15	25	15

Contrasts of interest might include contrasts involving differences in primer viscosity and differences in coating viscosity.

Treatment	1	2	3	4	5
Primer Viscosity	20	20	25	30	30
Coating Viscosity	15	25	15	25	15
Contrasts	c_1	c_2	c_3	c_4	c_5
1. Primer, 20 v 30	-1	-1	0	1	1
2. Primer, 20 v 25	-1	-1	2	0	0
3. Primer, 25 v 30	0	0	-2	1	1
4. Coating, 15 v 25	-2	3	-2	3	-2

These contrasts are simply four contrasts that might be considered. They are not the only contrasts that could be defined. In fact, they are far from the best contrasts for these data. They will be discussed more completely in subsequent examples in Chapter Five.

The value of the coefficients can be adjusted to allow for the comparison of different sized groups. This can be seen in contrasts 2, 3 and 4 in Example 4.2 above. In each case, the coefficients sum to zero, so they define a contrast, but the coefficients do not all have the same value. The comparison being made is defined by the signs of the contrasts, and the values can be anything as long as their sum is zero. In fact, changing the values will not change the comparison. Contrast 4 above could be expressed as { -4, 6, -4, 6, -4 }. Moreover, there is only an interpretative difference between a contrast and its negative. Contrast 4 above is equivalent to { 2, -3, 2, -3, 2 }, although the interpretation would have to be switched to reflect the change in signs.

EXERCISE 4.1: Contrasts for the Reflectivity Data:

Use the worksheet below to define the following contrasts for the Reflectivity Data.

1. Compare Concentration levels, 5 gm/ltr versus 10 gm/ltr.

2. Compare Temperatures, 75 versus 100.

3. Compare Temperatures, (75 and 100) versus 125.

4. Compare Temperatures, (75, 100 and 125) versus 150.

5. Compare Temperatures, (75, 100, 125 and 150) versus 175.

Subgroup	1	2	3	4	5	6	7	8	9	10
Concentration	5	5	5	5	5	10	10	10	10	10
Temperature	75	100	125	150	175	75	100	125	150	175

Contrasts

1. Conc. 5 vs 10	__	__	__	__	__	__	__	__	__	__
2. Temp. 75 vs 100	__	__	__	__	__	__	__	__	__	__
3. Temp. 75,100 vs 125	__	__	__	__	__	__	__	__	__	__
4. Temp. 75 to 125 vs 150	__	__	__	__	__	__	__	__	__	__
5. Temp. 75 to 150 vs 175	__	__	__	__	__	__	__	__	__	__

4.2 Evaluating Contrasts

Given k subgroups, a contrast defined upon the subgroup means, $\{ \mu_1, \mu_2, \ldots, \mu_j, \ldots, \mu_k \}$, by the coefficients $\{ c_1, c_2, \ldots, c_j, \ldots, c_k \}$, will have the form

$$C = \sum_{j=1}^{k} c_j \mu_j$$

and the estimate of this contrast will be

$$\hat{C} = \sum_{j=1}^{k} c_j \bar{X}_j$$

where the values $\{ \bar{X}_1, \bar{X}_2, \ldots, \bar{X}_k \}$ are the averages of the k subgroups. (Note: In some books the estimated contrast is defined in terms of the subgroup totals instead of the subgroup averages, which affects all of the formulas connected with contrasts.)

Thus, the ***estimated value*** for any contrast will be found by multiplying each subgroup average by its corresponding coefficient and summing the resulting values.

EXAMPLE 4.3: Estimated Contrast Values for the Lab Comparison Data:

Lab:	*A*	*B*	*C*	*D*	
Averages:	*56.52*	*59.66*	*61.12*	*61.96*	$\hat{C}$
Contrast	c_1	c_2	c_3	c_4	
1. A vs B	*1*	*-1*	*0*	*0*	*-3.14*
2. C vs D	*0*	*0*	*1*	*-1*	*-0.84*
3. A,B vs C,D	*1*	*1*	*-1*	*-1*	*-6.90*

These values for $\hat{C}$ *are found as follows:*

$\hat{C}_1$ = 1 (56.52) + (-1) (59.66) + 0 (61.12) + 0 (61.96) = -3.14

$\hat{C}_2$ = 0 (56.52) + 0 (59.66) + 1 (61.12) + (-1) (61.96) = -0.84

$\hat{C}_3$ = 1 (56.52) + 1 (59.66) + (-1) (61.12) + (-1) (61.96) = -6.90

When calculating the estimated value for a contrast there are some tricks that can simplify the work and reduce errors. Perhaps the most common error in calculating the estimated value for a contrast is using the wrong sign in the accumulation. If a term is added when it should be subtracted it will have a profound effect upon the result. The simplest way to avoid this mistake is to make two passes through the data. Begin by adding all of the terms that are positive, and then go through and subtract off all those terms that are negative. This removes one of the sources of confusion when performing the calculations, and therefore reduces errors.

The second trick to calculating estimated values for contrasts requires a calculator with at least k memory registers. By storing the subgroup averages in the memory registers, one does not have to repeatedly key in these values. When using this technique, it is usually best to make only one pass through the data instead of two as outlined above. However, since one is freed from having to enter the values of the averages, one can concentrate upon using the correct sign in the accumulation.

One other common error in calculating estimated values for contrasts is a failure to notice a coefficient value other than 1 or -1. With the effort required to get the right average and the right sign, it is easy to overlook the value of the coefficient. Combined with the fact that most coefficients will have a value of ±1 for convenience, the effort required to bring all the elements together make it inevitable that this mistake will occur from time to time.

EXAMPLE 4.4: Estimated Contrasts for the Coating Weight Data:

Treatment	*1*	*2*	*3*	*4*	*5*	
Primer Viscosity	*20*	*20*	*25*	*30*	*30*	
Coating Viscosity	*15*	*25*	*15*	*25*	*15*	
AVERAGES	*252.5*	*320*	*240*	*307.5*	*262.5*	
Contrasts						$\hat{C}$
1. Primer, 20 v 30	*-1*	*-1*	*0*	*1*	*1*	*-2.5*
2. Primer, 20 v 25	*-1*	*-1*	*2*	*0*	*0*	*-92.5*
3. Primer, 25 v 30	*0*	*0*	*-2*	*1*	*1*	*90*
4. Coating, 15 v 25	*-2*	*3*	*-2*	*3*	*-2*	*372.5*

EXERCISE 4.2: Estimated Contrasts for the Reflectivity Data:

Find the estimated value for the following contrasts:

Subgroup	1	2	3	4	5	6	7	8	9	10	
Concentration	5	5	5	5	5	10	10	10	10	10	
Temperature	75	100	125	150	175	75	100	125	150	175	
Averages	$36.\bar{6}$	$34.\bar{6}$	$31.\bar{3}$	$23.\bar{6}$	$19.\bar{6}$	$41.\bar{6}$	$39.\bar{6}$	$36.\bar{3}$	$40.\bar{6}$	33.0	$\hat{C}$
Contrasts											
1. 5 vs 10	1	1	1	1	1	-1	-1	-1	-1	-1	______
2. 75 vs 100	1	-1	0	0	0	1	-1	0	0	0	______
3. 75,100 vs 125	1	1	-2	0	0	1	1	-2	0	0	______
4. 75 to 125 vs 150	1	1	1	-3	0	1	1	1	-3	0	______
5. 75 to 150 vs 175	1	1	1	1	-4	1	1	1	1	-4	______

4.3 F-Ratios for Contrasts

Estimated contrast values that are close to zero are said to be nonsignificant. Estimated contrast values that are far from zero (in either direction) are said to be significant. This distinction between significant and nonsignificant contrasts will depend upon the F-ratios for each contrast. Of course, in order to compute an F-ratio for a contrast there must be a Mean Square for that contrast and a value for the degrees of freedom.

Since contrasts are comparisons between two groups of subgroup means, *every contrast will have exactly one degree of freedom, $\nu = 1$.* Because of this, the Mean Square for a contrast will be identical to the Sum of Squares for that contrast:

$$MS(C) = SS(C)$$

The F-ratio for an estimated contrast will have 1 and k(n-1) degrees of freedom and will be equal to

$$\text{F-ratio} = \frac{MS(C)}{MSW} = \frac{SS(C)}{MSW}$$

Thus, one must find the Sum of Squares for a contrast before one can check a given contrast for significance.

The ***Sum of Squares for a contrast*** will be obtained by squaring the estimate of that contrast and dividing by an adjustment term.

$$SS(C) = \frac{(\hat{C})^2}{\text{Adjustment Term}}$$

The value of the ***adjustment term*** will depend upon the *coefficients of the contrast* and the *number of observations per subgroup.* In its most general form, the adjustment term can handle unequal subgroup sizes. This general form is

$$\text{Adjustment Term} = \sum_{j=1}^{k} \left(\frac{c_j^2}{n_j} \right)$$

When the subgroups are all the same size, the n_j values will all equal n, and the adjustment term will simplify to

$$\text{Adjustment Term} = \frac{1}{n} \sum_{j=1}^{k} c_j^2$$

EXAMPLE 4.5: *F-Ratios for the Lab Comparison Data:*

Each of the four labs had n = 5, so the adjustment term for the first contrast is

$$\frac{1}{5}(1 + 1 + 0 + 0) = 0.4.$$

The adjustment term for the second contrast is

$$\frac{1}{5}(0 + 0 + 1 + 1) = 0.4.$$

And the adjustment term for the third contrast is

$$\frac{1}{5}(1 + 1 + 1 + 1) = 0.8.$$

Squaring $\hat{C}$ and dividing by the adjustment term gives SS(C) .

Lab:	A	B	C	D			
Averages:	56.52	59.66	61.12	61.96	$\hat{C}$	SS(C)	F-ratio
Contrast	c_1	c_2	c_3	c_4			
1. A vs B	1	-1	0	0	-3.14	24.65	11.47*
2. C vs D	0	0	1	-1	-0.84	1.76	0.82
3. A,B vs C,D	1	1	-1	-1	-6.90	59.51	27.68*

The MSW value for these data is 2.15, with 16 d.f. (See Example 4.1, p.112)
The F-ratios are the SS(C) values divided by the MSW value.
Each of these F-ratios has 1 and 16 degrees of freedom.
The 0.05 critical value for the F-distribution with 1 and 16 d.f. is 4.49.

Thus, Contrast 1 and Contrast 3 are found to be significant.
An interpretation of these contrasts will be given later.

The reader may have noticed that the F-tests in the previous example used the 95th percentile rather than the 90th percentile as in previous chapters. This switch in the method of obtaining a decision rule is due to the simultaneous examination of several contrasts.

With the ANOM procedures, the Tukey Post Hoc Test, and the One-Way ANOVA the α-level specified was the theoretical α-level for the whole procedure (the overall α-level). When examining contrasts the stated α-level is the theoretical value for each individual F-ratio. Thus, when p different F-ratios are examined simultaneously, a theoretical overall α-level may be computed according to the Bonferroni Inequality:

$$1-(1-\alpha)^p \leq \alpha_{\text{overall}} \leq p\alpha.$$

Once again the reader is reminded that in practice the α-level does nothing more than quantify the way in which the experimenter selects his decision rule for separating signals from the noise, but since we are using the probability model as a guide for this choice, the theoretical relationship summarized by this inequality does suggest that the overall procedure is still *exploratory* even when each of the individual decision rules is based upon the more conservative $\alpha = 0.05$.

This practice of using the more conservative decision rules when examining a set of contrasts will be used in this text. This is merely a convention. The analysis is still exploratory, and the only confirmation of a real cause and effect relationship will still be the nontrivial replication of results. There is certainly no intent to imply that one can give a stated "confidence level" or "significance level" for the results of any experiment. Such theoretical values have no well-defined meaning in practice.

EXAMPLE 4.6: F-Ratios for the Coating Weight Data:

From page 94 we see that $k = 5$, $n = 4$, and $MSW = 628.33$ with $\nu = 15$

Treatment	1	2	3	4	5				
Primer Viscosity	20	20	25	30	30				
Coating Viscosity	15	25	15	25	15				
AVERAGES	252.5	320	240	307.5	262.5	$\hat{C}$	Adj. Term	SS(C)	F-ratio
Contrasts	c_1	c_2	c_3	c_4	c_5				
1. Primer, 20 v 30	-1	-1	0	1	1	-2.5	1.0	6.25	0.01
2. Primer, 20 v 25	-1	-1	2	0	0	-92.5	1.5	5704.2	9.08*
3. Primer, 25 v 30	0	0	-2	1	1	90	1.5	5400.0	8.59*
4. Coating, 15 v 25	-2	3	-2	3	-2	372.5	7.5	18500.8	29.4*

The adjustment terms are:

$$\frac{(1 + 1 + 0 + 1 + 1)}{4} = 1.0$$

$$\frac{(1 + 1 + 4 + 0 + 0)}{4} = 1.5$$

$$\frac{(0 + 0 + 4 + 1 + 1)}{4} = 1.5$$

$$\frac{(4 + 9 + 4 + 9 + 4)}{4} = 7.5$$

The 0.05 critical value for the F-distribution with 1 and 15 d.f. is 4.54. Therefore, Contrast 1 is said to be nonsignificant, while Contrast 2, Contrast 3, and Contrast 4 are said to be significant.

EXERCISE 4.3: F-Ratios for the Reflectivity Data:

There were three observations per subgroup, and MSW = 14.00, with 20 d.f. Find the F-ratios for the 5 contrasts below.

Subgroup	1	2	3	4	5	6	7	8	9	10	
Concentration	5	5	5	5	5	10	10	10	10	10	
Temperature	75	100	125	150	175	75	100	125	150	175	
Reflectivities:	35	31	30	28	19	38	36	39	35	30	
	39	37	31	20	18	46	44	32	47	38	
	36	36	33	23	22	41	39	38	40	31	
Averages	$36.\bar{6}$	$34.\bar{6}$	$31.\bar{3}$	$23.\bar{6}$	$19.\bar{6}$	$41.\bar{6}$	$39.\bar{6}$	$36.\bar{3}$	$40.\bar{6}$	33.0	$\hat{C}$
Contrasts											
1. 5 vs 10	1	1	1	1	1	-1	-1	-1	-1	-1	$-45.\bar{3}$
2. 75 vs 100	1	-1	0	0	0	1	-1	0	0	0	4.0
3. 75,100 vs 125	1	1	-2	0	0	1	1	-2	0	0	$17.\bar{3}$
4. 75 to 125 vs 150	1	1	1	-3	0	1	1	1	-3	0	$27.\bar{3}$
5. 75 to 150 vs 175	1	1	1	1	-4	1	1	1	1	-4	74.0

Contrasts	Adj. Term	SS(C)	F-ratio
1. 5 vs 10	_____	_________	________
2. 75 vs 100	_____	_________	________
3. 75,100 vs 125	_____	_________	________
4. 75 to 125 vs 150	_____	_________	________
5. 75 to 150 vs 175	_____	_________	________

4.4 Interpreting Contrasts

Those contrasts which are "significant" are the best candidates for signals within the data. Therefore, the experimenter will interpret a significant contrast as if it represents a real, detectable difference between the averages of the two groups being compared. Of course, the only confirmation of such a real difference would be its persistence in future studies. But, given that a contrast is thought to represent a real difference, one will want to estimate the size of this difference. The ***Estimated Contrast Effect*** will estimate the difference between the averages of the two groups compared by the contrast.

Since a contrast is a linear combination of the subgroup averages, one common symbol for contrasts is the letter *"l."* This symbol will be used here with the contrast effect. Denote the ***Estimated Contrast Effect*** by the symbol $\hat{l}$. This value is found by dividing the *estimated contrast,* $\hat{C}$, by the *sum of the positive coefficients for that contrast.*

$$\hat{l} = \frac{\hat{C}}{\text{sum of positive } c_j\text{'s}} .$$

In essence, $\hat{l}$ = the difference between the averages of the two groups being compared. If the "groups" consist of single subgroup averages, then $\hat{l}$ = the difference between these two averages. When the two "groups" consist of collections of subgroups, $\hat{l}$ = the difference between the average of the first group and the average of the second group. The sign of $\hat{l}$ will indicate which group has the greatest average response.

EXAMPLE 4.7: Interpreting Significant Contrasts for the Lab Comparison Data:

Lab:	*A*	*B*	*C*	*D*		
Averages:	*56.52*	*59.66*	*61.12*	*61.96*		
Contrast					$\hat{C}$	$\hat{l}$
1. A vs B	*1*	*-1*	*0*	*0*	*-3.14*	*-3.14*
2. C vs D	*0*	*0*	*1*	*-1*	*-0.84*	
3. A,B vs C,D	*1*	*1*	*-1*	*-1*	*-6.90*	*-3.45*

The first contrast indicates a detectable difference between Lab A and Lab B: Lab B averages 3.14 units higher than Lab A on this test.

There is insufficient evidence to detect any difference that might exist between the external labs.

The third contrast indicates a detectable difference between the internal labs and the external labs: The external labs average 3.45 units higher than the internal labs on this test. Notice that this is 3.45 units higher than the average *of lab A and lab B.*

EXAMPLE 4.8: Interpreting Significant Contrasts for the Coating Weight Data:

Treatment	1	2	3	4	5		
Primer Viscosity	20	20	25	30	30		
Coating Viscosity	15	25	15	25	15		
AVERAGES	252.5	320	240	307.5	262.5		
Contrasts						$\hat{C}$	$\hat{l}$
1. *Primer, 20 v 30*	-1	-1	0	1	1	-2.5	
2. *Primer, 20 v 25*	-1	-1	2	0	0	-92.5	-46.25
3. *Primer, 25 v 30*	0	0	-2	1	1	90	45
4. *Coating, 15 v 25*	-2	3	-2	3	-2	372.5	62.08

There is insufficient evidence to detect a difference in coating weight between Primer viscosities of 20 and 30.
Contrast 2 suggests that the coating weights average 46 grams higher when the Primer viscosity is 20 instead of 25.
Contrast 3 suggests that the coating weights average 45 grams higher when the Primer viscosity is 30 instead of 25.
Contrast 4 suggests that the coating weight averages 62 grams higher when the Coating viscosity is 25 instead of 15.
As always, one must reconcile the interpretations of these contrasts with prior knowledge about the process and previous experience with the process. Results that contradict prior knowledge will always require further investigation before they will be accepted.

4.5 Working With Contrasts

There are three basic steps for working with contrasts:

(1) define the contrasts,

(2) find $\hat{C}$, SS(C), and the F-ratio for the contrast, and

(3) find $\hat{l}$ for significant contrasts and interpret in the language of the problem.

The details of each of these basic steps are summarized on page 340 of the Appendix under the heading "Working With Contrasts." The reader should examine this review sheet now so as to be able to use it effectively later.

Contrasts are part of the foundation of the Analysis of Variance. They are used throughout the remainder of this book. This was not done because contrasts represent the best way of performing ANOVA computations. When it comes to computation, contrasts are very pedestrian. Many other techniques provide easier computations. But, given today's software and calculators, few will ever need to do the computations by hand. Rather, contrasts are used throughout the remainder of this book because they provide a powerful unifying theme for the many different facets of ANOVA. Through an understanding of contrasts the reasons behind many different calculations become obvious, and the logic of different approaches becomes clear.

EXERCISE 4.4: Contrasts for the Steel Parts Data:

Twenty-four treatments were studied for their effect upon the length of some parts. Each treatment consisted of different combinations of three factors.
The type of Heat Treatment had two levels in this study (W and L).
The Machine on which the parts were processed had four levels (A, B, C, and D).
The Time of Day had three levels (8 am, 11 am, and 3 pm).
Four parts were produced under each treatment combination, giving a total of 96 parts.

Subgroup	1	2	3	4	5	6	7	8	9	10	11	12
Heat Treat	W	W	W	W	W	W	W	W	W	W	W	W
Machine	A	B	C	D	A	B	C	D	A	B	C	D
Time	8	8	8	8	11	11	11	11	3	3	3	3
Contrasts:												
Contrast 1	1	1	1	1	1	1	1	1	1	1	1	1
Contrast 2	1	0	-1	0	1	0	-1	0	1	0	-1	0
Contrast 3	0	1	0	-1	0	1	0	-1	0	1	0	-1
Contrast 4	1	-1	1	-1	1	-1	1	-1	1	-1	1	-1
Contrast 5	1	1	1	1	-1	-1	-1	-1	0	0	0	0
Contrast 6	1	1	1	1	1	1	1	1	-2	-2	-2	-2
Averages	4.75	6.5	1.75	5.5	2.25	6.75	1.5	8.25	6.0	7.75	2.0	6.75

Subgroup	13	14	15	16	17	18	19	20	21	22	23	24
Heat Treat	L	L	L	L	L	L	L	L	L	L	L	L
Machine	A	B	C	D	A	B	C	D	A	B	C	D
Time	8	8	8	8	11	11	11	11	3	3	3	3
Contrasts:												
Contrast 1	-1	-1	-1	-1	-1	-1	-1	-1	-1	-1	-1	-1
Contrast 2	1	0	-1	0	1	0	-1	0	1	0	-1	0
Contrast 3	0	1	0	-1	0	1	0	-1	0	1	0	-1
Contrast 4	1	-1	1	-1	1	-1	1	-1	1	-1	1	-1
Contrast 5	1	1	1	1	-1	-1	-1	-1	0	0	0	0
Contrast 6	1	1	1	1	1	1	1	1	-2	-2	-2	-2
Averages	2.75	4.5	0.0	4.5	0.75	3.5	0.5	5.5	4.0	6.25	-0.5	3.5

MSW is 6.2153, with 72 d.f.

(a) Identify the comparison made by each of these six contrasts.
(b) Find the F-ratio for each of these six contrasts.
(c) Interpret any significant contrasts found.

	$\hat{C}$	Adj. Term	SS(C)	F-ratio
Contrast 1				
Contrast 2				
Contrast 3				
Contrast 4				
Contrast 5				
Contrast 6				

CHAPTER FIVE

MULTIFACTOR ANOVA

The Analysis of Variance for multifactor studies will be presented in this chapter. This is done with the intent of creating a foundation for understanding Fractional Factorial Studies and Screening Designs, rather than presenting the many facets of multifactor studies. To this end, contrasts are used rather than the traditional approach. This use of contrasts with multifactor studies is somewhat pedestrian, but since contrasts will be an important part of subsequent topics, their use with Multifactor ANOVA will facilitate these later topics. Moreover, with the availability of software for ANOVA computations, it seems more important for a text to create insight than to give the most efficient computational formulas, and contrasts offer this insight.

5.1 Partitioning the Between-Subgroup Sum of Squares

The Analysis of Variance is based upon successive partitions of the Total Sum of Squares. The first division is that seen in the One-Way ANOVA:

$$\text{Total SS} = \text{Between SS} + \text{Within SS}.$$

The very term "one-way" refers to this one division of Total SS into two portions. The next step in this process is the division of the Between-Subgroup SS into identifiable components. Of course this division will depend upon the structure present in the data. The number of factors and the relationships between the factor levels in the study will determine just what subdivisions of the Between-Subgroup SS will be of interest. At the same time, the degrees of freedom will determine how many independent subdivisions can be made, since each subdivision will require at least one degree of freedom. Thus, the Between-Subgroup SS can be divided into at most (k-1) independent portions.

Special sets of contrasts will accomplish this subdivision of the Between-Subgroup SS.

EXAMPLE 5.1: A Decomposition for the Lab Comparison Data:

Lab	*A*	*B*	*C*	*D*
Values	55.9	58.7	60.7	62.7
	56.1	61.4	60.3	64.5
	57.3	60.9	60.9	63.1
	55.2	59.1	61.4	59.2
	58.1	58.2	62.3	60.3
Averages	56.52	59.66	61.12	61.96
s^2	1.352	1.983	0.592	4.668

One-Way ANOVA

Source	*SS*	*DF*	*MS*
Between	85.9255	3	28.64
Within	34.38	16	2.15
Total	120.3055	19	

The F-ratio is 13.33, which exceeds the 0.05 critical value of 3.24.

The contrasts defined in Chapter 4 have the following SS(C) values: (See Example 4.5, p.118)

Lab:	*A*	*B*	*C*	*D*	$\hat{C}$	SS(C)
Contrast	c_1	c_2	c_3	c_4		
A vs B	1	-1	0	0	-3.14	24.649
C vs D	0	0	1	-1	-0.84	1.764
A,B vs C,D	1	1	-1	-1	-6.90	59.5125
						85.9255

The sum of these SS(C) values is 85.9255, which equals the Between-Subgroup SS.

Thus, these three contrasts decompose the Between-Subgroup SS into single degree-of-freedom components, and the ANOVA table can be re-written as:

Source	*SS*	*DF*	*MS*	*F*
A vs B	24.65	1	24.65	11.47
C vs D	1.76	1	1.76	0.82
A,B vs C,D	59.51	1	59.51	27.68
Within	34.38	16	2.15	
Total	120.30	19		

The decomposition of the Between-Subgroup Sum of Squares in this example is due to a special property of the *set* of contrasts. Without this special property the sums of squares for the contrasts would not have added up to equal the Sum of Squares Between. With this special property, you will always get a decomposition of the Between-Subgroup Sum of Squares. A set of contrasts with this property is said to be a ***maximal mutually orthogonal set of contrasts***. The following sections will explain and illustrate just what this phrase means.

5.2 Orthogonal Contrasts

The term "orthogonal" is a mathematical expression for the generalized notion of perpendicularity (which defines one long word with a longer one). For contrasts, the notion of orthogonality is fairly simple. Contrasts that are "orthogonal" will make comparisons that are independent of each other. In addition, the sums of squares for orthogonal contrasts will display the same type of independence that is displayed by the Between-Subgroup SS and the Within-Subgroup SS.

Two contrasts, defined on the same set of k subgroup means by

$$C = \sum_{j=1}^{k} c_j \mu_j$$

and

$$D = \sum_{j=1}^{k} d_j \mu_j$$

will be said to be ***orthogonal*** if and only if

$$\sum_{j=1}^{k} c_j d_j = 0.$$

In English, in order for two contrasts to be orthogonal, the products of their corresponding coefficients must sum to zero.*

If the sum of the products is zero, then either the products are all zero, or else the sum of the positive products is canceled by the sum of the negative products. In the latter case, the set of products $\{ c_1 d_1, \ldots, c_k d_k \}$ will define a contrast. It may be a new contrast, or it may be the positive or negative version of a contrast that has already been defined, but it will define a contrast. This result leads to a second way of defining the property of orthogonality.

Two contrasts, such as C and D above, will be said to be orthogonal if and only if their product is either a contrast or a vector of zero coefficients. (The notion of a product used here is that of vector multiplication where corresponding coefficients are multiplied together.)

The Reflectivity Data will be used to illustrate how one can check to see if two contrasts are orthogonal.

---

* $\sum_{j=1}^{k} \frac{c_j d_j}{n_j} = 0$ when the k subgroups are not all the same size.

EXAMPLE 5.2: *Orthogonal Contrasts for the Reflectivity Data:*

The contrasts defined earlier for the Reflectivity Data are given below.

Subgroup	1	2	3	4	5	6	7	8	9	10	
Concentration	5	5	5	5	5	10	10	10	10	10	
Temperature	75	100	125	150	175	75	100	125	150	175	
Averages	36.67	34.67	31.33	23.67	19.67	41.67	39.67	36.33	40.67	33.0	$\hat{C}$
Contrasts											
1. *5 vs 10*	1	1	1	1	1	-1	-1	-1	-1	-1	-45.33
2. *75 vs 100*	1	-1	0	0	0	1	-1	0	0	0	4.00
3. *75,100 vs 125*	1	1	-2	0	0	1	1	-2	0	0	17.33
4. *75 to 125 vs 150*	1	1	1	-3	0	1	1	1	-3	0	27.33
5. *75 to 150 vs 175*	1	1	1	1	-4	1	1	1	1	-4	74.00

To see if a pair of contrasts is orthogonal, multiply their corresponding coefficients together, and examine the result. If the result is a string of 0 coefficients, or if the result defines a contrast, then the original pair of contrasts is orthogonal.

											Orthogonal?
1. x 2. gives	1	-1	0	0	0	-1	1	0	0	0	*Yes*
1. x 3. gives	1	1	-2	0	0	-1	-1	2	0	0	*Yes*
1. x 4. gives	1	1	1	-3	0	-1	-1	-1	3	0	*Yes*
1. x 5. gives	1	1	1	1	-4	-1	-1	-1	-1	4	*Yes*

Thus, Contrast 1 is orthogonal to Contrasts 2, 3, 4 and 5.

2. x 3. gives	1	-1	0	0	0	1	-1	0	0	0	*Yes*
2. x 4. gives	1	-1	0	0	0	1	-1	0	0	0	*Yes*
2. x 5. gives	1	-1	0	0	0	1	-1	0	0	0	*Yes*

So, Contrast 2 is orthogonal to Contrasts 3, 4, and 5.

3. x 4. gives	1	1	-2	0	0	1	1	-2	0	0	*Yes*
3. x 5. gives	1	1	-2	0	0	1	1	-2	0	0	*Yes*
4. x 5. gives	1	1	1	-3	0	1	1	1	-3	0	*Yes*

So, Contrast 3 is orthogonal to Contrasts 4 and 5,
and Contrast 4 is orthogonal to Contrast 5.

EXAMPLE 5.3: Orthogonal and Nonorthogonal Contrasts for the Coating Weight Data:

The contrasts defined earlier for the Coating Weight Data are given below.

Treatment	1	2	3	4	5	
Primer Viscosity	20	20	25	30	30	
Coating Viscosity	15	25	15	25	15	
Contrasts						
1. PV, 20 v 30	-1	-1	0	1	1	
2. PV, 20 v 25	-1	-1	2	0	0	
3. PV, 25 v 30	0	0	-2	1	1	
4. CV, 15 v 25	-2	3	-2	3	-2	
						Orthogonal?
Product of 1. x 2.	1	1	0	0	0	No
Product of 1. x 3.	0	0	0	1	1	No
Product of 1. x 4.	2	-3	0	3	-2	Yes
Product of 2. x 3.	0	0	-4	0	0	No
Product of 2. x 4.	2	-3	-4	0	0	No
Product of 3. x 4.	0	0	4	3	-2	No

The reader should go over these examples until he is satisfied that he knows how to distinguish orthogonal pairs of contrasts from nonorthogonal pairs.

5.3 Sets of Mutually Orthogonal Contrasts

When every pair of contrasts in a set is an orthogonal pair, the set is said to be a ***mutually orthogonal set of contrasts***. The set of contrasts { C_1, C_2, C_3, C_4, C_5 } for the Reflectivity Data forms a mutually orthogonal set. This follows from that fact that in Example 5.2 each possible product was explicitly examined, and each contrast was found to be orthogonal to every other contrast.

On the other hand, the four contrasts for the Coating Weight Data shown in Example 5.3 do *not* form a mutually orthogonal set.

EXERCISE 5.1: Checking for Orthogonal Contrasts in the Tensile Strength Data:
The structure of the data for the Tensile Strength Data is shown below.
Define the contrasts indicated below, and check for orthogonality.

Subgroup	1	2	3	4	5	6	7	8	
Catalyst	1%	1%	1%	1%	2%	2%	2%	2%	
Time	2m	2m	6m	6m	2m	2m	6m	6m	
Temp	100	120	100	120	100	120	100	120	
Contrasts:									
1. Catalyst 1% v 2%	__	--	--	--	--	--	--	--	
2. Time 2m v 6 m	--	--	--	--	--	--	--	--	
3. Temp 100 v 120	--	--	--	--	--	--	--	--	
Are contrasts 1, 2, and 3 orthogonal to one another?									Orthogonal?
4. = 1. x 2.	--	--	--	--	--	--	--	--	-----
5. = 1. x 3.	--	--	--	--	--	--	--	--	-----
6. = 2. x 3.	--	--	--	--	--	--	--	--	-----

Do contrasts 1, 2, and 3 form a mutually orthogonal set?

At present, the only way to identify a mutually orthogonal set is by listing all possible products between contrasts in the set. This approach will become very tedious as the number of contrasts increases. Fortunately there are some shortcuts that will save a lot of this work. The first shortcut concerns sets of contrasts that have a special structure.

Given a mutually orthogonal set of contrasts { $C_1, C_2, \ldots, C_j$ } we may generally add products of these contrasts to this set without destroying the mutual orthogonality of the expanded set. That is, if { $C_1, C_2, \ldots, C_j$ } is mutually orthogonal, then the set defined by

$$\{ C_1, C_2, \ldots, C_j, C_1 x C_2, C_1 x C_3, \ldots, C_1 x C_j, \text{etc.} \}$$

will usually be a mutually orthogonal set.*

Likewise, in a multifactor study, if two factors are fully crossed, then the contrasts for one factor will be orthogonal to those for the other factor.

These shortcuts are very useful in identifying mutually orthogonal sets.

EXAMPLE 5.4: Mutually Orthogonal Sets for the Reflectivity Data:

Subgroup	1	2	3	4	5	6	7	8	9	10	
Concentration	5	5	5	5	5	10	10	10	10	10	
Temperature	75	100	125	150	175	75	100	125	150	175	
Averages	36.67	34.67	31.33	23.67	19.67	41.67	39.67	36.33	40.67	33.0	$\hat{C}$
Contrasts											
1. 5 vs 10	1	1	1	1	1	-1	-1	-1	-1	-1	-45.33
2. 75 vs 100	1	-1	0	0	0	1	-1	0	0	0	4.00
3. 75,100 vs 125	1	1	-2	0	0	1	1	-2	0	0	17.33
4. 75 to 125 vs 150	1	1	1	-3	0	1	1	1	-3	0	27.33
5. 75 to 150 vs 175	1	1	1	1	-4	1	1	1	1	-4	74.00
											Orthogonal?
6. 1. x 2. gives	1	-1	0	0	0	-1	1	0	0	0	Yes
7. 1. x 3. gives	1	1	-2	0	0	-1	-1	2	0	0	Yes
8. 1. x 4. gives	1	1	1	-3	0	-1	-1	-1	3	0	Yes
9. 1. x 5. gives	1	1	1	1	-4	-1	-1	-1	-1	4	Yes

The first five contrasts form a mutually orthogonal set.
Therefore, the set of contrasts { $C_1, C_2, \ldots, C_9$ } forms a mutually orthogonal set.

If we had to verify this by enumeration, there would be 36 pairs of contrasts to check.

EXERCISE 5.2: Sets of Contrasts for the Tensile Strength Data:

From the Tensile Strength Data, Exercise 5.1, does the set of contrasts { C_1, C_2, C_3, C_4 } form a mutually orthogonal set?

Are there larger mutually orthogonal sets that can be defined for these data?

* The exception would come when some contrasts have unequal weightings on one or both sides of their comparison, i.e., C = (1, 2, 1, -3, -1, 0) instead of C = (2, 2, 2, -3, -3, 0).

5.4 Maximal Mutually Orthogonal Sets of Contrasts

Given a mutually orthogonal set of contrasts defined on k subgroup means, the maximum number of contrasts that can be defined in that set of contrasts is equal to (k-1), which, by no accident, is exactly the number of degrees of freedom for the set of k subgroup averages. In other words, the maximum size for a mutually orthogonal set of contrasts defined on a set of subgroup means is equal to the Between-Subgroup Degrees of Freedom for that set of means. For k subgroups, this value will be (k-1). Mutually orthogonal sets of contrasts with fewer than (k-1) contrasts can have additional contrasts defined. A set with exactly (k-1) contrasts will be said to be a ***maximal mutually orthogonal set of contrasts.***

As shown in Example 5.1, p.126, a maximal mutually orthogonal set of contrasts will decompose the Between-Subgroup Sum of Squares into single degree of freedom components, and these (k-1) components will be independent of each other. This means that each component is uncontaminated by the others. If one contrast represents the effect of Factor A, and another contrast represents the effect of Factor B, then each of these two effects are estimated *without interference* from the other factor effect. This feature is one that will allow for very powerful experiments with very few observations.

EXERCISE 5.3: Question about the Reflectivity Data:

The set of contrasts defined as { $C_1, C_2, \dots, C_9$ } in Example 5.4 is a mutually orthogonal set. Is it a maximal mutually orthogonal set?

EXERCISE 5.4: Question about the Tensile Strength Data:

Is the set consisting of all six contrasts defined in Exercise 5.1 a mutually orthogonal set, and if so, then is it a maximal mutually orthogonal set?

Given that a maximal mutually orthogonal set will decompose the Between-Subgroup Sum of Squares, how can this information be put to use? A thorough answer to this question is the goal of the rest of this book. A brief answer will be given below.

The most obvious way that any set of contrasts can be used with ANOVA is in making specific comparisons. If these comparisons can be made within the context of a mutually orthogonal set of contrasts, then the comparisons will be made independently of each other. A maximal mutually orthogonal set of contrasts is a collection of such independent comparisons.

EXAMPLE 5.5: Single Degree of Freedom ANOVA for the Gauge Study Data:

Parts 1, 2 and 3 were produced on machine D on three different days, while parts 4 and 5 were produced on machine E on two different days.

A maximal mutually orthogonal set of contrasts for these data is given below. Contrast 1 represents the Operator Main Effect. Contrasts 2, 3, 4, and 5 represent the Parts Main Effect. (The Machine Main Effect is part of the Parts Main Effect.) Contrasts 6, 7, 8, and 9 represent the Operator by Parts Interaction Effect.

Subgroup	1	2	3	4	5	6	7	8	9	10		
Operator	A	A	A	A	A	B	B	B	B	B		
Part	1	2	3	4	5	1	2	3	4	5		
(machine)	D	D	D	E	E	D	D	D	E	E		
Values	20	20	25	50	45	20	15	15	45	35		
	15	25	25	50	40	20	10	10	20	40		
Averages	17.5	22.5	25.0	50.0	42.5	20.0	12.5	12.5	32.5	37.5		
Contrasts:											$\hat{C}$	*adj.tm.*
Contrast 1	1	1	1	1	1	-1	-1	-1	-1	-1	42.5	5
Contrast 2	1	-1	0	0	0	1	-1	0	0	0	2.5	2
Contrast 3	1	1	-2	0	0	1	1	-2	0	0	-2.5	6
Contrast 4	0	0	0	1	-1	0	0	0	1	-1	2.5	2
Contrast 5	2	2	2	-3	-3	2	2	2	-3	-3	-267.5	30
Contrast 6	1	-1	0	0	0	-1	1	0	0	0	-12.5	2
Contrast 7	1	1	-2	0	0	-1	-1	2	0	0	-17.5	6
Contrast 8	0	0	0	1	-1	0	0	0	-1	1	12.5	2
Contrast 9	2	2	2	-3	-3	-2	-2	-2	3	3	-27.5	30

The Single Degree of Freedom ANOVA table is

SOURCE	SS	DF	MS	F
Contrast 1	361.2500	1	361.25	9.32*
Contrast 2	3.1250	1	3.125	0.08
Contrast 3	1.0417	1	1.04	0.02
Contrast 4	3.1250	1	3.125	0.08
Contrast 5	2385.2083	1	2385.2	61.55*
Contrast 6	78.1250	1	78.125	2.01
Contrast 7	51.0417	1	51.04	1.32
Contrast 8	78.1250	1	78.125	2.01
Contrast 9	25.2083	1	25.21	0.65
Within	387.5000	10	38.75	
Total	3373.7500	19		

$F_{.95}(1,10) = 4.96$.
Contrasts 1 and 5 appear to represent signals.
This ANOVA Table is much more specific than the One-Way ANOVA on p.93.

For Contrast 1, $\hat{l}$ *= 8.5 ten-thousandths, which is the average difference between the two operator's measurements.*

For Contrast 5, $\hat{l}$ *= -22.3 ten-thousandths, which is the average difference between the parts from the two machines.*

Within each machine, there are no detectable day to day differences.

The use of a 0.05 α-level in the preceding example is due to the Bonferroni Inequality given in Chapter Four. The (theoretical) overall α-level for the set of 9 F-tests is computed to be somewhere between the values

$$1-(1-0.05)^9 = 0.37 \leq \text{overall } \alpha \leq 9(0.05) = 0.45.$$

Thus, while each F-test is performed using a 0.05 decision rule, the computed overall α-level for the set of F-tests is in the vicinity of 0.40, making this analysis much more of an exploratory analysis than it first appears to be.

EXERCISE 5.5: Estimated Contrast Values for the Reflectivity Data:

The One-Way ANOVA table for these data is

SOURCE	SS	DF	MS	F
Between	1403.8667	9	155.98	11.14*
Within	280.0000	20	14.0	
Total	1683.8667	29		

Find the missing values for $\hat{C}$ in the following table.

Subgroup	1	2	3	4	5	6	7	8	9	10	
Concentration	5	5	5	5	5	10	10	10	10	10	
Temperature	75	100	125	150	175	75	100	125	150	175	
Averages	$36.\bar{6}$	$34.\bar{6}$	$31.\bar{3}$	$23.\bar{6}$	$19.\bar{6}$	$41.\bar{6}$	$39.\bar{6}$	$36.\bar{3}$	$40.\bar{6}$	33.0	$\hat{C}$
Contrasts											
1. 5 vs 10	1	1	1	1	1	-1	-1	-1	-1	-1	$-45.\bar{3}$
2. 75 vs 100	1	-1	0	0	0	1	-1	0	0	0	4.0
3. 75,100 vs 125	1	1	-2	0	0	1	1	-2	0	0	$17.\bar{3}$
4. 75 to 125 vs 150	1	1	1	-3	0	1	1	1	-3	0	$27.\bar{3}$
5. 75 to 150 vs 175	1	1	1	1	-4	1	1	1	1	-4	74.0
6. 1. x 2. gives	1	-1	0	0	0	-1	1	0	0	0	____
7. 1. x 3. gives	1	1	-2	0	0	-1	-1	2	0	0	____
8. 1. x 4. gives	1	1	1	-3	0	-1	-1	-1	3	0	____
9. 1. x 5. gives	1	1	1	1	-4	-1	-1	-1	-1	4	____

EXERCISE 5.6: Single Degree of Freedom ANOVA for the Reflectivity Data:

Find the adjustment terms (remember, k = 10, n = 3).
Find the SS(C) values for each of the contrasts.
Fill in the ANOVA table below.

SOURCE	SS	DF	MS	F
Contrast 1	$616.533\overline{3}$	1	616.533	44.0
Contrast 2	12.0000	1	12.000	0.86
Contrast 3	$75.111\overline{1}$	1	75.111	5.37
Contrast 4	$93.388\overline{8}$	1	93.389	6.67
Contrast 5	410.7000	1	410.700	29.33
Contrast 6	--------	---	-------	-----
Contrast 7	--------	---	-------	-----
Contrast 8	--------	---	-------	-----
Contrast 9	--------	---	-------	-----
Within	--------	---	-------	
Total	$1683.866\overline{6}$	29		

Consider the maximal mutually orthogonal set of contrasts for the Reflectivity Data shown above. Contrast 1 makes the only interesting comparison between the two levels of Concentration. Contrasts 2 through 5 make comparisons between the different levels of temperature. Contrasts such as these, that make direct comparisons between the levels of *one* of the factors in a multifactor study, are traditionally said to represent ***main effects***.

Contrasts 6 through 9 above are products of the Concentration Contrast with one of the Temperature Contrasts. Contrasts such as these, that are expressed as the product of two or more main effect contrasts, are traditionally said to represent ***interaction effects***.

Two factors are said to interact when the effect of one factor upon the response changes depending on which level of the other factor is present. In the Reflectivity Data, the effect of temperature is moderated by the level of concentration. While in general, higher temperatures yield lower reflectivities, there is an increase in reflectivity as temperature goes from 125° to 150° at concentrations of 10 gms/ltr. (See Figure 2.11, p.74.) When two factors interact, it is impossible to talk about their effects upon the response separately. They must be considered in concert.

EXERCISE 5.7: Estimated Contrast Values for the Tensile Strength Data:

The Tensile Strength Data is a special type of fully crossed multifactor study; each factor has only two levels. This type of fully crossed study allows the inclusion of the maximum number of factors with the minimum number of subgroups.

The contrasts you defined in Exercise 5.1, p.130, should be similar to the following:

Subgroup	1	2	3	4	5	6	7	8	
Catalyst	1%	1%	1%	1%	2%	2%	2%	2%	
Time	2m	2m	6m	6m	2m	2m	6m	6m	
Temp	100	120	100	120	100	120	100	120	
Contrasts:									$\hat{C}$
1. Catalyst 1% v 2%	-1	-1	-1	-1	1	1	1	1	______
2. Time 2m v 6 m	-1	-1	1	1	-1	-1	1	1	______
3.Temp 100 v 120	-1	1	-1	1	-1	1	-1	1	______
4. = 1. x 2.	1	1	-1	-1	-1	-1	1	1	______
5. = 1. x 3.	1	-1	1	-1	-1	1	-1	1	______
6. = 2. x 3.	1	-1	-1	1	1	-1	-1	1	______

These contrasts form a mutually orthogonal set which is *not* maximal.
The missing contrast represents the three-factor interaction.
It can be found by multiplying 1 x 6, or 2 x 5, or 3 x 4.

7. = 1. x 2. x 3.	-1	1	1	-1	1	-1	-1	1	______

The subgroup averages for these data are:

998	1160	1151	1152	1184	1104	1194	983

Find the estimated contrast values.

EXERCISE 5.8: Single Degree of Freedom ANOVA for the Tensile Strength Data:

The Tensile Strength Data consists of k = 8 subgroups of size n = 5.

Since each of the contrasts in Exercise 5.7 contains 8 coefficients = ±1, the adjustment term will be the same for each contrast.

Find the adjustment term for the contrasts shown in Exercise 5.7.

Find the Single Degree of Freedom ANOVA table for the Tensile Strength Data.

SOURCE	SS	DF	MS	F
Contrast 1	_________	___	_______	_____
Contrast 2	_________	___	_______	_____
Contrast 3	_________	___	_______	_____
Contrast 4	_________	___	_______	_____
Contrast 5	_________	___	_______	_____
Contrast 6	_________	___	_______	_____
Contrast 7	_________	___	_______	_____
Within	24,870.0	32	777.1875	
Total	259,477.5	39		

This Single Degree of Freedom ANOVA is also called a Factorial ANOVA since each contrast uniquely represents a factor in the experiment or the interaction of factors in the experiment.

5.5 Factorial ANOVA

The traditional way of presenting the results of a multifactor study is a ***factorial ANOVA table.*** It is the purpose of this section to show just what the values in a factorial ANOVA table represent. Since the reader should now have a basic appreciation for contrasts, the factorial ANOVA will be presented from this perspective.

The traditional factorial ANOVA table exists only when the multifactor study is balanced and fully crossed. A study is said to be ***balanced*** when each subgroup has the same number of observations. In other words, equal subgroup sizes = balanced. A study is said to be ***fully crossed*** when each level of each factor in the study occurs in conjunction with every level of every other factor. That is, every combination of factor levels is present in the study. If Factor A has 4 levels, and Factor B has 3 levels, and Factor C has 2 levels, then a fully crossed study would have to have 4 x 3 x 2 = 24 subgroups, and each subgroup would have to represent a unique combination of factor levels. Anything less is called a partially crossed study.

The Tensile Strength Data, the Reflectivity Data, and the Steel Parts Data all represent balanced and fully crossed studies. The Gauge Study Data is balanced and fully crossed only in a special sense.

The Coating Weight Data, when considered as a factorial study (see Example 4.2, p.113) is balanced but only partially crossed. In order for it to be a fully crossed study, four observations would need to be obtained with Primer Viscosity at 25 and Coating Viscosity at 25.

Given a balanced and fully crossed study, the factorial ANOVA table may be obtained from the Single Degree of Freedom ANOVA table. The SS for the main effect contrasts for Factor A will be summed to get the Sum of Squares due to Factor A, SSA. This will be repeated for each factor in turn. Next all the SS for the contrasts derived from the products of a Factor A contrast and a Factor B contrast will be summed to get the SS for the AB interaction, SSAB. This is repeated for all pairs of factors in the study. This same process is then repeated for three-factor interactions, etc. The degrees of freedom are "pooled" in the same manner, and the result is the factorial ANOVA.

EXAMPLE 5.6: Factorial ANOVA for the Gauge Study Data:

Orthogonal contrasts for the Gauge Study Data were given in Example 5.5, p.133.
Contrast 1 represents the Operator Main Effect.
Contrasts 2, 3, 4, and 5 represent the Parts Main Effect.
Contrasts 6, 7, 8, and 9 represent the Operator by Parts Interaction Effect.
So the Factorial ANOVA table is

SOURCE	SS	DF	MS	F
Operators	361.2500	1	361.25	9.32*
Parts	2392.5000	4	598.125	15.4*
Oper. x Part	232.5000	4	58.125	1.50
Within	387.5000	10	38.75	
Total	3373.7500	19		

SS(Parts) = 3.125 + 1.0417 + 3.125 + 2385.2083 = 2392.5000
Parts d.f. = 1 + 1 + 1 + 1 = 4

SS(OxP) = 78.125 + 51.0417 + 78.125 + 25.2083 = 232.5000, with 4 d.f.

EXERCISE 5.9: Factorial ANOVA Table for the Reflectivity Data:

Find the values for the factorial ANOVA using the values from Exercise 5.6.

SOURCE	SS	DF	MS	F
Concentration	___________	___	_________	_______
Temperature	___________	___	_________	_______
Conc. x Temp	___________	___	_________	_______
Within	___________	___	_________	
Total	1683.8667	29		

5.6 The Invariance of Factorial ANOVA

As stated earlier, it is traditional to present the results of a balanced and fully crossed multifactor study in terms of the Factorial ANOVA Table. Yet, when compared to the Single Degree of Freedom ANOVA, the Factorial ANOVA Table can be much less specific. The F-ratios for a given main effect or a given interaction are only slightly more specific than the F-ratio for the One-Way ANOVA. If the ratio is significant, then that main effect or interaction can be said to cause detectable differences in the response variable, but no information is provided about just which levels cause which differences.

In Example 5.6, p.139, the Factorial ANOVA for the Gauge Study Data shows significant differences due to Operators and significant differences due to Parts, but no detectable interaction effects. Since there were only two operators in the study, the significant difference due to operators is fairly specific, but with five parts, the significant difference due to parts is very nonspecific. The contrasts in Example 5.5, p.133 were much more revealing. Two contrasts were significant. One represented the difference between the two operators, and the other represented the difference between the parts produced on the two different machines.

So, given this difference in specificity between the Factorial ANOVA and the Single Degree of Freedom ANOVA, why is the Factorial ANOVA the traditional choice? It is because the Factorial ANOVA *will remain the same* even though a different maximal mutually orthogonal set of contrasts may be used. This invariance of the Factorial ANOVA means that it is not tied down to just one set of comparisons. It is the same for all sets of orthogonal comparisons. This will be illustrated using the Reflectivity Data.

EXAMPLE 5.7: A Second Maximal Mutually Orthogonal Set for the Reflectivity Data:

A new maximal mutually orthogonal set of contrasts is defined below.

Subgroup	*1*	*2*	*3*	*4*	*5*	*6*	*7*	*8*	*9*	*10*	
Concentration	*5*	*5*	*5*	*5*	*5*	*10*	*10*	*10*	*10*	*10*	
Temperature	*75*	*100*	*125*	*150*	*175*	*75*	*100*	*125*	*150*	*175*	
Averages	*36.67*	*34.67*	*31.33*	*23.67*	*19.67*	*41.67*	*39.67*	*36.33*	*40.67*	*33.0*	$\hat{C}$
Contrasts											
11. 5 vs 10	*1*	*1*	*1*	*1*	*1*	*-1*	*-1*	*-1*	*-1*	*-1*	*-45.333*
12. 75 vs 100	*1*	*-1*	*0*	*0*	*0*	*1*	*-1*	*0*	*0*	*0*	*4.000*
13. 150 vs 175	*0*	*0*	*0*	*1*	*-1*	*0*	*0*	*0*	*1*	*-1*	*11.667*
14. 75,100v150,175	*1*	*1*	*0*	*-1*	*-1*	*1*	*1*	*0*	*-1*	*-1*	*35.667*
15. 125 vs others	*1*	*1*	*-4*	*1*	*1*	*1*	*1*	*-4*	*1*	*1*	*-1.000*
16. 11. x 12. gives	*1*	*-1*	*0*	*0*	*0*	*-1*	*1*	*0*	*0*	*0*	*0*
17. 11. x 13. gives	*0*	*0*	*0*	*1*	*-1*	*0*	*0*	*0*	*-1*	*1*	*-3.667*
18. 11. x 14. gives	*1*	*1*	*0*	*-1*	*-1*	*-1*	*-1*	*0*	*1*	*1*	*20.333*
19. 11. x 15. gives	*1*	*1*	*-4*	*1*	*1*	*-1*	*-1*	*4*	*-1*	*-1*	*-20.333*

While three of these contrasts (numbers 11, 12, and 16) were in the other set, six are new.

The Single Degree of Freedom ANOVA Table is:

SOURCE	SS	DF	MS	F
Contrast 11	616.5332	1	616.53	44.04**
Contrast 12	12.0000	1	12.00	0.86
Contrast 13	102.0833	1	102.08	7.29**
Contrast 14	477.0417	1	477.04	34.07**
Contrast 15	0.0750	1	0.075	0.01
Contrast 16	0.0000	1	0.000	0.0
Contrast 17	10.0833	1	10.08	0.72
Contrast 18	155.0417	1	155.04	11.07**
Contrast 19	31.0083	1	31.01	2.21
Within	280.0000	20	14.00	
Total	1683.8665	29		

These values are different from those you found in Exercise 5.6, p.135.

EXAMPLE 5.8: The Factorial ANOVA for the Reflectivity Data:

Based on the contrasts in the preceding Example, the Factorial ANOVA is:

SOURCE	SS	DF	MS	F
Concentration	616.5332	1	616.53	44.04**
Temperature	591.200	4	147.80	10.56**
Interaction	196.1333	4	49.03	3.50**
Within	280.000	20	14.00	
Total	1683.8665	29		

Except for roundoff, this should match the table you found in Exercise 5.9, p.139.

Thus, the Factorial ANOVA Table for a balanced and fully crossed study will remain the same regardless of the maximal mutually orthogonal set of contrasts defined. It is more specific than the One-Way ANOVA, but is not tied to any particular set of comparisons.

A special case exists when all the factors have only two levels each. Here the Single Degree of Freedom ANOVA and the Factorial ANOVA are identical since each main effect and each interaction effect have only one degree of freedom. This implies that there is effectively *only one* maximal mutually orthogonal set of contrasts that can be defined for data such as the Tensile Strength Data. Situations such as this will be exploited later.

5.7 Multifactor Studies with Subgroups of Size One

Because the number of subgroups required for a fully crossed study is equal to the product of the number of levels of each of the factors, it is very easy to end up with huge values for k with only a few factors. With multiple observations per subgroup, the total amount of data needed can quickly become overwhelming. One of the steps taken to remedy this problem is the use of only one observation per subgroup. However, when n = 1, there is no way to compute a Within-Subgroup Estimate of the Variance.

With one observation per subgroup, the Within-Subgroup row disappears from the ANOVA table, and SS Total = SS Between. Contrasts can still be used to decompose the Between-Subgroup Sum of Squares, and each contrast will still have one degree of freedom, but there will be no honest F-ratio for the evaluation of the contrasts. In such a case we can either resort to Pooling or Plotting in an attempt to separate signals from noise. ***Pooling*** occurs when some of the contrasts are combined in order to obtain a Mean Square Error term. ***Plotting*** consists of using one of several graphic techniques for gauging the relative signal strengths in the various contrasts.

Before discussing either of these approaches, it will be helpful to have an example with which to work.

<u>*EXAMPLE 5.9:*</u> <u>*The Adhesive Strength Data:*</u>

An experiment was run to compare two formulations of a particular adhesive compound. Thus, Formulation defines one factor with two levels. In addition to the two formulations, three different Primer Viscosities and three different Adhesive Viscosities were used in the study. A fully crossed design would have required 2x3x3 = 18 subgroups. Since certain combinations of Primer Viscosity and Adhesive Viscosity were not included in the study only 12 subgroups were obtained.

One part was produced and tested to rupture for each subgroup.

The response was the force in pounds required to separate the two components. These values were measured and recorded to the nearest 10 pounds (i.e. a value of 253 is equal to a force of 2530 pounds).

In spite of the partial crossing of two of the factors in this study, it was possible to obtain a maximal, mutually orthogonal set of contrasts to use with the data. The orthogonality of these contrasts guarantees that there is no cross-contamination between the contrasts. They will completely decompose the Between-Subgroup Sum of Squares.

(Some of the tricks for defining contrasts with partially crossed layouts are described in Sections 5.13 and 5.14. Other tricks are beyond the scope of this text. Fortunately, when an experiment is planned in advance, one can usually obtain a maximal, mutually orthogonal set of contrasts from a book of tables, so that it is not absolutely essential for the beginner to master the techniques for

defining contrasts.)

Subgroup	1	2	3	4	5	6	7	8	9	10	11	12
Formula	1	1	1	1	1	1	2	2	2	2	2	2
Primer Visc.	L	L	L	M	H	H	L	L	L	M	H	H
Adhes. Visc.	L	M	H	M	L	H	L	M	H	M	L	H
Responses	253	273	233	178	187	147	263	199	213	203	203	156
<u>*Contrasts:*</u>												
1. Formula	-1	-1	-1	-1	-1	-1	1	1	1	1	1	1
2. Primer LvH	-1	0	-1	0	1	1	-1	0	-1	0	1	1
3. Primer LvM	0	-1	0	1	0	0	0	-1	0	1	0	0
4. Adhesive LvH	-1	0	1	0	-1	1	-1	0	1	0	-1	1
5. = 2. x 4.	1	0	-1	0	-1	1	1	0	-1	0	-1	1
6. Adh. L,H v M *	1	-2	1	-2	1	1	1	-2	1	-2	1	1
7. = 1. x 2.	1	0	1	0	-1	-1	-1	0	-1	0	1	1
8. = 1. x 3.	0	1	0	-1	0	0	0	-1	0	1	0	0
9. = 1. x 4.	1	0	-1	0	1	-1	-1	0	1	0	-1	1
10. = 1. x 5.	-1	0	1	0	1	-1	1	0	-1	0	-1	1
11. = 1. x 6.	-1	2	-1	2	-1	-1	1	-2	1	-2	1	1

The Estimated Contrast Values and Single Degree of Freedom ANOVA Table are:

$\hat{C}$	*adj. term*	*SOURCE*	*SS*	*DF*
-34	12	*Contrast 1.*	96.333	1
-269	8	*Contrast 2.*	9,045.125	1
-91	4	*Contrast 3.*	2,070.250	1
-157	8	*Contrast 4.*	3,081.125	1
-17	8	*Contrast 5.*	36.125	1
-51	24	*Contrast 6.*	108.375	1
35	8	*Contrast 7.*	153.125	1
99	4	*Contrast 8.*	2,450.250	1
-37	8	*Contrast 9.*	171.125	1
23	8	*Contrast 10.*	66.125	1
113	24	*Contrast 11.*	<u>532.042</u>	<u>1</u>
		Total	17,810.000	11

These values will be used to illustrate both Pooling and Plotting techniques.

* Contrast 6 is a complex contrast selected in order to get a maximal orthogonal set of contrasts. It actually represents more than merely Adhesive Viscosity.

5.8 A Priori Pooling

There are two types of pooling that can be used. The first type of pooling is ***A Priori* Pooling**. *A Priori* Pooling occurs when certain contrasts are chosen for the pooling *before* the data are collected. These selected contrasts will then make up the error term for the analysis. An example of this might be the choice of all interaction contrasts for pooling. As long as this choice is made prior to the calculation of the sums of squares it can be said to be *A Priori* Pooling. With *A Priori* Pooling it is customary to use the critical points from the regular F-distribution (Table F). (The justification of the use of these critical points is the objectivity of the choice regarding which sums of squares were pooled.)

EXAMPLE 5.10: A Priori Pooling for the Adhesive Strength Data:

Say that it is decided, a priori, *to pool all interaction contrasts. Including Contrast 5, this gives a total of 6 d.f. for the mean square error (MSE) term.*

SOURCE	*SS*	*DF*	*Pooled*	*F-Ratios*
Contrast 1.	*96.333*	*1*		*0.17*
Contrast 2.	*9,045.125*	*1*		*15.9**
Contrast 3.	*2,070.250*	*1*		*3.64*
Contrast 4.	*3,081.125*	*1*		*5.42**
Contrast 5.	*36.125*	*1*	*Yes*	
Contrast 6.	*108.375*	*1*		*0.19*
Contrast 7.	*153.125*	*1*	*Yes*	
Contrast 8.	*2,450.250*	*1*	*Yes*	
Contrast 9.	*171.125*	*1*	*Yes*	
Contrast 10	*66.125*	*1*	*Yes*	
Contrast 11.	*532.042*	*1*	*Yes*	
Total	*17,810.000*	*11*		

Six Interaction SS(C) values are pooled (a priori) to obtain an Error Term:

Error Term	*3408.792*	*6*	*MSE = 568.1*

This MSE term may be used to form an F-ratio for each of the other five contrasts. These F-ratios would then be compared with an F critical point. With $\alpha = .10$ *and 1 and 6 d.f. this value is 3.78: thus, Contrasts 2 and 4 are said to be significant.*

Recall that each SS(C) value is a Method Three (Between Subgroup) Estimate of V(X). Method Three Estimates of V(X) are subject to being inflated whenever a signal exists, but otherwise they do estimate the variance of the distribution of X. (See the examples for Data Set One.) The idea of *A Priori* Pooling is to combine some Method Three Estimates of V(X) to use in lieu of the missing Method Two (Within Subgroup) Estimate. Of course the experimenter is supposed to pool those SS(C) values which, in his or her judgment, are least likely to be inflated by the presence of signals. When this is done successfully, the pooled MSE value will yield reasonable F-ratios. The danger of *A Priori* Pooling is that one may pool a sum of squares which is inflated by the presence of some signal and thereby deflate all of the computed F-ratios. This problem is illustrated in Example 5.10: Contrast 8 was pooled, but it had the third largest sum of squares, and the MSE value was inflated by its inclusion.

In Example 5.10 the experimenter is immediately tempted to remove Contrast 8 from the pooling and recompute the MSE value. Unfortunately, any revision of the MSE value based upon the values of the SS(C) terms will also change the distributions involved into conditional distributions, and one may no longer use critical values from the ordinary F-distribution. (As soon as the choice of what to pool is dependent upon the values of the sums of squares, the objectivity of the choice is lost.) This is the difference between *A Priori* Pooling and *Post Hoc* Pooling. With *Post Hoc* Pooling the pooling is subjective.

EXERCISE 5.10: A Priori Pooling for the Tensile Strength Data:

The Tensile Strength Data had multiple observations per subgroup, so that Pooling was not necessary. However, for the sake of practice, use the Contrast Sums of Squares given below as if the data consisted of 8 subgroups of size n=1.

Contrast	SS(C)	F-ratios
1	10	
2	722.5	
3	10,240	
4	40,960	pool
5	128,822.5	pool
6	53,290	pool
7	562.5	pool

(a) Assume that the interaction contrasts were selected for pooling prior to the collection of the data: Pool the four interaction sums of squares (Contrasts 4, 5, 6, and 7) to obtain a MSE term:

(b) Use this MSE term to obtain an F-ratio for the three main effect contrasts.

(c) Use the regular F Distribution (with 1 and 4 d.f.) to evaluate these ratios.

5.9 Post Hoc Pooling

Any pooling that is decided upon *after* the sums of squares have been calculated is said to be *Post Hoc* Pooling. This distinction between *A Priori* Pooling and *Post Hoc* Pooling is important for the following reason: Once the sums of squares have been computed, it will always be logical to choose the smallest SS(C) values to represent the background variation. (Recall that, in the absence of any signals, each SS(C) value is a Method Three estimate of V(X). Therefore, the larger SS(C) values are the most likely to be inflated by signals, while the smaller SS(C) values are more likely to be estimates of the background noise.) However, when the smallest SS(C) values are pooled to form a MSE term, the ratios obtained are subject to being inflated, and **it is no longer appropriate to use the critical values from the regular F-distribution**.

With *A Priori* Pooling the choice of what to pool into MSE is made independently of the SS(C) values, so the denominator of the F-ratios is independent of the numerators. With *Post Hoc* Pooling this independence is missing. The way that the SS(C) values affect the choice of what to pool creates a dependency between the MSE term and all of the possible numerators for F-type ratios. Moreover, this dependency exists even when the contrasts are orthogonal.

Therefore, with *Post Hoc* Pooling, a new set of critical values is needed. These are the critical values for a particular Modulus-Ratio Statistic, the Mod F-ratio.

Assume that a set of (k-1) orthogonal contrasts has sums of squares:

$$\{ SS(C_1), SS(C_2), \ldots, SS(C_{k-1}) \}$$

furthermore, assume that these sums of squares are arranged in order according to magnitude, and let the ranked SS(C) values be denoted by SS_i so that:

$$SS_1 \le SS_2 \le \ldots \le SS_{k-1}$$

and assume that the p smallest sums of squares are pooled to obtain a MSE term:

$$MSE_p = \frac{SS_1 + SS_2 + \ldots + SS_p}{p},$$

then one version of the Modulus-Ratio Statistic is:

$$\text{Mod F}(j, p, k\text{-}1) = \frac{SS_j}{MSE_p}, \quad \text{for any } j = (p+1), (p+2), \ldots, (k\text{-}1).$$

This Mod F-ratio will depend upon three parameters: j = rank of SS value in the numerator, p = number of values pooled in the denominator, and (k-1) = the number of SS(C) values for the set of orthogonal contrasts. Percentiles for several of these Mod F Distributions are given in Table G in the Appendix. For a more complete discussion of the Generalized Modulus-Ratio Statistic the reader is referred to Voss [1988].

If the Mod F-ratio for a given contrast is found to be in the upper tail of the appropriate Mod F distribution, then that contrast is likely to represent a signal rather than mere random variation. The four percentiles given in Table G allow any given ratio to be allocated to one of five regions in the Mod F distribution: (1) the lower half, (2) the upper half, (3) the upper quartile, (4) the upper ten percent, and (5) the upper five percent.

Since the Mod F distribution changes as the rank of the numerator changes (i.e., as j changes), one will have to look up a different set of percentiles for each ratio obtained. In particular:

$$\text{Compare } \frac{SS_{k-1}}{MSE_p} \text{ against Mod } F_{1-\alpha}(k\text{-}1, p, k\text{-}1)$$

$$\text{Compare } \frac{SS_{k-2}}{MSE_p} \text{ against Mod } F_{1-\alpha}(k\text{-}2, p, k\text{-}1)$$

$$\text{Compare } \frac{SS_{k-3}}{MSE_p} \text{ against Mod } F_{1-\alpha}(k\text{-}3, p, k\text{-}1)$$

and finally

$$\text{Compare } \frac{SS_{p+1}}{MSE_p} \text{ against Mod } F_{1-\alpha}(p\text{+}1, p, k\text{-}1).$$

EXAMPLE 5.11: Post Hoc Pooling for the Adhesive Strength Data:

SOURCE	*SS*	*DF*	*Pooled*	*Modulus Ratio*	*Rank*	*Mod F (j,5,11)*	*Ratio in*
Contrast 1.	*96.333*	*1*	*Yes*				
Contrast 2.	*9,045.125*	*1*		*98.3**	*j = 11*	$F_{.90} = 94.5$	*upper 10%*
Contrast 3.	*2,070.250*	*1*		*22.5**	*j = 8*	$F_{.90} = 19.6$	*upper 10%*
Contrast 4.	*3,081.125*	*1*		*33.5**	*j = 10*	$F_{.75} = 28$	*upper 25%*
Contrast 5.	*36.125*	*1*	*Yes*				
Contrast 6.	*108.375*	*1*	*Yes*				
Contrast 7.	*153.125*	*1*	*Yes*				
Contrast 8.	*2,450.250*	*1*		*26.6**	*j = 9*	$F_{.75} = 17.7$	*upper 25%*
Contrast 9.	*171.125*	*1*		*1.86*	*j = 6*	$F_{.50} = 3.37$	*lower half*
Contrast 10	*66.125*	*1*	*Yes*				
Contrast 11.	*532.042*	*1*		*5.78*	*j = 7*	$F_{.50} = 4.95$	*upper half*
Total	*17,810.000*	*11*					

Here the ranked sums of squares are:
{ $SS_1 = 36.1$, $SS_2 = 66.1$, $SS_3 = 96.3$, $SS_4 = 108.4$, $SS_5 = 153.1$,
$SS_6 = 171.1$, $SS_7 = 532.0$, $SS_8 = 2070.$, $SS_9 = 2450.$, $SS_{10} = 3081.$, $SS_{11} = 9045.$ }

The five smallest SS(C) values are pooled (post hoc) to obtain an Error Term:

Error Term 460.083 5 *MSE* = 92.02

The values for j range from 11 down to 6, the value for p is 5, and the number of contrasts in the set of contrasts is k-1 = 11. Thus, we use the percentiles from the Mod F (j,5,11) distributions:

$$\frac{SS_{11}(C_2)}{MSE} = 98.3 \geq 94.5 \pm 4.9 = \text{Mod } F_{.90}(11, 5, 11)$$

$$\frac{SS_{10}(C_4)}{MSE} = 33.5 > 28.0 \pm 0.9 = \text{Mod } F_{.75}(10, 5, 11)$$

$$\frac{SS_9(C_8)}{MSE} = 26.6 > 17.7 \pm 0.5 = \text{Mod } F_{.75}(9, 5, 11)$$

$$\frac{SS_8(C_3)}{MSE} = 22.5 > 19.6 \pm 1.0 = \text{Mod } F_{.90}(8, 5, 11)$$

$$\frac{SS_7(C_{11})}{MSE} = 5.78 > 4.95 \pm 0.08 = \text{Mod } F_{.50}(7, 5, 11)$$

$$\frac{SS_6(C_9)}{MSE} = 1.86 < 3.37 \pm 0.05 = \text{Mod } F_{.50}(6, 5, 11)$$

Contrasts 2, 4, 8, and 3 have Mod F-ratios in the upper quartile, thus these four contrasts could represent potential signals. Contrast 11 has a Mod F-ratio which is merely in the upper half of its distribution, and Contrast 9 has a Mod F-ratio which is below the median, therefore, it is unlikely that these contrasts represent signals.

Thus, the critical values of the Mod F distribution given in Table G allow one to evaluate the ratios obtained by *Post Hoc* Pooling. As long as the SS(C) values pooled into the MSE term do not contain signals, the sensitivity of the Mod F-ratio will increase as the number of pooled values increases. The more terms used for MSE, the greater the sensitivity of the Mod F-ratio. However, since every SS(C) value is a Method Three Estimate of V(X), each SS(C) value has the potential of being inflated by some signal. When such inflated SS(C) values are pooled into the MSE term the sensitivity of the Mod F-ratio will decrease. Thus, the trick to using the Mod F-ratio is the appropriate choice of what to pool into the MSE term. Pool too many terms and the Mod F-ratio becomes a signal-to-signal ratio. Pool too few terms and the Mod F-ratio will be weak and insensitive. The Scree Plot described below will provide a picture which will help one choose just how many SS(C) values to pool into MSE.

Finally, with either pooling technique, there is always the possibility that the F-ratios will be deflated by the pooling of some real effects. With *A Priori* Pooling it is possible that significant contrasts may be pooled. With *Post Hoc* Pooling there is a chance that even the smallest sums of squares may represent some real effects. Because of these possibilities of deflated F-ratios, it is customary to use very large α-values with either type of pooling. This is especially true for exploratory studies since the error to avoid is the deletion of a real effect.

Thus, Pooling is essentially an act of desperation. It is an attempt to hang onto the semblance of an F-test even though an honest F-test is impossible. Unfortunately, both *A Priori* Pooling and *Post Hoc* Pooling are quite weak until a fair number of values are pooled into MSE, and the potential for contamination increases each time another sum of squares is pooled. This means that with the pooling approach the user is confronted with a choice between a shot in the dark (*A Priori* Pooling) or a subjective choice (*Post Hoc* Pooling).

EXERCISE 5.11 Post Hoc Pooling for the Tensile Strength Data:

For the sake of practice, use the Contrast Sums of Squares given below as if the data consisted of eight subgroups of size n = 1.

Contrast	SS(C)	Mod F-Ratios
1	10	
2	722.5	
3	10,240	
4	40,960	
5	128,822.5	
6	53,290	
7	562.5	

(a) Pool the three smallest Sums of Squares to obtain a MSE term:

(b) Use this MSE term to obtain the Mod F-ratios for the other contrasts.

(c) Use the Mod F(j, 3, 7) Percentiles to evaluate these ratios:

(d) Which Contrasts appear to represent signals?

The inability to get an honest F-test when n = 1 is one reason that people have resorted to plotting techniques. With the advantage of a graphic display of the results it is generally easier to obtain a consensus on just what conclusions the data will support than it is with numerical techniques.

Two different plots will be discussed here: the Scree Plot and the Normal Plot.

5.10 Scree Plots

A ***Scree Plot*** provides a graphic tool for analyzing the sums of squares for orthogonal contrasts. While the idea of a Scree Plot was suggested to this author, in a different context, by John Philpot, its application to the Single Degree of Freedom ANOVA is debuted here.

The word "scree" describes the rubble at the bottom of a cliff. The Scree Plot plots the SS(C) values in descending order of magnitude, and connects these points to form the profile of a cliff. The rubble at the bottom of the cliff will represent background noise, while the cliff represents any signals which may be present in the data. Moreover, the height of any cliff above the rubble will indicate the relative strength of the signal.

The basic premise of the Scree Plot is the following. Every SS(C) value is a Method Three Estimate of V(X), and at least some of these Method Three Estimates should be inflated by any signals which are present. Any Method Three Estimates which are not so inflated may be taken to be estimates of V(X). Thus, when the SS(C) values are plotted in descending order of magnitude, any SS(C) values which happen to be inflated by a signal will stand out above the other SS(C) values, resulting in a plot that looks like the profile of a cliff with rubble at the bottom.

Therefore, a Scree Plot is, in effect, a Pareto Chart for the SS(C) values. However, because of the way the Scree Plot is interpreted, the Scree Plot will use lines to connect the SS(C) points instead of the stair-step bar graph of the Pareto Chart.

By comparing the slope of the "rubble pile" with the height of the "cliff" portion of the Scree Plot one can make a visual comparison of the background noise level to the potential signal strength. Likewise, the rubble will best define the contrasts which might well be pooled into a MSE term for the Mod F-ratio.

The Scree Plot for the Adhesive Strength Data shows Contrast 2 to be the strongest effect by far. Contrasts 4, 8, and 3 form a small plateau that may or may not turn out to be significant in the long run, while the remainder of the contrasts have SS(C) values that appear to be part of the rubble.

When using a Scree Plot the objective is to find the "elbow" of the curve. Points above the elbow will be likely to indicate a signal, while those below the elbow will, most likely, be noise. Of course, some Scree Plots will not have a nice elbow, and the choice becomes somewhat subjective. When this happens, the "slope" of the rubble pile is a useful guide to where the cliff begins.

Intermediate plateaus may contain signals or they may be all noise. The way they are treated will depend upon how many strong signals are present. With many strong signals, additional weak signals are of little interest. With few strong signals, additional weak signals are of some interest.

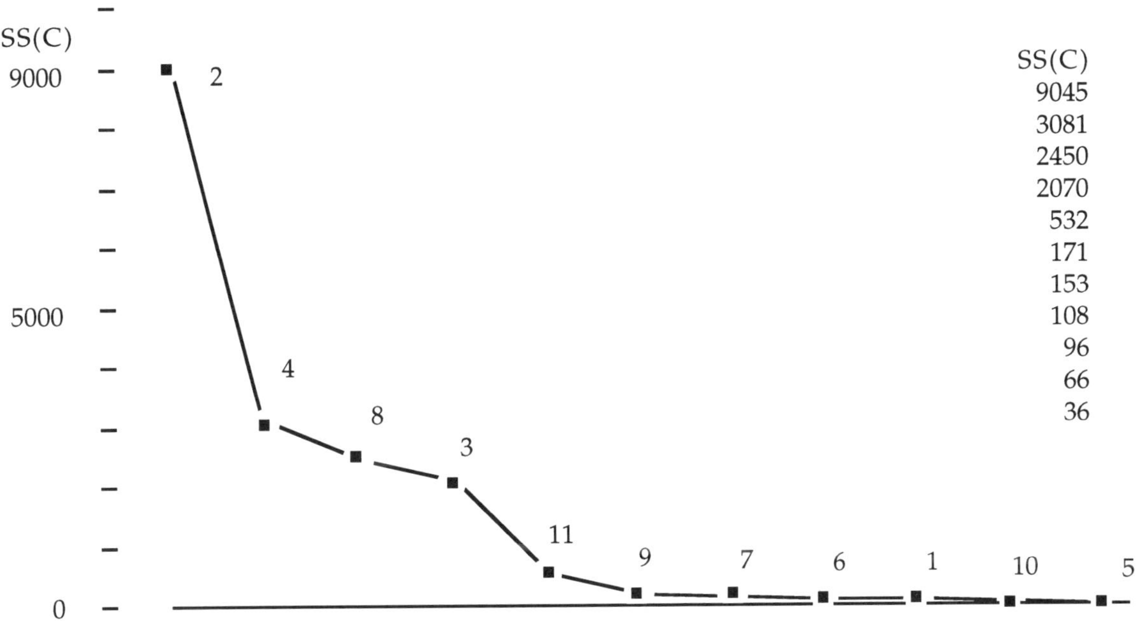

Figure 5.1: Scree Plot for the Adhesive Strength Data:

Note that there is no hint of any probabilistic argument with a Scree Plot. Those contrasts with SS(C) values which stand out above the rubble are simply the best candidates for signals. The only way that we can ever confirm that a contrast does or does not represent a real effect is by the replication of results. If a subsequent experiment using the same factors shows the same contrasts high on the Scree Plot, then these contrasts are likely to represent real effects. If a contrast is high on the Scree Plot in one study, but does not show up on the cliff portion in a subsequent study, then it is unlikely that it represents a real effect (or at least it is unlikely that it represents a very strong effect). This approach uses the traditional confirmatory tool of the scientific method. It does not require any assumptions, nor does it require the use of critical values and probability theory. It simply rests on the nontrivial replication of results from one study to another. The Scree Plot identifies the contrasts of interest; the persistence of a given effect from one study to another confirms that it is a real effect rather than an accidental effect.

Figure 5.2 shows a nonsignificant Scree Plot for 27 contrasts. The purpose of including this plot is to show how Scree Plots with many values may show a low cliff even when there is no signal in the data. This is why it is important to evaluate the height of the "cliff" relative to the "slope" of the rubble. If the cliff does not stand out dramatically above the rubble, there may be no signal present.

Of course, for those who are addicted to using F-ratios, one may use the Mod F ratio to check a Scree Plot for significance. For the data in Figure 5.2, pool the 10 smallest SS(C) values to get MSE = 20.522. The first two Mod F ratios are:

$$\frac{SS_{27}}{MSE} = 26.3 < \text{Mod } F_{.50}(27, 10, 27) \qquad \frac{SS_{26}}{MSE} = 19.9 < \text{Mod } F_{.50}(26, 10, 27)$$

Since the two largest Mod F ratios fail to exceed the median of their respective Mod F distributions, testing is discontinued and the Scree Plot is judged to be nonsignificant.

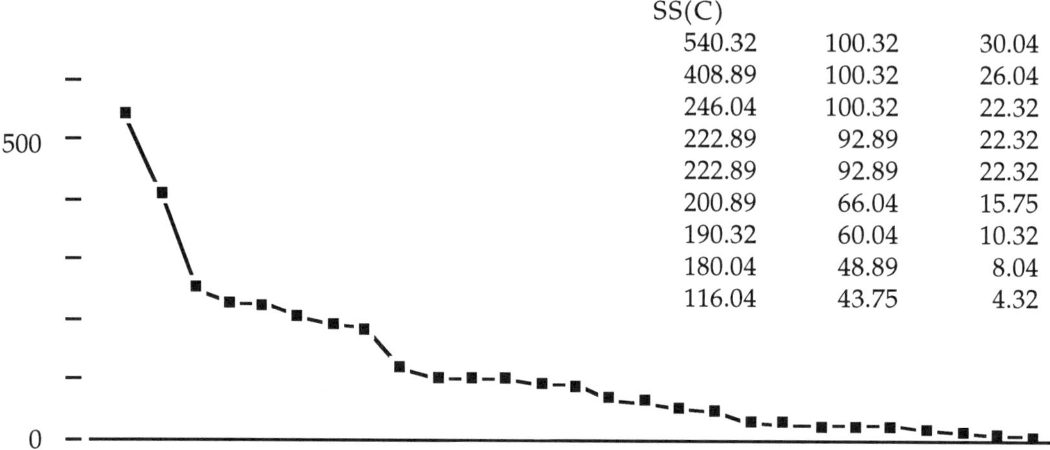

Figure 5.2: A Nonsignificant Scree Plot for 27 Contrasts

When the Scree Plot appears to be "significant," yet the Mod F ratio does not indicate any significance, it will generally be best to depend upon the Scree Plot. In particular, this author has observed the following: An initial study had a Scree Plot with three effects standing out above the others (say A, B and C), however all of the Mod F-ratios were below their median values. The follow-up study had a Scree Plot which was in agreement with the first study (A, B, and C standing out from the rest), but once again none of the Mod F-ratios were above the median values. The replication of the Scree Plot results from one study to the other was a much stronger positive confirmation than the negative results from the Mod F-ratio. In this case, the Scree Plot results were used as the basis for action. Subsequent results proved this to be the correct decision. (The simplest explanation of how such inconsistencies can occur is for the MSE term to be contaminated by signals.)

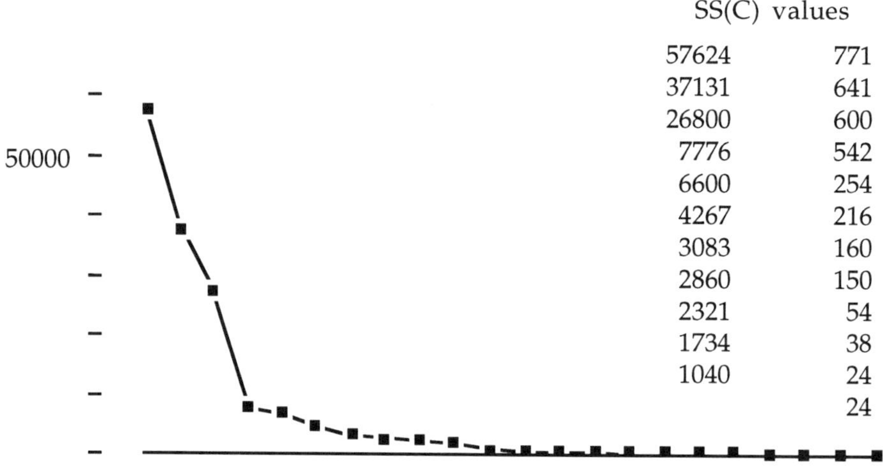

Figure 5.3: A Significant Scree Plot for 23 Contrasts

Figure 5.3 shows a significant Scree Plot for 23 contrasts. The height of the cliff relative to the slope of the rubble is much greater than in Figure 5.2. Clearly, the three contrasts with the largest SS(C) values stand out as potential signals, and the remainder of the contrasts do not contain any strong signals.

The Mod F-ratios also indicate significance for these data: pooling the 10 smallest sums of squares gives MSE = 206.2, so the Mod F-ratios are:

$$\frac{SS_{23}}{MSE} = 279.5 > 143 \pm 13 = \text{Mod } F_{.95}(23, 10, 23), \text{ upper five percent}$$

$$\frac{SS_{22}}{MSE} = 180.1 > 86 \pm 8 = \text{Mod } F_{.95}(22, 10, 23), \text{ upper five percent}$$

$$\frac{SS_{21}}{MSE} = 130.0 > 61 \pm 6 = \text{Mod } F_{.95}(21, 10, 23), \text{ upper five percent}$$

$$\frac{SS_{20}}{MSE} = 37.71 > 27 \pm 1 = \text{Mod } F_{.75}(20, 10, 23), \text{ upper quartile}$$

$$\frac{SS_{19}}{MSE} = 32.01 \approx 31.2 \pm 2.7 = \text{Mod } F_{.90}(19, 10, 23), \text{ upper ten percent}$$

$$\frac{SS_{18}}{MSE} = 20.69 > 17.1 \pm 0.4 = \text{Mod } F_{.75}(18, 10, 23), \text{ upper quartile}$$

$$\frac{SS_{17}}{MSE} = 14.95 \approx 14.7 \pm 0.6 = \text{Mod } F_{.75}(17, 10, 23), \text{ upper quartile}$$

$$\frac{SS_{16}}{MSE} = 13.87 > 11.8 \pm 0.5 = \text{Mod } F_{.75}(16, 10, 23), \text{ upper quartile}$$

$$\frac{SS_{15}}{MSE} = 11.26 > 9.92 \pm 0.11 = \text{Mod } F_{.75}(15, 10, 23), \text{ upper quartile}$$

$$\frac{SS_{14}}{MSE} = 8.409 > 8.00 \pm 0.21 = \text{Mod } F_{.75}(14, 10, 23), \text{ upper quartile}$$

$$\frac{SS_{13}}{MSE} = 5.044 > 4.81 \pm 0.18 = \text{Mod } F_{.50}(13, 10, 23), \text{ upper half}$$

$$\frac{SS_{12}}{MSE} = 3.739 < 3.95 \pm 0.10 = \text{Mod } F_{.50}(12, 10, 23), \text{ lower half}$$

$$\frac{SS_{11}}{MSE} = 3.109 < 3.26 \pm 0.07 = \text{Mod } F_{.50}(11, 10, 23), \text{ lower half}$$

So the three largest SS(C) values are quite likely to represent signals, and the seven next largest may represent weaker signals.

Thus, the Scree Plot and the Mod F ratios make the same essential comparisons. While the Mod F ratio does this numerically, the Scree Plot does it graphically. Both techniques are *post hoc* procedures. With the Mod F ratios the subjectivity comes in choosing which SS(C) values to pool. With the Scree Plot the subjectivity comes in choosing where to draw the line between the cliff and the rubble. The Scree Plot will generally do a better job of communicating the results than the Mod F ratios. Moreover, the Scree Plot still indicates relative strengths even when all contrasts represent a signal. These characteristics give the Scree Plot a greater utility than the Mod F approach. The Scree Plot should be used routinely, regardless of whether the Mod F ratios are used or not.

The two preceding figures point out the importance of plotting *all* of the SS(C) values and of drawing the horizontal axis for the value of SS(C) = 0. Without all of the rubble, one cannot estimate the slope of the rubble, and without the horizontal reference line, one cannot easily discern the relative height of the cliff portion of the curve.

The Scree Plot must have the points equally spaced on the horizontal axis for proper interpretation. Moreover, the vertical axis must be given without any discontinuities or breaks. The creation of a Scree Plot is an exercise in plotting the relative sizes of the SS(C) values. The interpretation depends upon the overall shape more than it does on having the individual values plotted exactly. Thus, the integrity of the scales on the two axes is more important than the exactness of the plotting.

Bar graphs are not as effective as the connected dots for the Scree Plot because of the way the curve is interpreted.

Since the curve is interpreted relative to itself, the choice of a vertical scale is of little consequence, as long as the curve is large enough to be easily seen.

5.11 Normal Probability Plots

Like Scree Plots, Normal Probability Plots also make a relative comparison. However, instead of plotting the SS(C) values, a Normal Probability Plot plots the Estimated Contrast Effects, $\hat{l}$. These effects are computed for every contrast, arranged in rank order, and plotted versus the appropriate percentages on Normal Probability Paper.

Normal Probability Paper has the property that a random sample drawn from a normal distribution will yield a straight line, more or less, when plotted against the appropriate percentages. One axis of Normal Probability Paper has a nonlinear scale marked in percentages ranging from 0.1 to 99.9 (or proportions of 0.001 to 0.999). The other axis is a simple linear scale.

For a set of (k-1) orthogonal contrasts, the Normal Probability Plot is obtained by plotting the (k-1) Estimated Contrast Effects, in numerical order on the linear scale, against the points

$$\left\{ \frac{1}{2(k-1)}, \frac{3}{2(k-1)}, \frac{5}{2(k-1)}, \ldots, \frac{2(k-1)-1}{2(k-1)} \right\}$$

on the nonlinear scale.

Consider a set of (k-1) = 5 Estimated Contrast Effects, and let the ranked values be denoted by:

$$\{ \hat{l}_{(1)}, \hat{l}_{(2)}, \hat{l}_{(3)}, \hat{l}_{(4)}, \hat{l}_{(5)} \}$$

Then the points to be plotted on the Normal Probability Plot would be:

$$(\hat{l}_{(1)}, 0.10), (\hat{l}_{(2)}, 0.30), (\hat{l}_{(3)}, 0.50), (\hat{l}_{(4)}, 0.70)(\hat{l}_{(5)}, 0.90)$$

where the first value in each pair would be plotted on the linear scale, while the second value in each pair would be plotted on the nonlinear scale. (If this doesn't make sense to you, stick to the Scree Plot.)

<u>EXAMPLE 5.12:</u> <u>Normal Probability Plot for Adhesive Strength Data:</u>
The Estimated Contrast Effects for the 11 contrasts in Example 5.9 are shown below, along with the appropriate proportions and percentiles for plotting on Normal Probability Paper.

Contrast	*Estimated Effect*	*proportion*	*percentile*
Contrast 2	*-67.25*	*1/22*	*4.5%*
Contrast 3	*-45.50*	*3/22*	*13.6%*
Contrast 4	*-39.25*	*5/22*	*22.7%*
Contrast 6	*-6.38*	*7/22*	*31.8%*
Contrast 1	*-5.67*	*9/22*	*40.9%*
Contrast 5	*-4.25*	*11/22*	*50.0%*
Contrast 11	*3.0*	*13/22*	*59.1%*
Contrast 10	*5.75*	*15/22*	*68.2%*
Contrast 7	*8.75*	*17/22*	*77.3%*
Contrast 9	*9.25*	*19/22*	*86.4%*
Contrast 8	*49.5*	*21/22*	*95.4%*

The Normal Probability Plot for these data is shown in Figure 5.4.

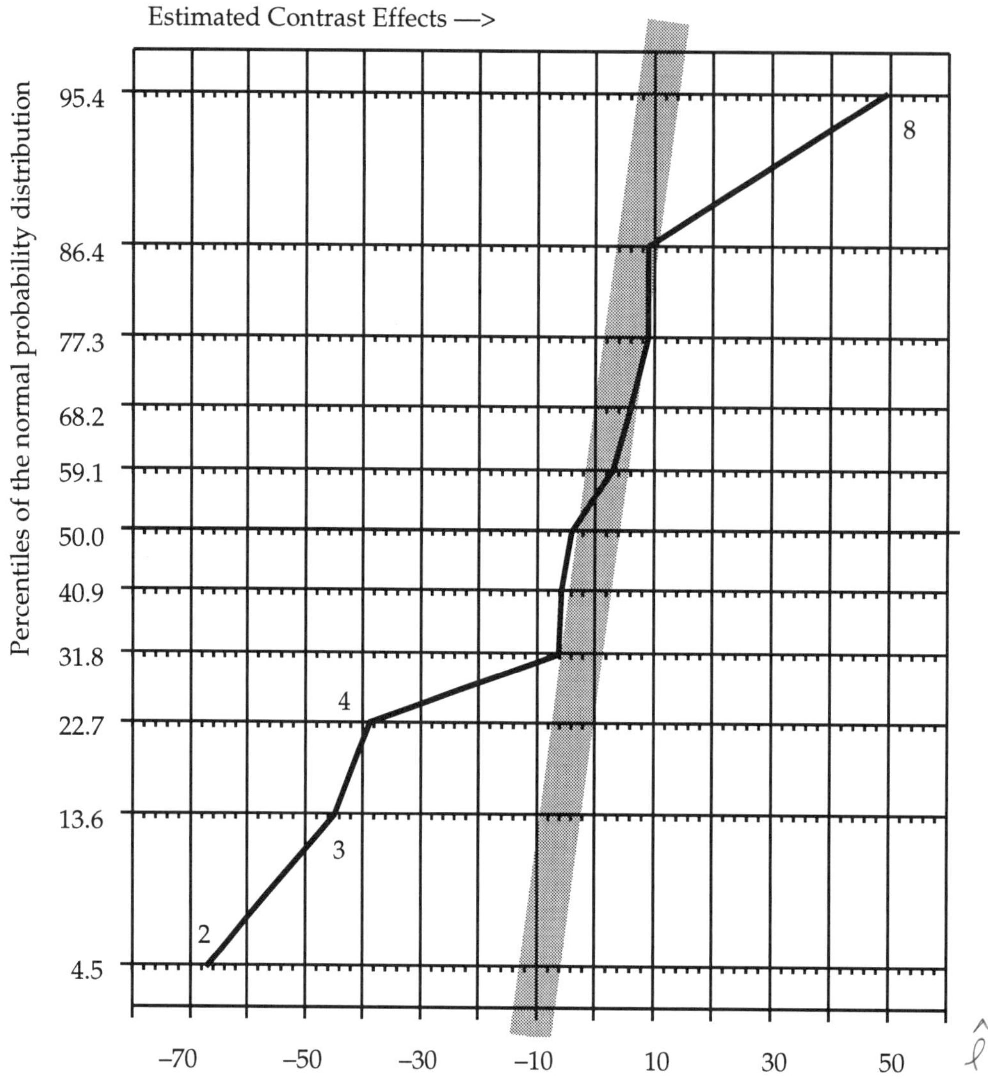

Figure 5.4: Normal Probability Plot for the Adhesive Strength Data

In interpreting this plot, it is the points which deviate from the straight line which represent signals. Here the Estimated Effects for Contrasts 2, 3, 4, and 8 appear to stand out as potential signals.

If all of the contrasts represent nothing but random variation (if there are no real differences between the subgroup averages) then the Estimated Contrast Effects should form a reasonably straight line on the Normal Probability Plot. However, if some contrasts represent real differences rather than random variation, then these contrasts should fall to the lower left or the upper right of the straight line formed by the other contrasts. Thus, the key to interpreting a Normal Probability Plot is to visualize the straight line defined by those contrasts which are most likely to represent random variation.

Since this straight line is supposed to represent the background noise, emphasis is given to those contrast effect values which are close to zero. Generally one will define the reference line by using the steepest portion of the curve in the neighborhood of a contrast effect of zero. While trying to decide upon how to construct a line which fits the steepest portion of the plot in the vicinity of zero it is useful to know that the line should pass close to the point defined by a contrast effect of 0.0 and the 50th percentile on the probability scale. Therefore, one may force a line through this point and pivot it until it is parallel to the steepest portion of the curve.

Once the steep, central portion has been used to visualize a straight line, contrast effects which deviate from the vicinity of this straight line to an appreciable degree are likely to represent signals.

In deciding just which contrasts represent possible signals, one will consider departures to the left of the line only if they correspond to small percentiles (≤50%). Likewise, departures to the right will be considered only if they correspond to large percentiles (≥50%).

A departure to the left which corresponds to a high percentage will simply serve to suggest the need of a steeper straight line. Likewise for departures to the right which correspond to a small percentage.

Occasionally the Normal Probability Plots are turned on their side with the percentiles on a horizontal (nonlinear) scale and the effects on a vertical, linear scale. The reader is therefore warned to read the labels on the axes of any computer generated Normal Probability Plots.

Some special forms for Normal Probability Plots are given in the Appendix, pages 355–361. These forms already have the percentiles selected for various numbers of contrasts. This greatly simplifies the job of manually preparing a Normal Probability Plot: One arranges the Estimated Contrast Effect values in ascending order, chooses a scale for the horizontal axis which will allow these values to be plotted on the graph, and plots the points in an ascending manner on the plot, one point on each horizontal line. (The smallest value goes on the lowest horizontal line, the second smallest value goes on the second lowest horizontal line, etc.)

Another version of the Normal Probability Plot is called a Quantile-Quantile Plot (Johnson and Tukey, 1987). Here the contrast effects are plotted on the horizontal-axis, and the corresponding quantiles for a normal probability distribution are plotted on the vertical axis. These quantiles are obtained from the formula:

$$Q(i) = 4.91 \left\{ \left[\frac{i-.375}{(k-1)+.25}\right]^{0.14} - \left\{ 1 - \left[\frac{i-.375}{(k-1)+.25}\right] \right\}^{0.14} \right\}$$

where i is the rank of the contrast and (k-1) is the total number of contrasts being plotted. The values plotted are therefore:

$$(\hat{l}_{(i)}, Q(i)) \qquad \text{for } i = 1, 2, 3, \ldots, (k-1).$$

This approach does not require special graph paper, but it does require considerable computation.

EXAMPLE 5.13: Normal Probability Plot for Data of Figure 5.2:

The Estimated Contrast Effects which correspond to the Sums of Squares shown in Figure 5.2 are:

-8.79	-5.36	-3.79	-1.79	0.79	1.79	2.50	3.07	5.07
-7.64	-5.21	-3.64	-1.79	1.07	1.93	2.64	3.79	5.64
-5.64	-4.07	-3.64	-1.50	1.21	2.07	2.93	3.79	5.93

The Normal Probability Plot for these Estimated Contrast Effects is shown in Figure 5.5. Once again, there is no real evidence of any signals in these data.

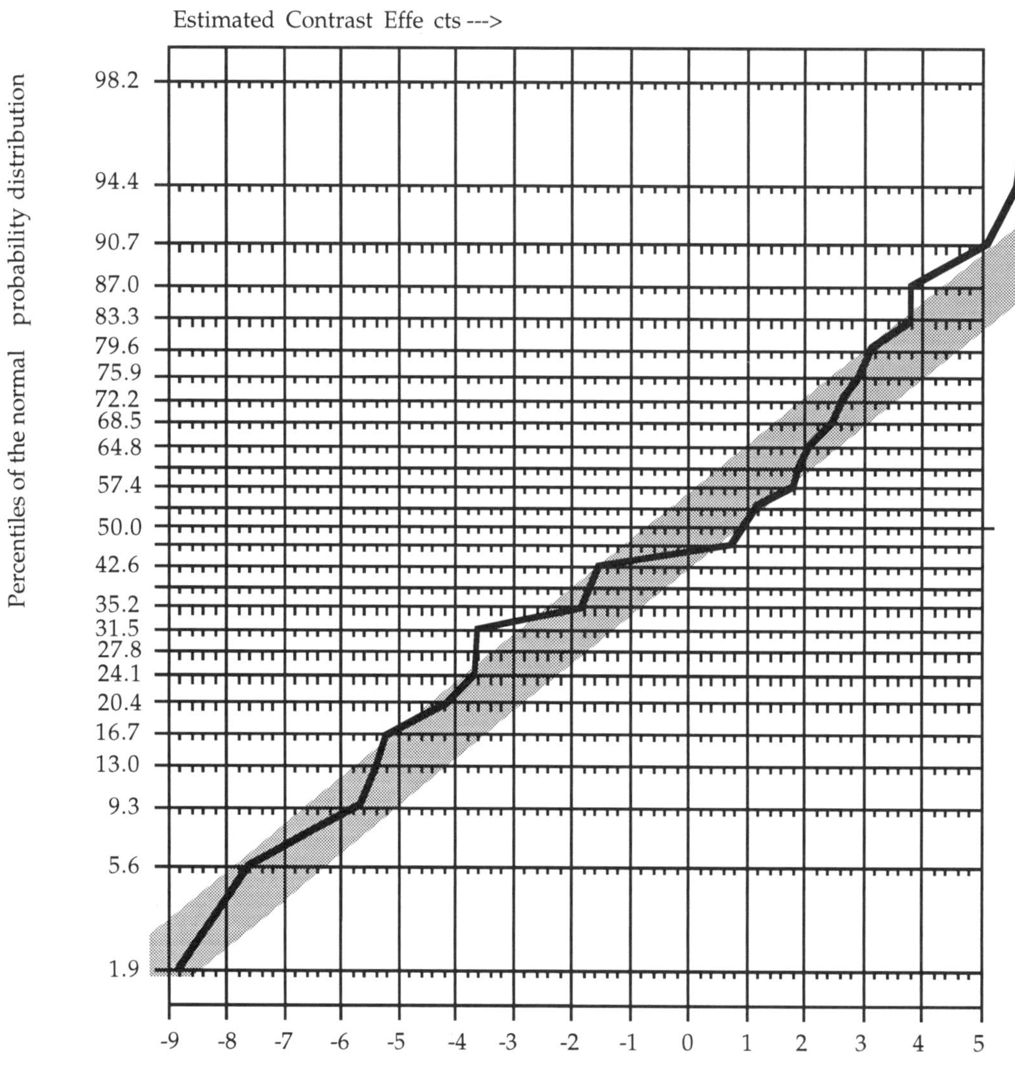

Figure 5.5: Normal Probability Plot for Data of Figure 5.2

EXERCISE 5.12: Plots for Tensile Strength Data:

Use the form on the following page to plot both the Scree Plot and the Normal Probability Plot for the Tensile Strength Data.

The Contrasts, Sums of Squares, and Estimated Effects are:

Contrast	SS(C)	$\hat{l}$
1	10	1
2	722.5	8.5
3	10,240	-32
4	40,960	-64
5	128,822.5	-113.5
6	53,290	-73
7	562.5	7.5

Comparing the two Pooling techniques with the two Plotting techniques above, the advantages of the Plotting techniques should be apparent. While *A Priori* Pooling is a shot in the dark, the other techniques are subjective. *Post Hoc* Pooling involves the subjective choice of which Sums of Squares to pool; the Scree Plot involves the subjective choice of where to draw the line between the cliff portion and the rubble; and the Normal Probability Plot involves the subjective decision of where and how to locate the reference line. The advantage of the plots is the power they possess to communicate the results to others quickly and efficiently.

The purpose of analysis is insight, rather than numbers, and the simplest technique which will provide that insight is the technique to use. These Pooling and Plotting techniques are among the simplest available for analyzing data when there is only one observation per subgroup. The student may use any of these techniques with the assurance that they will all perform with approximately the same sensitivity and robustness.

There are two other plotting techniques which the student might encounter: Daniel's Half-Normal Plots, and Box, Hunter, and Hunter's technique for plotting points against a Student's-t Distribution. These techniques present essentially the same information as the Scree Plot and the Normal Probability Plot.

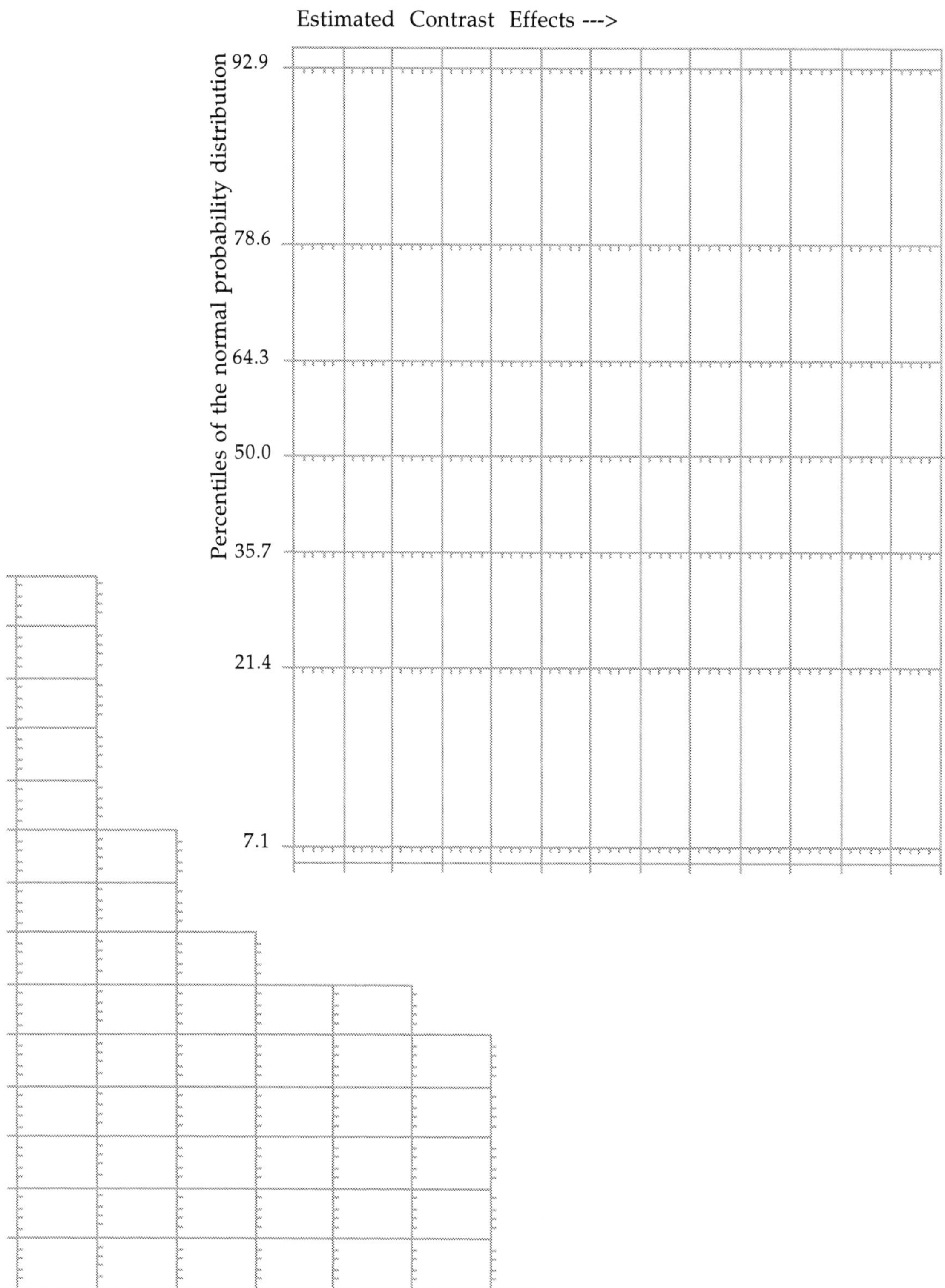

Scree Plot and Normal Probability Plot for Tensile Strength Data

5.12 Interpreting the Results: Response Plots and ANOM Plots

The Pooling and Plotting techniques of the previous section are concerned with the separation of signals and noise. They have as their objective the identification of potential signals which may be present in the data. Once these potential signals have been identified, they should be used to organize the data into a coherent graph. Those Factors associated with the significant contrasts will be included on the list of potentially active factors. The responses, or the subgroup averages, will then be plotted as a function of these potentially active factors, and the graph will be interpreted in the context of the data. One may use either an ANOM type plot, or a Response Plot, or both for this interpretation of the results of the experimental study. Both of these plots will be illustrated using the Adhesive Strength Data of Example 5.9, p.143.

EXAMPLE 5.14: *ANOM Plot for Adhesive Strength Data:*

Although the subgroup size is one, we may still plot the responses in an ANOM type plot.
The Scree Plot suggests that Contrasts 2, 3, 4, and 8 might represent signals.
Contrasts 2 and 3 involve the Primer Viscosity.
Contrast 4 involves the Adhesive Viscosity.
And Contrast 8 is an interaction contrast for Primer Viscosity and Formulation.
Therefore, all three factors apparently have an impact upon the adhesive force.
Thus, all three factors will be used for the ANOM Plot for these data.
(If some factors did not appear to have an impact, the subgroups could be combined to eliminate those factors and simplify the ANOM Plot.)
The Scree Plot assures us that at least some of the differences shown on the following plot are likely to be real effects.

Subgroup	1	2	3	4	5	6	7	8	9	10	11	12
Formula	1	1	1	1	1	1	2	2	2	2	2	2
Primer Visc.	L	L	L	M	H	H	L	L	L	M	H	H
Adhes. Visc.	L	M	H	M	L	H	L	M	H	M	L	H
Responses	253	273	233	178	187	147	263	199	213	203	203	156

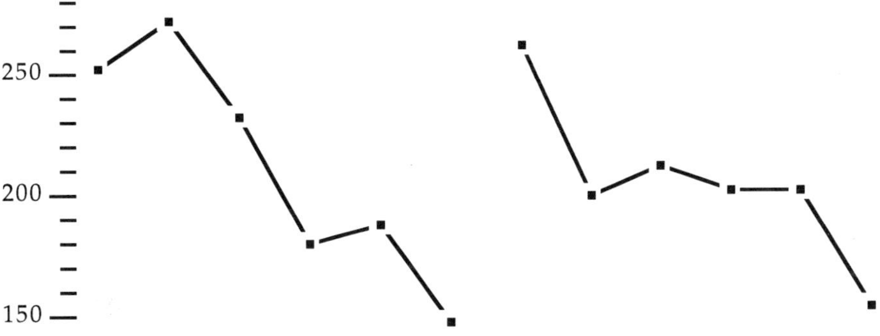

With subgroups of size one, there is no objective way to obtain ANOM decision limits. However, since this ANOM Plot is drawn for interpretation purposes, rather than as the primary analysis, the lack of limits is of little consequence. The Scree Plot assures the user that all three factors in the study have an impact upon the adhesive strength. The ANOM Plot is intended to clarify just what the effects of these factors may be. The picture may not be completely developed, but these data define the broad outlines of how these factors affect adhesive strength.

A Response Plot is a different way of plotting the data as a function of the factor levels. In a Response Plot each axis will represent a factor, and positions along each axis will correspond to different levels of each factor. With this scheme, each possible combination of factor levels will be represented by a point in the space defined by the Factor Axes. The responses obtained in conjunction with a given combination of factor levels will be written on the Response Plot at the point defined by the combination of factor levels. (That is, if a response of 42 is obtained for A high, B low, then the value of 42 is inserted into the Response Plot at the point defined by A high and B low.)

This technique results in a different perspective than the ANOM Plot. Together these two plots will usually provide the needed insight on what to do next.

EXAMPLE 5.15: *Response Plot for Adhesive Strength Data:*

The Scree Plot suggested that Contrasts 2, 3, 4, and 8 might represent signals.
Contrasts 2 and 3 involve the Primer Viscosity.
Contrast 4 involves the Adhesive Viscosity.
And Contrast 8 is an interaction contrast for Primer Viscosity and Formulation.
Therefore, all three factors will be used to construct a Response Plot for these data.
The partial crossing will be seen by the points in the factor space which do not have any values.

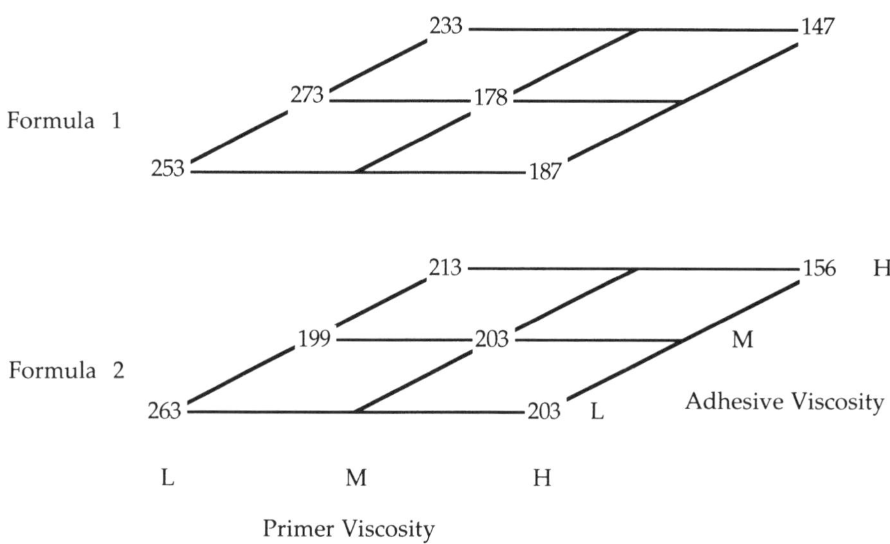

Clearly, the best Adhesion Strength occurs when both Primer and Adhesive Viscosities are low. Moreover, Formula 2 appears to be rather insensitive to Primer or Adhesive Viscosity except at Low/Low and High/High combinations.

Response Plots are useful in examining the factor space. These plots help the experimenter to define regions of greatest interest for future work, and combinations of factor levels that may, in the interim, work better than the current practice.

Since the purpose of analysis is insight, the analysis of experimental data is not complete without either an ANOM type plot, a Response Plot, or some other graphic presentation of the data. Much of the preliminary analysis will be focused on just how to actually construct these graphs. Often one or more of the factors in the study may be deleted from the final graph. When this happens the resulting simplification will make the graph much more powerful and easy to interpret. When no factor drops out, as happens with the Adhesive Strength Data, the knowledge that all the factors are important is essential to a proper interpretation of the data. The beginnings of understanding involve a knowledge of just which factors affect the response. Include too many, and complexity results. Include too few, and all your conclusions will be biased.

Use the Subgroup Averages shown on page 106 to construct a Response Plot for the Tensile Strength Data.

5.13 Partially Crossed Multifactor Studies

When a multifactor study is balanced and fully crossed, the contrasts for one factor will automatically be orthogonal to those of other factors. When the study is unbalanced or partially crossed, contrasts for different factors are usually nonorthogonal. This lack of orthogonality results in the contamination of the contrasts. For example, a contrast on Factor A may be contaminated by the effect of Factor B. Such contamination is called ***confounding.*** An example of partial confounding is provided by the contrasts defined earlier for the Coating Weight Data.

EXAMPLE 5.16: Partially Crossed Factors for the Coating Weight Data:

The Primer Viscosity, PV, occurs at three levels, while the Coating Viscosity, CV, occurs at two levels. A fully crossed design would require 3 x 2 = 6 subgroups.

Since there are only five subgroups in the study, Primer Viscosity is partially crossed with Coating Viscosity.

Of the four contrasts defined below, only Contrast 1 and Contrast 4 are orthogonal. All other pairings were shown to be nonorthogonal in Example 5.3, p.129. The F-ratios shown below were found in Example 4.6, p.119.

Treatment	1	2	3	4	5	
PV	20	20	25	30	30	
CV	15	25	15	25	15	
Averages	252.5	320	240	307.5	262.5	
Contrasts						*F-Ratio*
1. PV, 20 v 30	-1	-1	0	1	1	0.01
2. PV, 20 v 25	-1	-1	2	0	0	9.08*
3. PV, 25 v 30	0	0	-2	1	1	8.59*
4. CV, 15 v 25	-2	3	-2	3	-2	29.4*

Contrasts 2, 3, and 4 were found to be significant.

The Estimated Effect for Contrast 2 is $\hat{l} = -92.5 / 2 = -46.25$. *This would suggest that the coating weights average 46 grams* more *when the Primer Viscosity is 20 than when it is 25.*

The Estimated Effect for Contrast 3 is $\hat{l} = 90 / 2 = 45$, *which suggests that the coating weight averages 45 grams* more *when the Primer Viscosity is 30 than when it is 25.*

Thus, there would appear to be a strong dip in Coating Weight when the Primer Viscosity is equal to 25 units.

The Estimated Effect for Contrast 4 is $\hat{l} = 372.5 / 6 = 62.1$, *which suggests that the coating weight averages 62 grams* more *when the Coating Viscosity is 25 than when it is 15.*

While these three results seem fairly straightforward, the lack of orthogonality between the contrasts means that they are partially confounded, so that one significant effect can contaminate several contrasts, making the interpretation of the contrasts difficult.

A clearer idea of just how partial confounding can muddy the interpretation of the results is made possible by using a Response Plot for the Coating Weight Data.

EXAMPLE 5.17: Response Plot for the Coating Weight Data:

The Response Plot for the Coating Weight Data is shown below: Each subgroup is represented by the subgroup average in this response plot, and each subgroup occurs at a different point in the Factor Space defined by the two Factors.

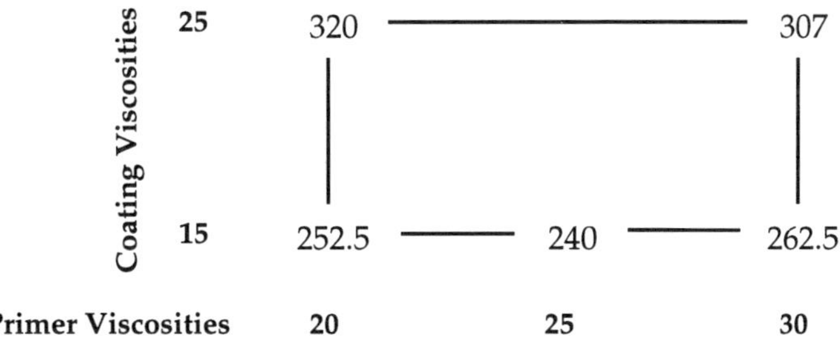

The Response Plot may be used to display the comparisons being made by the different contrasts. The circles will indicate those subgroups with one sign, the squares will denote those having the opposite sign, and the arrows will show the pairings used by the contrast.

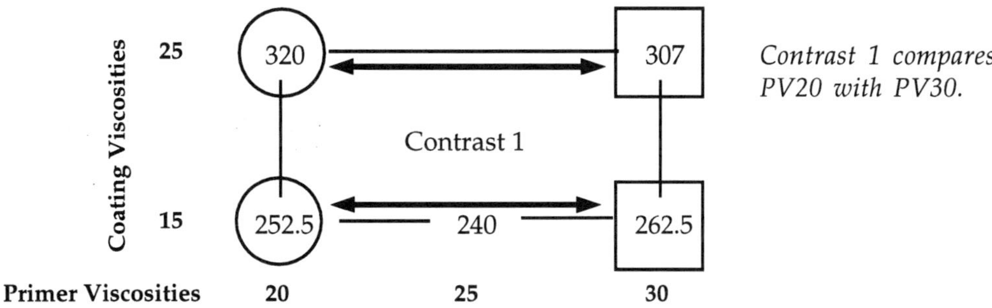

Contrasts	*Treatment*	*1*	*2*	*3*	*4*	*5*
	PV	*20*	*20*	*25*	*30*	*30*
	CV	*15*	*25*	*15*	*25*	*15*
Contrasts	***AVERAGES***	***252.5***	***320***	***240***	***307.5***	***262.5***
1. PV, 20 v 30		*-1*	*-1*	*0*	*1*	*1*
2. PV, 20 v 25		*-1*	*-1*	*2*	*0*	*0*
3. PV, 25 v 30		*0*	*0*	*-2*	*1*	*1*
4. CV, 15 v 25		*-2*	*3*	*-2*	*3*	*-2*

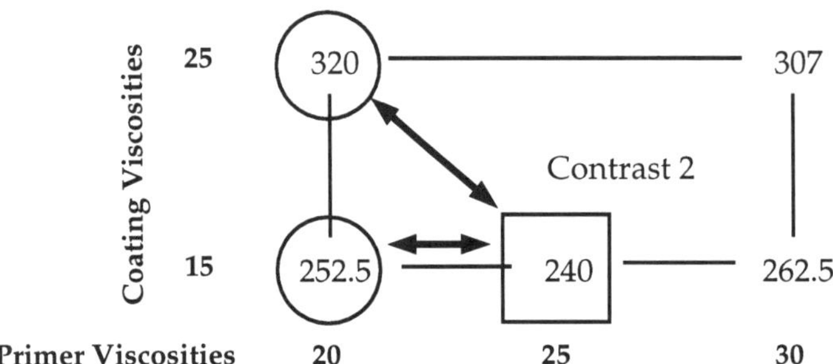

Contrast 2 compares PV20 with PV25, but it uses a pairing that allows CV to vary.

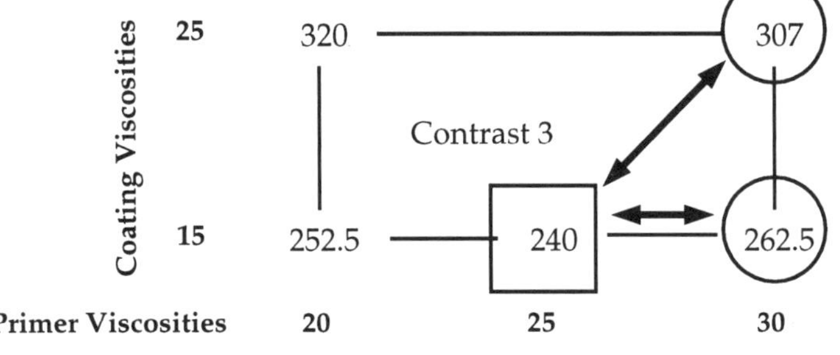

Contrast 3 compares PV25 with PV30, but it also uses a pairing that allows CV to vary.

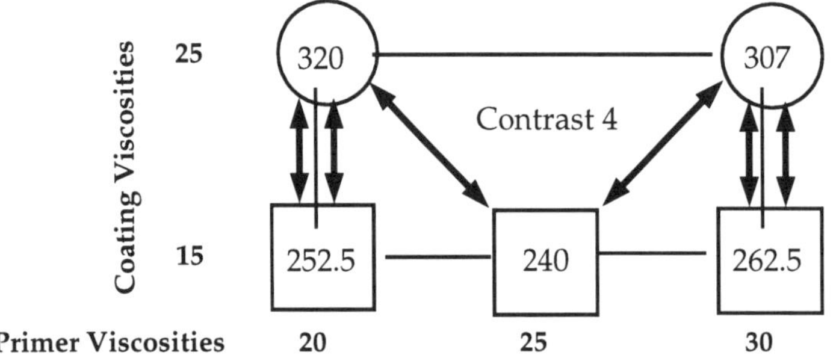

Contrast 4 compares CV15 with CV25, and it includes the same diagonal arrows seen in Contrast 2 and Contrast 3.

Since Contrast 2 and Contrast 3 each share the one diagonal arrow with Contrast 4, there is some cross-contamination between these contrasts. Contrast 2 actually compares PV20 with PV25 and CV25 with CV15. Likewise, Contrast 3 actually compares PV25 with PV30 and CV15 with CV25. Given the size of the Coating Viscosity Effect, and the inconsistent results when Contrasts 2 and 3 are interpreted as a Primer Viscosity Effect, it is most likely that the significance of Contrast 2 and Contrast 3 is due to contamination rather than a real effect on the part of the Primer Viscosity.

Once the nature of this partial confounding is clear, the obvious interpretation of the Coating Weight Data is that there is no detectable effect due to Primer Viscosity:

the whole story is told by the Coating Viscosity. (This may be seen by defining a new contrast on Primer Viscosity which uses only the CV15 data.)

Thus, diagonal comparisons will, in general, result in partial confounding. Partial confounding will contaminate the results and muddy the interpretation. Therefore, when defining contrasts, one should avoid the use of "diagonal" comparisons.

With partially crossed studies, it is the definition of the contrasts that will determine just what will be discovered from the data. If the contrasts are cross contaminated by considerable amounts of partial or total confounding, then the results will, of necessity be muddy and unclear. If the contrasts minimize the cross contamination, then the results will be clearer and easier to interpret.

Balanced and fully crossed studies are of interest because they permit orthogonal contrasts. In fact, the great efficiency with which information can be extracted from the data of a fully crossed study is primarily due to the inherent orthogonality of a fully crossed layout.

When faced with a partially crossed layout, the clearest interpretation of the data will be based on that analysis which minimizes the cross contamination between the various factors. There are several ways of defining such sets of contrasts, most of which are rather involved mathematically, but there is one approach which is simple enough to perform by hand. This approach is outlined in the next section. It should be of special interest for those faced with the task of snooping through databases.

5.14 Analyzing Messy Data

Interpretable contrasts can often be defined even when the data are very messy. The principle is to make up each contrast with balanced pairs of subgroups or other small contrasts that make sense. By pairing that which is to be compared while holding all other factors constant within each pairing one may even end up with orthogonal contrasts. But even for those contrasts which are nonorthogonal, the low degree of cross-contamination will still result in contrasts which are easy to interpret. Since results are seldom useful if they cannot be effectively communicated to others, this emphasis on interpretability is appropriate in all but the most sophisticated environments.

This method of constructing contrasts will be illustrated with a modified version of the Steel Parts Data.

EXAMPLE 5.18: Contrasts for Partially Crossed Version of the Steel Parts Data: Six of the 24 subgroups were deleted to represent missing treatment combinations such as might occur with data coming from a data base or from an experiment where certain treatment combinations are impossible. Moreover, only the first observation of each subgroup was retained in order to simulate one observation per treatment combination.

Subgroup	1	2	3	4	5	6	7	8	9	10	11	12	13	14	15	16	17	18	19	20	21	22	23	24
Heat Treat	W	W	W	W	W		W		W		W	W	L		L		L	L	L	L	L	L	L	
Machine	A	B	C	D	A		C		A		C	D	A		C		A	B	C	D	A	B	C	
Time	8	8	8	8	11		11		3		3	3	8		8		11	11	11	11	3	3	3	
Lengths:	6	7	2	6	3		2		6		2	8	1		0		1	4	0	6	3	7	0	

1. *The contrast for comparing the two levels of Heat Treatment is formed as follows:*

Heat Treat	W	W	W	W	W		W		W		W	W	L		L		L	L	L	L	L	L	L	
Machine	A	B	C	D	A		C		A		C	D	A		C		A	B	C	D	A	B	C	
Time	8	8	8	8	11		11		3		3	3	8		8		11	11	11	11	3	3	3	
	1	....	....	....	....	....	....	....	....	....	....	....	-1											
			1	....	....	....	....	....	....	....	....	....	....	....	-1											
					1	....	....	....	....	....	....	....	....	....	....	....	-1							
							1	....	....	....	....	....	....	....	....	....	....	-1						
									1	....	....	....	....	....	....	....	....	....	....	-1				
											1	....	....	....	....	....	....	....	....	....	....	-1		

Six pairwise comparisons are combined into one contrast that uses 12 of the 18 subgroups:

	1	2	3	4	5	6	7	8	9	10	11	12	13	14	15	16	17	18	19	20	21	22	23	24
1. Heat Treat	1	0	1	0	1		1		1		1	0	-1		-1		-1	0	-1	0	-1	0	-1	

2. *A contrast for comparing Machine A with Machine C is formed below:*

Heat Treat	W	W	W	W	W	W	W	W	W	L	L	L	L	L	L	L	L	L
Machine	A	B	C	D	A	C	A	C	D	A	C	A	B	C	D	A	B	C
Time	8	8	8	8	11	11	3	3	3	8	8	11	11	11	11	3	3	3
	1	....	-1															
					1	-1												
							1	-1										
										1	-1							
												1	....	-1				
																1	....	-1

Six pairwise comparisons are combined into one contrast:

2. Mach. AvC	1	0	-1	0	1	-1	1	-1	0	1	-1	1	0	-1	0	1	0	-1

3. *A contrast for comparing Machine B with Machine D is found below:*

Heat Treat	W	W	W	W	W	W	W	W	W	L	L	L	L	L	L	L	L	L
Machine	A	B	C	D	A	C	A	C	D	A	C	A	B	C	D	A	B	C
Time	8	8	8	8	11	11	3	3	3	8	8	11	11	11	11	3	3	3
		1	....	-1														
													1	....	-1			

Two pairwise comparisons are combined into one contrast:

3. Mach. BvD	0	1	0	-1	0	0	0	0	0	0	0	0	1	0	-1	0	0	0

4. *A contrast for comparing the average of Machines A and C with that of B and D is found by combining two smaller contrasts. Each smaller contrast holds Heat Treat. and Time of Day constant while making its comparison:*

Heat Treat	W	W	W	W	W	W	W	W	W	L	L	L	L	L	L	L	L	L
Machine	A	B	C	D	A	C	A	C	D	A	C	A	B	C	D	A	B	C
Time	8	8	8	8	11	11	3	3	3	8	8	11	11	11	11	3	3	3
4. AC v BD	1	-1	1	-1	0	0	0	0	0	0	0	1	-1	1	-1	0	0	0

5. *A contrast comparing 8 a.m. versus 11 a.m. is found below:*

Heat Treat	W	W	W	W	W	W	W	W	W	L	L	L	L	L	L	L	L	L
Machine	A	B	C	D	A	C	A	C	D	A	C	A	B	C	D	A	B	C
Time	8	8	8	8	11	11	3	3	3	8	8	11	11	11	11	3	3	3
	1	.....	....	.....	-1													
			1	....	....	-1												
										1	....	-1						
											1	....	....	-1				

Four pairwise comparisons are combined into one contrast:

5. Time 8v11	1	0	1	0	-1	-1	0	0	0	1	1	-1	0	-1	0	0	0	0

6. A contrast comparing 8 and 11 a.m. with 3 p.m. is found below:

Heat Treat	W	W	W	W	W	W	W	W	W	L	L	L	L	L	L	L	L	L
Machine	A	B	C	D	A	C	A	C	D	A	C	A	B	C	D	A	B	C
Time	8	8	8	8	11	11	3	3	3	8	8	11	11	11	11	3	3	3
	1	.....	....	.....	1	.....	-2											
			1	....	.....	1	....	-2										
										1	....	1	....	.....	....	-2		
											1	....	.....	1	....	.....	....	-2

Four smaller contrasts are combined into one contrast:

6. Time 8,11v3	1	0	1	0	1	1	-2	-2	0	1	1	1	0	1	0	-2	0	-2

These six contrasts are given below, along with their sums of squares.

Subgroup	1	2	3	4	5	6	7	8	9	10	11	12	13	14	15	16	17	18	19	20	21	22	23	24
Heat Treat	W	W	W	W	W		W		W		W	W	L		L		L	L	L	L	L	L	L	
Machine	A	B	C	D	A		C		A		C	D	A		C		A	B	C	D	A	B	C	
Time	8	8	8	8	11		11		3		3	3	8		8		11	11	11	11	3	3	3	
Lengths:	6	7	2	6	3		2		6		2	8	1		0		1	4	0	6	3	7	0	
Contrasts:																								
1. Heat Treat	1		1		1		1		1		1		-1		-1		-1		-1		-1		-1	
2. Mach. AvC	1		-1		1		-1		1		-1		1		-1		1		-1		1		-1	
3. Mach. BvD		1		-1														1		-1				
4. AC v BD	1	-1	1	-1													1	-1	1	-1				
5. Time 8v11	1		1		-1		-1						1		1		-1		-1					
6. Time 8,11v3	1		1		1		1		-2		-2		1		1		1		1		-2		-2	

Contrast 4 is not orthogonal to Contrast 6, otherwise this would be a mutually orthogonal set of contrasts. Given this high degree of orthogonality, a Scree Plot was used to compare the SS(C) values.

Contrast:	$\hat{C}$	*Adj. Term*	*SS(C)*
1. Heat	16	12	21.33
2. Mach. AvC	14	12	16.33
3. Mach. BvD	-1	4	0.25
4. AC v BD	-14	8	24.50
5. Time 8 v 11	3	8	1.125
6. Time 8,11 v 3	-7	24	2.04

Contrasts 4, 1 and 2 stand out. The partial confounding between Contrasts 4 and 6 would have made these contrasts hard to interpret if both had turned out to have large SS(C) values. Since only Contrast 4 has a large SS(C), it is unlikely that the contamination is appreciable.

The estimated effect of Heat Treatment is $\hat{l} = 2.67$.

The estimated effect of the difference between Machines A and C is $\hat{l} = 2.33$

The estimated difference between the average of Machines A and C and the average of Machines B and D is $\hat{l} = -3.5$.

Using all 96 observations from the complete data set, the corresponding comparisons were found to be significant, and the estimated effects were, respectively, 2.0, 2.5, and –3.6.

This approach for finding contrasts for a set of partially crossed factors will allow one to make sense of the data when other approaches fail. Nonorthogonal contrasts will result in partially confounded effects, and this can complicate the interpretation of the results, but with the help of response plots, it is often possible to construct coherent descriptions of the effects of at least some of the factors.

With irregular partial crossings the efficiency with which the information is extracted from the data will drop as the crossing becomes more partial. In extreme cases, the contrasts will become very sparse, each one using only a small number of the total number of subgroups. Thus, even though contrasts provide a way to extract information from irregular partial crossings, it is better to avoid such partial crossings whenever possible.

The ability to salvage messy data by defining contrasts in the manner described above is no excuse for deliberately collecting experimental data in an unplanned manner. The procedure outlined above is intended to be used primarily for snooping around in historical data.

EXERCISE 5.13 Analysis of Messy Data: 24 Subgroups from Data Set Eleven:

Twenty-four of the 128 subgroups in Data Set Eleven, p.341, were selected by the toss of a die. In addition the amount of data recorded for each subgroup was also determined by another toss of the die. Thus, the data on the next page represent a random selection of subgroups and a random selection of subgroup sizes from the 256 values listed in Data Set Eleven. Contrasts will be used to salvage what information may be obtainable from these data. The factors are labeled as A, B, C, D, E, F, and G, the response variable is labeled Y.

In order to facilitate comparison of treatment combinations the "plus" and "minus" notation given in the table for Data Set Eleven is replaced with "1" for the low level and "2" for the high level of each factor. This will allow the subgroups to be placed in a numerical ordering, which will make it easier to find the appropriate pairings. Note that due to the size of this data set the subgroups are listed in ROWS, and the contrast coefficients for each Factor are written in COLUMNS.

EXERCISE 5.13 Continued — Analysis of Messy Data for Data Set Eleven:

The contrasts shown for Main Effects for A, B, C, D, E, and F were obtained by the pairing approach outlined above. For example, the contrast for A was found by pairing Runs 122 and 58, Runs 120 and 56, and Runs 107 and 43. No other pairings would hold the other factors constant while allowing Factor A to change levels.

(a) Find the contrast coefficients for the main effect for Factor G:

	Factor Levels		Data		Contrast Coefficients for Main Effects								
Run	A BCD	EFG	Y_1	Y_2	A	B	C	D	E	F	G	$\bar{Y}$	s^2
122	1 1 1 1	2 2 1	47.02	46.42	-1	0	0	0	0	0		46.72	0.18
120	1 1 1 2	1 1 1	46.67	47.65	-1	0	-1	0	0	-1		47.16	0.4802
118	1 1 1 2	1 2 1	47.75	49.40	0	0	0	0	0	1		48.575	1.3613
117	1 1 1 2	1 2 2	48.65		0	0	0	0	0	0		48.65	
107	1 1 2 1	2 1 2	48.41		-1	0	0	0	0	0		48.41	
104	1 1 2 2	1 1 1	54.17	54.59	0	0	1	0	0	0		54.38	0.0882
63	2 1 1 1	1 1 2	44.71	46.46	0	0	0	-1	-1	0		45.585	1.5313
59	2 1 1 1	2 1 2	46.67		0	0	-1	-1	1	0		46.67	
58	2 1 1 1	2 2 1	45.79		1	0	0	0	0	0		45.79	
56	2 1 1 2	1 1 1	45.14		1	-1	0	0	-1	0		45.14	
55	2 1 1 2	1 1 2	47.58	46.94	0	-1	-1	1	-1	0		47.26	0.2048
52	2 1 1 2	2 1 1	46.00	47.48	0	0	0	0	1	0		46.74	1.0952
51	2 1 1 2	2 1 2	46.16	47.75	0	0	0	1	1	0		46.955	1.2641
44	2 1 2 1	2 1 1	47.08		0	0	0	0	0	0		47.08	
43	2 1 2 1	2 1 2	48.64		1	0	1	0	0	0		48.64	
39	2 1 2 2	1 1 2	55.16	56.82	0	0	1	0	0	-1		55.99	1.3778
37	2 1 2 2	1 2 2	55.80	57.87	0	0	0	0	0	1		56.835	2.1424
29	2 2 1 1	1 2 2	47.70		0	0	0	0	-1	0		47.70	
25	2 2 1 1	2 2 2	47.01	47.29	0	0	0	0	1	0		47.15	0.0392
24	2 2 1 2	1 1 1	45.96		0	1	0	0	0	0		45.96	
23	2 2 1 2	1 1 2	45.90		0	1	0	0	0	0		45.90	
15	2 2 2 1	1 1 2	50.20		0	0	0	0	0	0		50.20	
14	2 2 2 1	1 2 1	50.27		0	0	0	0	0	0		50.27	
3	2 2 2 2	2 1 2	56.26		0	0	0	0	0	0		56.26	

EXERCISE 5.13 Continued — Analysis of Messy Data for Data Set Eleven:

(b) Use the eleven subgroups with n = 2 to find the MSW term for these data.
(c) Find the Estimated Contrast for Factor D
(d) Find the Adjustment Term for Factor D (Hint, use the form in Chapter Four, p.117).
(e) Find the Sum of Squares and the F-ratio for Factor D.
(f) Draw a Response Plot for the Factors which appear to affect the Response Variable.

Contrast	Estimated Contrast	Adjustment Term	Sums of Squares	F-Ratios
A	-2.72	5.0	1.4797	1.667
B	-0.54	3.5	0.0833	0.09
C	17.92	4.0	80.28	90.44
D				
E	1.83	5.5	0.609	0.69
F	2.26	2.0	2.554	2.88
G	3.91	8.0	1.911	2.15

5.15 The Problem of Varying One Factor At A Time

Much of the experimentation done in industry uses the One-At-A-Time approach. While there are several variations of this approach, this procedure may be summarized in the following manner. In order to study a set of several factors, the experimenter performs a sequence of runs using a given piece of equipment: between one run and the next the experimenter will change the level of one of the factors, and between the next pair of runs the experimenter will change the level of another one of the factors. By successively changing the levels of each of the factors, the experimenter will eventually get around to changing all of the factor levels.

This approach will allow the experimenter to study (k-1) factors using only k experimental runs. While the small number of experimental runs is appealing, the inefficiency of this design will result in the loss of valuable information. In particular, none of the main effects will be estimated very efficiently, and there will be absolutely no information about possible interaction effects obtained.

The following examples based upon the Tensile Strength Data will be used to illustrate two different One-At-A-Time strategies. The approach outlined in the introductory paragraph is summarized by Figure 5.6(a) and is used in Example 5.19. The approach shown in Figure 5.6(b) is a common variation on the One-At-A-Time approach. It will be used in Example 5.20.

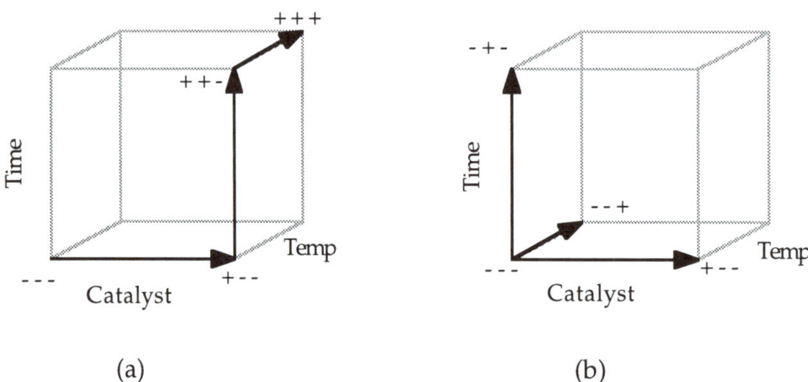

Figure 5.6: Different One-At-A-Time Strategies

One-At-A-Time experimentation will always have three serious drawbacks:

(1) there will be no way to detect the presence of interactions,
(2) the data will be used very inefficiently, and
(3) you could be misled about main effects.

The first problem is inherent with non-orthogonal partial crossings. The only way to detect an interaction effect between two factors is to systematically cross those two factors during the course of an experiment (using orthogonal partial crossings or fully crossed experiments). In the One-At-A-Time approach there is absolutely no crossing of the factors. In fact, the whole idea of this approach is to avoid any systematic crossing of the factors. Therefore, it should be no surprise to find that there is no way to detect interaction effects. If interaction effects are present in the data, they will confuse the interpretation of the results and may create inconsistencies from study to study.

EXAMPLE 5.19: One-At-A-Time Experiment for Tensile Strength Data: The three factors are still Catalyst, Time and Temperature, and the response variable is the tensile strength of a particular rubber compound. Two observations are obtained per subgroup, and a total of four subgroups are obtained. The response values shown are the last two response values listed in the corresponding subgroups in the complete data set, p.106.

Subgroup:	*1*	*5*	*7*	*8*
Catalyst	*1%*	*2%*	*2%*	*2%*
Time	*2m*	*2m*	*6m*	*6m*
Temp	*100°*	*100°*	*100°*	*120°*
Tensile	*995*	*1195*	*1200*	*980*
	1010	*1190*	*1210*	*990*
Averages	*1002.5*	*1192.5*	*1205*	*985*
s^2	*112.5*	*12.5*	*50*	*50*

Contrasts:						$\hat{C}$
	Catalyst	*-1*	*1*	*0*	*0*	*190*
	Time	*0*	*-1*	*1*	*0*	*12.5*
	Temp	*0*	*0*	*-1*	*1*	*-220*

(not a mutually orthogonal set of contrasts)

ANOVA Table:

Source	*SS*	*df*	*MS*	*F*
Catalyst	*36,100*	*1*	*36,100*	*642.**
Time	*156.25*	*1*	*156.25*	*2.78*
Temp	*48,400*	*1*	*48,400*	*860.**
Within	*225*	*4*	*56.25*	
	(non additive)			

We appear to have found the following effects:
As Catalyst goes from 1% to 2% the tensile strength goes up 190 units;
There is no detectable effect due to Time:
As Temperature goes from 100° to 120° the tensile strength goes down 220 units;

NONE OF THESE RESULTS ARE CORRECT (see pp.108 and 137).

EXAMPLE 5.20: Another One-At-A-Time Approach for the Tensile Strength Data:

Subgroup	1	5	3	2
Catalyst	1%	2%	1%	1%
Time	2m	2m	6m	2m
Temp	100°	100°	100°	120°
Tensile	995	1195	1125	1165
	1010	1190	1190	1170
Averages	1002.5	1192.5	1157.5	1167.5
s^2	112.5	12.5	2112.5	12.5

Contrasts:						$\hat{C}$
	Catalyst	-1	1	0	0	190
	Time	-1	0	1	0	155
	Temp	-1	0	0	1	165

(not a mutually orthogonal set of contrasts)

Source	SS	DF	MS	F
Catalyst	36100	1	36100	64.1*
Time	24025	1	24025	42.7*
Temp	27225	1	27225	48.4*
Within	2250	4	562.5	

(non-additive)

With this design we appear to have found the following effects:
As Catalyst goes from 1% to 2% the tensile strength goes up 190 units;
As Time goes from 2m to 6m the tensile strength goes up 155 units; and
As Temperature goes from 100° to 120° the tensile strength goes up 165 units.

These results contradict those obtained in Example 5.19.

ONCE AGAIN ALL OF THESE RESULTS ARE WRONG.

(compare these results with those on pp.108 and 137).

The inefficiency of the One-At-A-Time approach is seen in the fact that every contrast uses only two subgroups. Each and every main effect is estimated using only a fraction of the data collected. Moreover, as the number of factors increases, the proportion of the data used for each factor continues to drop. The One-At-A-Time approach guarantees that, regardless of the number of runs in the experiment, only two runs will be used for estimating each main effect.

Thus, if there are no interaction effects present, the One-At-A-Time approach may allow one to detect the larger main effects, but the analysis will be less revealing and less sensitive than is possible with screening designs such as those in Chapter Seven. Additionally, when strong interaction effects are present, the One-At-A-Time approach will yield results which are clouded and which may possibly be contradictory.

While contrasts allow one to analyze messy data and data from One-At-A-Time studies, it will always be much better to design the experiment to use an orthogonal partial crossing or a fully crossed layout (see Chapters 6 and 7).

One book on the market today suggests using two One-At-Time studies together. This approach will be illustrated using Data Set Eleven, p.341.

EXAMPLE 5.21 — Another Naive Approach to Multifactor Experiments:

The following 16 subgroups of size n=1 are drawn from the data given in Data Set Eleven. The Run numbers are those in the original data set.

Run No.	Treatment Combinations							Response Y_1	Contrast Coefficients for Main Effect Contrasts						
	A	B	C	D	E	F	G		A	B	C	D	E	F	G
1	+	+	+	+	+	+	+	56.94	1	1	1	1	1	1	1
128	–	–	–	–	–	–	–	46.31	–1	–1	–1	–1	–1	–1	–1
2	+	+	+	+	+	+	–	56.00	0	0	0	0	0	0	–1
127	–	–	–	–	–	–	+	44.64	0	0	0	0	0	0	1
3	+	+	+	+	+	–	+	56.26	0	0	0	0	0	–1	0
126	–	–	–	–	–	+	–	48.58	0	0	0	0	0	1	0
5	+	+	+	+	–	+	+	56.14	0	0	0	0	–1	0	0
124	–	–	–	–	+	–	–	43.71	0	0	0	0	1	0	0
9	+	+	+	–	+	+	+	50.90	0	0	0	–1	0	0	0
120	–	–	–	+	–	–	–	46.67	0	0	0	1	0	0	0
17	+	+	–	+	+	+	+	49.62	0	0	–1	0	0	0	0
112	–	–	+	–	–	–	–	49.90	0	0	1	0	0	0	0
33	+	–	+	+	+	+	+	56.81	0	–1	0	0	0	0	0
96	–	+	–	–	–	–	–	45.34	0	1	0	0	0	0	0
65	–	+	+	+	+	+	+	55.82	–1	0	0	0	0	0	0
64	+	–	–	–	–	–	–	46.25	1	0	0	0	0	0	0

The contrasts shown in columns above yield the following ANOVA Table:

Contrast	$\hat{C}$	SS(C)
A	1.06	0.28
B	–0.84	0.18
C	10.91	29.76
D	6.40	10.24
E	–1.80	0.81
F	2.95	2.18
G	–0.73	0.13

This approach has the rather dubious distinction of having collected sixteen observations while using only four observations for each contrast. Moreover, the first two observations are used repeatedly, while each of the other 14 observations is used only once. Thus, this approach utilizes the data inefficiently, provides no information about interaction effects, and may be misleading in the presence of strong interaction effects. Designs which require no more work, but which do not suffer from these drawbacks are given in Chapter Seven.

5.16 Nested Factors

In any multifactor experiment, each pair of factors may be either fully crossed, partially crossed, or nested. A pair of factors is said to be fully crossed if every possible combination of factor levels is present in the study. That is, if Factor A has three levels, and Factor B has five levels, then Factors A and B are fully crossed if and only if all 15 possible combinations are present in the study. If even one possible combination is missing, then the factors are said to be partially crossed. As more combinations are lost, the crossing becomes more partial, and the extraction of information can become severely limited.

Nesting is the opposite of crossing. Say Factor B and Factor A both have several levels present in the data. Factor B is said to be nested in Factor A if (1) it is impossible to cross Factor B and Factor A and (2) knowledge of the level of Factor B completely defines which level of Factor A is present. When Factor B is nested in Factor A, Factor B is said to be the "inferior factor," and Factor A is said to be the "superior factor."

An example of nested factors is provided by Example 5.5, p.133. There Parts are nested in Machine. Each part naturally and inevitably occurs in conjunction with one and only one machine—it is impossible to cross Parts with Machines. Parts 1, 2, and 3 came from Machine D, and Parts 4 and 5 came from Machine E: thus, Parts are nested in Machine. Parts are called the "inferior" or "nested" factor, and Machines is called the "superior" factor. When two factors are nested, knowledge of which level of the inferior factor is present will completely define which level of the superior factor is present.

When factors are nested they cannot interact. Factors may interact only if they are crossed (partially or fully).

The words inferior and superior denote a hierarchy, and therefore designs with nested factors are also called hierarchical designs.

If Factor B is nested in Factor A, then the data may be analyzed by using Factor B alone. Factor A simply exists as a partitioning of the levels of Factor B. This means that all contrasts on either Factor A or Factor B will look like contrasts defined for Factor B.

Of course, if Factor B is nested in Factor A, usually denoted as B(A), and if Factor B is fully crossed with Factor C, then Factor A is also fully crossed with Factor C. An example of this combination of crossed and nested factors is given in the next example.

EXAMPLE 5.22: Contrasts for Nested Factors:

An experiment is performed to evaluate the relative effectiveness of the old and a new method of operation. Since this change is proposed for company-wide adoption, the study is to be carried out in all plants. Thus, the factors in the study are (A) Plant (2 levels), (B) Machine (2 or 3 per plant), (C) Operator (2 or 3 per Machine), and (D) Method (old (O) and new (N)). The layout for this experimental study is shown below.

Subgroup Number:

1	2	3	4	5	6	7	8	9	10	11	12	13	14	15	16	17	18	19	20	21	22
Plant 1										Plant 2											
Machine 1				Machine 2						Machine 3				Machine 4				Machine 5			
Op.1		Op.2		Op.3		Op.4		Op.5		Op.6		Op.7		Op.8		Op.9		Op.10		Op.11	
O	N	O	N	O	N	O	N	O	N	O	N	O	N	O	N	O	N	O	N	O	N

Machines are nested in Plants: each machine is present in only one plant.

Operators are nested in Machines: each operator works with only one machine.

Operators are also nested in Plants

(therefore operators are said to be doubly nested).

Operators are fully crossed with Method: each operator uses each method, giving a total of 22 subgroups.

Since Operators are crossed with Method,

those factors in which Operators are nested are also crossed with Method.

A mutually orthogonal set of contrasts for these factors is shown below:

	Plant 1										Plant 2											
	Machine 1				Machine 2						Machine 3				Machine 4				Machine 5			
	Op.1		Op.2		Op.3		Op.4		Op.5		Op.6		Op.7		Op.8		Op.9		Op.10		Op.11	
	O	N	O	N	O	N	O	N	O	N	O	N	O	N	O	N	O	N	O	N	O	N
Plant	6	6	6	6	6	6	6	6	6	6	-5	-5	-5	-5	-5	-5	-5	-5	-5	-5	-5	-5
Mach(P)	3	3	3	3	-2	-2	-2	-2	-2	-2	0	0	0	0	0	0	0	0	0	0	0	0
Mach(P)	0	0	0	0	0	0	0	0	0	0	1	1	1	1	-1	-1	-1	-1	0	0	0	0
Mach(P)	0	0	0	0	0	0	0	0	0	0	1	1	1	1	1	1	1	1	-2	-2	-2	-2
O(PM)	1	1	-1	-1	0	0	0	0	0	0	0	0	0	0	0	0	0	0	0	0	0	0
O(PM)	0	0	0	0	1	1	-1	-1	0	0	0	0	0	0	0	0	0	0	0	0	0	0
O(PM)	0	0	0	0	1	1	1	1	-2	-2	0	0	0	0	0	0	0	0	0	0	0	0
O(PM)	0	0	0	0	0	0	0	0	0	0	1	1	-1	-1	0	0	0	0	0	0	0	0
O(PM)	0	0	0	0	0	0	0	0	0	0	0	0	0	0	1	1	-1	-1	0	0	0	0
O(PM)	0	0	0	0	0	0	0	0	0	0	0	0	0	0	0	0	0	0	1	1	-1	-1
Method	-1	1	-1	1	-1	1	-1	1	-1	1	-1	1	-1	1	-1	1	-1	1	-1	1	-1	1

The first 10 contrasts are essentially contrasts defined on the 11 operators. However, some of these contrasts represent differences between the levels of the two superior factors, plants and machines. Thus, while 11 operators should result in 10 degrees of freedom for operator differences, one of these degrees of freedom is used to compare the two plants. Another degree of freedom is used to compare the two machines within Plant 1, and two more degrees of freedom are used to compare the three machines within Plant 2. This leaves six degrees of freedom for comparing operators, and this is precisely enough to compare the operators within each nest (those operators using the same machine).

Contrast 5 compares the two operators using Machine 1 (within Plant 1). Contrasts 6 and 7 compare the three operators using Machine 2 (within Plant 1). Contrast 8 compares the two operators using Machine 3 (within Plant 2), Contrast 9 compares the two operators using Machine 4 (within Plant 2), and Contrast 10 compares the two operators using Machine 5 (within Plant 2).

The eleventh contrast is a contrast on the two methods, and the remaining 10 of the total of 21 contrasts will be the interaction contrasts defined by multiplying Contrasts 1 through 10 by Contrast 11. (Plants, Machines and Operators may interact with Methods since they are crossed with Methods.)

5.17 Using the Different Analysis Techniques

At this point the details have probably accumulated to the point that the reader may be confused about just when to use each analysis technique. This section is offered to help sort out when to use the different analysis techniques.

<u>One Factor At k Levels:</u>

Use One-Way ANOM.
May use One-Way ANOVA.
If subgroup size is constant, may use Tukey Post Hoc Test.
If all subgroups are of size n = 1, then plot histogram of responses and look for gaps or distinct clusters of response values.

<u>Multiple Factors, Fully Crossed:</u>

Use One-Way ANOM.
If Balanced, then get Main-Effect ANOM, and get Factorial ANOVA from computer if interactions appear to be likely.
May use ANOMR if interested.
If subgroup size is n = 1, then plot ANOM plot without decision limits, define orthogonal contrasts to obtain Single d.f. ANOVA, plot Scree Plot, and, if desired, plot the Normal Probability Plot.

<u>Multiple Factors, Completely Hierarchical:</u>

Use One-Way ANOM.
May use One-Way and Hierarchical ANOVA.
If Balanced, may use Tukey Post Hoc Test.
If n = 1, define orthogonal contrasts to obtain Single d.f. ANOVA, and find Scree Plot.

<u>Partially Hierarchical Designs:</u>

(Some pairs nested, other pairs fully crossed, no partial crossings of factors):
Use One-Way ANOM.
May collapse the hierarchical factors into one factor to obtain a design which appears to be fully crossed.
May be able to define a set of contrasts (possibly orthogonal) and use a Single Degree of Freedom ANOVA

Partially Crossed Designs:

If partial crossing is orthogonal, the orthogonal contrasts will be known:

use these contrasts to obtain the Single d.f. ANOVA and Scree Plot.

May plot One-Way ANOM if subgroup size exceeds n = 1.

If partial crossing is nonorthogonal, then data are said to be messy, and a salvage operation may be attempted using interpretable contrasts.

Whatever the analysis, the purpose is to filter out the noise so that you can concentrate on any signals present in the data. Once the potential signals have been identified, one should always summarize the findings with a plot of some kind. Typically these plots are either Response Plots or ANOM Plots.

UNTIL YOU HAVE SOME SORT OF PICTURE OF YOUR DATA, ORGANIZED IN SUCH A MANNER TO SHOW THE EFFECTS OF ANY SIGNALS PRESENT IN THE DATA, YOU HAVE NOT OBTAINED UNDERSTANDABLE AND USABLE RESULTS.

CHAPTER SIX

FRACTIONAL FACTORIAL DESIGNS

Beginning in this chapter a new convention will be used. Contrast coefficients will be written in columns rather than in rows. This is done to accommodate greater numbers of subgroups than can be listed across the page. By flipping the contrast coefficients into columns, all of the coefficients for a single contrast can be listed in one column, rather than in several rows. At the same time, subgroups will now be listed down the side of the tables, rather than across the top, and observations, averages, and ranges will be listed in columns instead of rows.

6.1 Redundancies of Full Factorial Designs

Special partially crossed multifactor layouts that are useful for preliminary studies are discussed in this chapter. The next section illustrates why such designs work, and the later sections describe specific characteristics of such designs. In order to have a basis for comparison, we begin with a fully crossed factorial study. The Factorial ANOVA for the complete version of the Steel Parts Data is given in the table below.

EXAMPLE 6.1: Factorial ANOVA for the Steel Parts Data:

SOURCE	*SS*	*DF*	*MS*	*F*
Heat Treatment	*100.042*	*1*	*100.042*	*16.1**
Machines	*393.416*	*3*	*131.139*	*21.1**
Time of Day	*12.896*	*2*	*6.448*	*1.04*
Heat x Machine	*1.542*	*3*	*0.514*	*0.08*
Heat x Time of Day	*1.645*	*2*	*0.823*	*0.13*
Machine x Time	*71.021*	*6*	*11.836*	*1.90*
Heat x Mach. x Time	*9.771*	*6*	*1.629*	*0.26*
Within	*447.500*	*72*	*6.215*	
Total	*1037.833*	*95*		

Only Heat Treatment and Machine main effects appear to be significant using decision rules based upon $\alpha = 0.05$. These two main effects are broken down into single degree of freedom components in the following table.

Contrast	$\hat{C}$	SS(C)	F	$\hat{l}$
Heat Treatment	24.5	100.042	16.1*	2.04
A vs C	15.25	77.521	12.47*	2.54
B vs D	1.25	0.521	0.08	
A,C vs B,D	-43.5	315.375	50.74*	-3.625

Thus, using all 96 observations, we find a detectable difference between Heat Treatments, a detectable difference between Machines A and C, and a detectable difference between the average of Machines A and C and the average of Machines B and D. No other main effects or interaction effects are found to be significant with these data.

The results noted above correspond exactly to the results noted in Example 5.18, p.170. Yet in Example 5.18 only 18 observations were used. A comparison between the estimated effects from the two studies is given in the following table.

TABLE 6.1 Estimated Contrast Effects for the Steel Parts Data

Contrast	Example 5.18 Partially Crossed 18 Observations	Example 6.1 Fully Crossed 96 Observations
Heat Treatment	2.67	2.04
A vs C	2.33	2.54
A,C vs B,D	-3.50	-3.625

Thus, using less than 20 percent of the data, the significant main effects were found and estimated in Example 5.18. Given this performance, what was the purpose of collecting a total of 96 observations? Why not use less data? One answer to this question concerns the way the data are used. Consider the degrees of freedom in the complete ANOVA table above. Six degrees of freedom were used for main effect contrasts. Seventeen degrees of freedom were used for interaction contrasts, and 72 degrees of freedom were used for estimation of MSW. The 72 degrees of freedom for MSW will provide a solid foundation for the F-ratios, which can be important when one is conducting a conservative analysis. However, in an exploratory study the discriminatory power of the F-ratio is not so crucial, and a large within-subgroup degrees of freedom is unnecessary. In effect, 72 of the 96 observations were obtained in order to improve the power of the analysis. In a preliminary analysis such power is not always necessary.

For the purposes of a preliminary or exploratory study, many observations could be eliminated by reducing the degrees of freedom for MSW. In the Steel Parts Data, subgroups of size $n = 2$ would give 24 degrees of freedom for MSW, but this would still require 48 observations in all. A further reduction in the number of observations is possible by dropping to one observation per subgroup. While cutting the number of observations down to 24, this will also eliminate the use of MSW for the F-ratios. This will usually not be a severe handicap in an exploratory study. With k observations, one degree of freedom is available for estimating the average, and k-1 degrees of freedom are available for a maximal mutually orthogonal set of contrasts. However, with the Steel Parts Data, only 6 of these 23 degrees of freedom are used for estimation of main effects. 11 d.f. are used for two-factor interactions, and 6 d.f. are used for contrasts involving three-factor interaction effects. If one is interested only in the main effects, or only in the main effects and the two-factor interactions, then 24 observations may still be excessive.

Thus, in any fully crossed and balanced multifactor study, many of the degrees of freedom are used to estimate within subgroup variation and high-order interactions. If the study is an exploratory one, the mean square within and the high-order interaction effects may well be of little interest. For this reason preliminary and exploratory studies will often be carried out with Fractional Factorial Designs.

6.2 Some 2^{k-p} Fractional Factorial Designs

Among the class of all partially crossed multifactor designs a special subset consists of those designs that represent a fraction of some fully crossed design. Some of these designs will be constructed in this section. Since most of the commonly used Fractional Factorial Designs are tabled, the actual methods of constructing these designs do not have to be mastered by the reader. The construction is shown here in order for the reader to see how these designs are related to Full Factorial Designs. A second objective of the following sections is the introduction of certain concepts and the common terminology associated with Fractional Factorial Designs.

To understand the relationship between a Fractional Factorial and a Full Factorial Design, it is best to begin with a Full Factorial Design, such as the 2^4 design given in the Flex Rate Data. (Four factors at two levels each require 2 x 2 x 2 x 2 = 2^4 subgroups for a fully crossed design.)

EXAMPLE 6.2: *The Flex Rate Data:*

Four factors, each at two levels, are studied in a Fully Crossed Design.
With one observation per combination, k = 16 and n = 1.
For brevity, the "subgroups" are called "Runs," since they are hardly "groups."
Factor A = Formulation (1 or 2)
Factor B = Cycle Time (15 sec or 25 sec)
Factor C = Pressure (300 psi or 375 psi)
Factor D = Temperature (110 or 130)
Response Y = The Flex Rate of the product in pounds per inch.

The Run Numbers, Factor Levels, and Response Values are given below:

Run	A	B	C	D	Response
1	1	15	300	110	71
2	2	15	300	110	73
3	1	25	300	110	74
4	2	25	300	110	75
5	1	15	375	110	77
6	2	15	375	110	77
7	1	25	375	110	78
8	2	25	375	110	80
9	1	15	300	130	71
10	2	15	300	130	72
11	1	25	300	130	74
12	2	25	300	130	74
13	1	15	375	130	77
14	2	15	375	130	77
15	1	25	375	130	76
16	2	25	375	130	78

The design of the study described in the Flex Rate Data is already rather efficient. Four factors are studied at two levels each with only sixteen observations. However, it will be used to obtain some fractional factorial designs. The sixteen runs of the Flex Rate Data allow for fifteen mutually orthogonal contrasts. To find these 15 contrasts begin with the contrasts for each of the four main effects, A, B, C, and D. Since A, B, C, and D are fully crossed, their contrasts will be orthogonal, and the interaction contrasts may be obtained from these four main effect contrasts by appropriate multiplications.

EXAMPLE 6.3: *Contrasts for the Flex Rate Data:*

The labels at the top of each column denote the main effect or interaction effect represented by the contrast coefficients in that column.

Run	*A*	*B*	*AB*	*C*	*AC*	*BC*	*ABC*	*D*	*AD*	*BD*	*ABD*	*CD*	*ACD*	*BCD*	*ABCD*	*Y*
1	-	-	+	-	+	+	-	-	+	+	-	+	-	-	+	71
2	+	-	-	-	-	+	+	-	-	+	+	+	+	-	-	73
3	-	+	-	-	+	-	+	-	+	-	+	+	-	+	-	74
4	+	+	+	-	-	-	-	-	-	-	-	+	+	+	+	75
5	-	-	+	+	-	-	+	-	+	+	-	-	+	+	-	77
6	+	-	-	+	+	-	-	-	-	+	+	-	-	+	+	77
7	-	+	-	+	-	+	-	-	+	-	+	-	+	-	+	78
8	+	+	+	+	+	+	+	-	-	-	-	-	-	-	-	80
9	-	-	+	-	+	+	-	+	-	-	+	-	+	+	-	71
10	+	-	-	-	-	+	+	+	+	-	-	-	-	+	+	72
11	-	+	-	-	+	-	+	+	-	+	-	-	+	-	+	74
12	+	+	+	-	-	-	-	+	+	+	+	-	-	-	-	74
13	-	-	+	+	-	-	+	+	-	-	+	+	-	-	+	77
14	+	-	-	+	+	-	-	+	+	-	-	+	+	-	-	77
15	-	+	-	+	-	+	-	+	-	+	-	+	-	+	-	76
16	+	+	+	+	+	+	+	+	+	+	+	+	+	+	+	78

A new convention is introduced in this array. Since all of the contrast coefficients are either +1 or -1, the value of the coefficient has been dropped for simplicity, and only the *sign* is printed.

A one-half replicate of the design for the Flex Rate Data can be obtained by choosing any one contrast in the table above and selecting only those runs that correspond to a plus sign in that contrast. Since the contrast used in this manner will be lost to the analysis, it is customary to choose the contrast for one of the higher order interactions.

EXAMPLE 6.4: *A One-Half Replicate for the Flex Rate Data:*

The ABC interaction contrast was used to define the following one-half replicate of the design above.

Run	*A*	*B*	*AB*	*C*	*AC*	*BC*	*ABC*	*D*	*AD*	*BD*	*ABD*	*CD*	*ACD*	*BCD*	*ABCD*	*Y*
2	+	-	-	-	-	+	+	-	-	+	+	+	+	-	-	*73*
3	-	+	-	-	+	-	+	-	+	-	+	+	-	+	-	*74*
5	-	-	+	+	-	-	+	-	+	+	-	-	+	+	-	*77*
8	+	+	+	+	+	+	+	-	-	-	-	-	-	-	-	*80*
10	+	-	-	-	-	+	+	+	+	-	-	-	-	+	+	*72*
11	-	+	-	-	+	-	+	+	-	+	-	-	+	-	+	*74*
13	-	-	+	+	-	-	+	+	-	-	+	+	-	-	+	*77*
16	+	+	+	+	+	+	+	+	+	+	+	+	+	+	+	*78*

This one-half replicate has only eight runs or subgroups of size one. While the ABC column is filled with plus signs, the rest of the columns define contrasts. Thus, there are fourteen contrasts left. But these fourteen contrasts are not unique. First of all, with only eight runs or subgroups, there can be no more than seven contrasts in a mutually orthogonal set. This means that there must be some confounding between the contrasts above. Inspection of the table shows that pairs of contrasts are ***totally confounded***.

The contrasts for A and for BC are identical. The main effect for A is said to be totally confounded with the BC interaction effect. Likewise, main effect B is totally confounded with AC. Main effect C is totally confounded with AB. Main effect D is totally confounded with ABCD. The AD interaction effect is totally confounded with BCD. The BD interaction effect is totally confounded with ACD, and ABD is totally confounded with CD. This leaves a set of seven mutually orthogonal contrasts, and each contrast represents the sum of the two confounded effects. The two effect names are called ***aliases*** for each of the contrasts. Any One-Half Replicate defined by a contrast in the fully crossed design will have two totally confounded effects represented by each contrast.

The ABC interaction "contrast" consists of a string of +1 coefficients. The terminology for this condition is "The ABC Interaction Effect is confounded with the mean." The origin of this expression is the fact that if one used these "coefficients" to compute the estimated contrast effect for ABC one would obtain the Grand Average. Whenever an effect is said to be confounded with the mean, that effect will have been used to define a split of the orthogonal array.

EXAMPLE 6.5: A One-Half Replicate for the Flex Rate Data:

Removing the redundant contrasts, the One-Half Replicate looks like:

Run	A BC	B AC	AB C	D ABCD	AD BCD	BD ACD	ABD CD	Y
2	+	-	-	-	-	+	+	73
3	-	+	-	-	+	-	+	74
5	-	-	+	-	+	+	-	77
8	+	+	+	-	-	-	-	80
10	+	-	-	+	+	-	-	72
11	-	+	-	+	-	+	-	74
13	-	-	+	+	-	-	+	77
16	+	+	+	+	+	+	+	78

This One-Half Replicate can be said to be a 2^{4-1} design. This notation indicates a One-Half Replicate (2^{-1}) of a sixteen run design (2^4) involving four factors at two levels each.

This One-Half Replicate could be split to obtain a One-Quarter Replicate using the same technique. Those runs that correspond to the plus signs in one of the contrasts could be taken as a One-Quarter Replicate.

EXAMPLE 6.6: The First One-Quarter Replicate for the Flex Rate Data:

Using the ABD+CD contrast to define a split yields the following:

Run	A BC	B AC	AB C	D ABCD	AD BCD	BD ACD	ABD CD	Y
2	+	-	-	-	-	+	+	73
3	-	+	-	-	+	-	+	74
13	-	-	+	+	-	-	+	77
16	+	+	+	+	+	+	+	78

Which, with the confounding, collapses down to the following.

Run	A BC BD ACD	B AC AD BCD	AB C D ABCD	Y
2	+	-	-	73
3	-	+	-	74
13	-	-	+	77
16	+	+	+	78

This design can be denoted as a 2^{4-2} design. While there are only 2^2 runs, there are four factors, and the design is a One-Quarter Replicate design.

The degree of confounding in a Fractional Factorial Design will depend upon the amount of fractionation. One-Quarter Replicates will have four main effects or interaction effects represented by each contrast. In the One-Quarter Replicate shown above there are twelve effects represented by the three contrasts. The fully crossed design had 15 effects. Where are the other three effects? They have been used to generate this One-Quarter Replicate, and cannot be estimated.

If one used the minus signs instead of the plus signs in defining a split, one would obtain another fraction of the original design. Returning to the One-Half Replicate shown in Example 6.5, the negative half of the ABD+CD contrast will define the following One-Quarter Replicate:

EXAMPLE 6.7: A Second One-Quarter Replicate for the Flex Rate Data:

The confounding pattern is not as obvious here as it was before.

Run	*A*	*B*	*AB*	*D*	*AD*	*BD*	*ABD*	*Y*
	BC	*AC*	*C*	*ABCD*	*BCD*	*ACD*	*CD*	
5	-	-	+	-	+	+	-	77
8	+	+	+	-	-	-	-	80
10	+	-	-	+	+	-	-	72
11	-	+	-	+	-	+	-	74

A + BC is confounded with the negative of the BD + ACD contrast.
B + AC is confounded with the negative of the AD + BCD contrast.
AB + C is confounded with the negative of the D + ABCD contrast.

Run	*A*	*B*	*AB*	*Y*
	BC	*AC*	*C*	
	-BD	*-AD*	*-D*	
	-ACD	*-BCD*	*-ABCD*	
5	-	-	+	77
8	+	+	+	80
10	+	-	-	72
11	-	+	-	74

The same approach can be used with the One-Half Replicate defined by the negative signs of the ABC contrast. These two One-Quarter Replicates are given below:

EXAMPLE 6.8: A Third and a Fourth One-Quarter Replicate for the Flex Rate Data:

Using the negative signs of the ABC contrast yields a different One-Half Replicate:

Run	*A*	*B*	*AB*	*C*	*AC*	*BC*	*ABC*	*D*	*AD*	*BD*	*ABD*	*CD*	*ACD*	*BCD*	*ABCD*	*Y*
1	-	-	+	-	+	+	-	-	+	+	-	+	-	-	+	71
4	+	+	+	-	-	-	-	-	-	-	-	+	+	+	+	75
6	+	-	-	+	+	-	-	-	-	+	+	-	-	+	+	77
7	-	+	-	+	-	+	-	-	+	-	+	-	+	-	+	78
9	-	-	+	-	+	+	-	+	-	-	+	-	+	+	-	71
12	+	+	+	-	-	-	-	+	+	+	+	-	-	-	-	74
14	+	-	-	+	+	-	-	+	+	-	-	+	+	-	-	77
15	-	+	-	+	-	+	-	+	-	+	-	+	-	+	-	76

The reader should compare this array with that shown in Example 6.4, p.188.

Once again, the confounding is a negative confounding.

When the confounded pairs are combined, this One-Half Replicate becomes:

Run	*A* -*BC*	*B* -*AC*	*AB* -*C*	*D* -*ABCD*	*AD* -*BCD*	*BD* -*ACD*	*ABD* -*CD*	*Y*
1	-	-	+	-	+	+	-	71
4	+	+	+	-	-	-	-	75
6	+	-	-	-	-	+	+	77
7	-	+	-	-	+	-	+	78
9	-	-	+	+	-	-	+	71
12	+	+	+	+	+	+	+	74
14	+	-	-	+	+	-	-	77
15	-	+	-	+	-	+	-	76

The positive portion of the ABD contrast can be used to define a third One-Quarter Replicate:

Run	*A* -*BC* *BD* -*ACD*	*B* -*AC* *AD* -*BCD*	*AB* -*C* *D* -*ABCD*	*Y*
6	+	-	-	77
7	-	+	-	78
9	-	-	+	71
12	+	+	+	74

While the negative portion of the ABD contrast can be used to define a fourth One-Quarter Replicate:

Run	A -BC -BD ACD	B -AC -AD BCD	AB -C -D ABCD	Y
1	-	-	+	71
4	+	+	+	75
14	+	-	-	77
15	-	+	-	76

Thus, by using various contrasts to define splits of a design, a fractional design can be created. For each split there will be some corresponding confounding of the effects. While inspection served in the cases above, there is a simple method of finding the aliases for a particular contrast that will be described in Section 6.4.

6.3 Using Fractional Factorial Designs

The One-Quarter Replicates given in section two of this chapter will serve to illustrate how Fractional Factorial Designs can be used to discover the basic relationships between the factors and the response variable. We begin by analyzing the data for the First One-Quarter Replicate.

EXAMPLE 6.9: The First One-Quarter Replicate of the Flex Rate Data:

Run	A BC BD ACD	B AC AD BCD	AB C D ABCD	Y
2	+	-	-	73
3	-	+	-	74
13	-	-	+	77
16	+	+	+	78

The sums of squares for the three contrasts are:

Contrast	$\hat{C}$	SS(C)	$\hat{l}$
A + BC + BD + ACD	0	0	
B + AC + AD + BCD	2	1.0	
AB + C + D + ABCD	8	16.0	4.0

A comparison of the SS(C) values for this rather minimal example suggests that the AB contrast is the most interesting. This might be the AB interaction, the C main effect, the D main effect, or the ABCD interaction. Based on these data alone it is impossible to tell which.

The only way to untangle this confounding is by means of a follow-up experiment. This was done using the Second One-Quarter Replicate for the Flex Rate Data, shown in Example 6.7, p.190.

EXAMPLE 6.10: The Second One-Quarter Replicate of the Flex Rate Data:

Run	*A* *BC* *-BD* *-ACD*	*B* *AC* *-AD* *-BCD*	*AB* *C* *-D* *-ABCD*	*Y*
5	-	-	+	77
8	+	+	+	80
10	+	-	-	72
11	-	+	-	74

The sums of squares for the three contrasts are:

Contrast	$\hat{C}$	SS(C)	$\hat{l}$
A + BC - BD - ACD	1	0.25	
B + AC - AD - BCD	5	6.25	
AB + C - D - ABCD	11	30.25	5.5

Once again the SS for the AB contrast is by far the largest SS. This effect can be said to be due to the AB interaction, the C main effect, the negative of the D main effect, or the negative of the ABCD interaction. Combining the two studies we can write the alias equations:

$$4.0 = AB + C + D + ABCD \text{ effect}$$

and

$$5.5 = AB + C - D - ABCD \text{ effect}$$

which makes it unlikely that either the D main effect or the ABCD interaction have a detectable effect upon the response variable. (These two effects changed signs but the estimated effect did not change sign.) Thus, this effect has been narrowed down to either the AB interaction or the C main effect or the sum of these two effects. Yet another experiment will be required to decide between these options.

EXAMPLE 6.11: The Third One-Quarter Replicate of the Flex Rate Data:

Run	*A* *-BC* *BD* *-ACD*	*B* *-AC* *AD* *-BCD*	*AB* *-C* *D* *-ABCD*	*Y*
6	+	-	-	77
7	-	+	-	78
9	-	-	+	71
12	+	+	+	74

The sums of squares for the three contrasts are:

Contrast	$\hat{C}$	SS(C)	$\hat{l}$
A	2	1.0	
B	4	4.0	
AB	-10	25.00	-5

Yet again, the AB contrast has the largest SS. This time, this contrast represents the AB interaction, the negative of the C main effect, the D main effect, and the negative of the ABCD interaction. Having eliminated the D main effect and the ABCD interaction, and given the sign change, it should be apparent that this contrast represents the main effect for Factor C. The alias equations from the three studies are:

$$4.0 = AB + C + D + ABCD$$

$$5.5 = AB + C - D - ABCD$$

and

$$-5.0 = AB - C + D - ABCD.$$

Combining these three results, an estimate of the effect of Factor C is:

$$\hat{l} = (1/3)\,[\,4.0 + 5.5 - (-5.0)\,] = 4.833.$$

As Factor C goes from its low level to its high level, the response variable increases, on the average, about 5 units. Moreover, this effect dominates all others among these four factors.

EXERCISE 6.1: Interpreting Contrast B for the Flex Rate Data:

The estimated effects of Contrast B in the three One-Quarter Replicates given in Examples 6.9, 6.10, and 6.11 are, respectively, 1.0, 2.5, and 2.0. Write the three alias equations for Contrast B.

Do these values estimate the main effect for Factor B, the interaction of A and C, the interaction of A and D, or the interaction of B, C and D?

Get a combined value for the estimate of this effect.

There are two distinct advantages to the approach illustrated above. First of all, the replication of the same pattern of responses from study to study confirms that the effects that stand out above the rubble are indeed real effects instead of random variation. This is the most reliable indicator of real effects. ***When it comes to separating real effects from random variation, no probabilistic procedure (such as F-tests) can approach the reliability of the nontrivial replication of results.*** By getting the same pattern three times (even though the treatment combinations present in each One-Quarter Replicate were different), one can be sure that the effect represented by contrast AB is indeed real.

Secondly, one can discover what is happening without performing the complete 2^4 experiment. Admittedly, in the example shown there is only a slight difference between the 16 runs of the fully crossed design and the 12 runs of the three One-Quarter Replicates, but this is primarily due to the small size of the fully crossed design. In other situations, the use of Fractional Factorials can reduce the number of runs by orders of magnitude.

These two characteristics of the use of a sequence of Fractional Factorials make this approach most appropriate for industrial experimentation. Individually, each Fractional Factorial Design reveals only a portion of the overall pattern, but collectively, in sequence, they reveal the whole pattern, and they will usually do this long before the total number of runs approaches the number of runs needed for a fully crossed factorial design. Moreover, when using a sequence of small studies with one observation per treatment combination, it is unnecessary to randomize the order of the treatment combinations in the study. This will usually make the study much easier to perform. This point will be discussed more completely in the next chapter.

Another illustration of the use of the sequential approach to multifactor studies is provided by the following examples. The data in these examples comes from Data Set Eleven in the Appendix. In the complete data set there are 128 subgroups of size n = 2, involving seven factors at two levels each, for a total of 256 observations. Instead of analyzing the complete set of data, two One-Sixteenth Replicates will be used to "sample" the data. Each of these Fractional Factorial Designs will have only 8 runs, and each will use only the first observation listed for the subgroups selected.

A One-Sixteenth Replicate of the Full Factorial Design for Data Set Eleven is a 2^{7-4} design. Such a design will require only 8 runs, but each contrast will have 16 aliases. However, if we restrict our attention to the main effects and the two-factor interactions, there will be only four aliases per contrast. The particular 2^{7-4} designs used are Fractional Factorial Designs that are given in the next chapter. The Run numbers shown identify the subgroup in the table for Data Set Eleven in the Appendix, p.341. Example 6.12 uses a Reflected 8-Run Plackett-Burman Design, while Example 6.13 uses a Basic 8-Run Plackett-Burman Design.

EXAMPLE 6.12: A Fractional Factorial Study for Data Set Eleven:

A One-Sixteenth Replicate of a 2^7 study is performed to examine seven factors. The contrasts, aliases and responses are given below.

Run	A	B	C	D	E	F	G	Response
	BC	AC	AB	AE	AD	AG	AF	
	DE	DF	DG	BF	BG	BD	BE	
	FG	EG	EF	CG	CF	CE	CD	
								Y_1
1	+	+	+	+	+	+	+	56.94
16	+	+	+	-	-	-	-	48.34
61	+	-	-	-	-	+	+	46.11
52	+	-	-	+	+	-	-	46.00
103	-	-	+	+	-	-	+	56.60
106	-	-	+	-	+	+	-	50.39
91	-	+	-	-	+	-	+	45.10
86	-	+	-	+	-	+	-	49.47

The additional 12 aliases for each contrast involve three-factor and higher-order interactions.

The estimated contrasts and the SS(C) values for these data are:

Contrast	$\hat{C}$	SS(C)	$\hat{l}$
A	-4.17	2.1736	
B	0.75	0.0703	
C	25.59	81.8560	6.4
D	19.07	45.4581	4.8
E	-2.09	0.5460	
F	6.87	5.8996	1.7
G	10.55	13.9128	2.6

The Scree Plot suggests that Contrasts C, D and possibly G stand out above the rubble. Just what these contrasts represent is still not known. They could be main effects, or they could be interactions. But we can focus our attention on the effects represented by these four contrasts.

The way to untangle the confounding is to run a different Fractional Factorial study. If the right Fractional Factorial is used, the confounding pattern will complement the one above, and it will be possible to separate the main effects from the interaction effects.

EXAMPLE 6.13: A Second Fractional Factorial for Data Set Eleven:

The contrasts, aliases, and responses are shown below:

Run	*A*	*B*	*C*	*D*	*E*	*F*	*G*	*Response*
	-BC	*-AC*	*-AB*	*-AE*	*-AD*	*-AG*	*-AF*	
	-DE	*-DF*	*-DG*	*-BF*	*-BG*	*-BD*	*-BE*	
	-FG	*-EG*	*-EF*	*-CG*	*-CF*	*-CE*	*-CD*	
								Y_1
128	-	-	-	-	-	-	-	*46.31*
113	-	-	-	+	+	+	+	*47.53*
68	-	+	+	+	+	-	-	*53.75*
77	-	+	+	-	-	+	+	*50.81*
26	+	+	-	-	+	+	-	*47.29*
23	+	+	-	+	-	-	+	*45.90*
38	+	-	+	+	-	+	-	*57.91*
43	+	-	+	-	+	-	+	*48.64*

The estimated contrasts and the SS(C) values for these data are:

Contrast	$\hat{C}$	*SS(C)*	$\hat{l}$
A	1.34	0.2245	
B	-2.64	0.8712	
C	24.08	72.4808	6.0
D	12.04	18.1202	3.0
E	-3.72	1.7298	
F	8.94	9.9905	2.2
G	-12.38	19.1581	-3.1

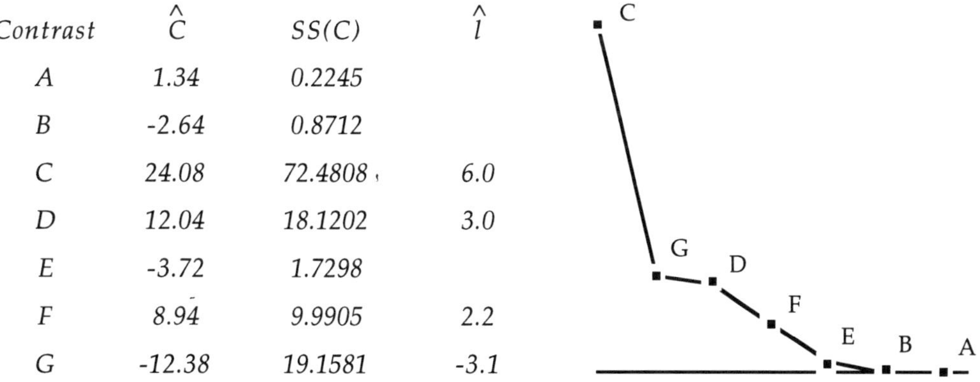

The Scree Plot suggests that contrasts C, G, D and possibly F represent signals within these data. Here F is more clearly indicated than it was before. If F had been part of the rubble, it is unlikely that it would have appeared in the same relative position in both studies.

Combining the different confounding patterns, we have two alias equations for Contrast C:

$$C + AB + DG + EF = 6.4$$

and

$$C - AB - DG - EF = 6.0$$

which suggest that Contrast C represents the main effect for Factor C.

The two alias equations for Contrast D are:

$$D + AE + BF + CG = 4.8$$

and

$$D - AE - BF - CG = 3.0$$

which suggest that Contrast D represents the main effect for Factor D.

The two alias equations for Contrast F are:

$$F + AG + BD + CE = 1.7$$

and

$$F - AG - BD - CE = 2.2$$

which suggest that Contrast F represents the main effect for Factor F.

The two alias equations for Contrast G are:

$$G + AF + BE + CD = 2.6$$

and

$$G - AF - BE - CD = -3.1$$

which suggest that Contrast G represents an interaction effect. Since neither Factors A nor B were found to have an impact, the best bet is the CD interaction, but this is essentially a guess.

Thus, with sixteen observations, the seven factors and 21 two-factor interactions have been narrowed down to three main effects and one of three possible two-factor interactions. A fully crossed factorial study involving three factors requires only 8 subgroups, so if further confirmation is needed, it can be had with an additional 8 or 16 observations. And only 6 to 12 percent of the total amount of data shown in Data Set Eleven was needed to make these discoveries.

It might also be instructive to compare that which has been discovered with these 16 runs with that which was discovered in Example 5.21, p.177. The Doubled One-At-A-Time Approach of Mr. Bhote used 16 runs to study the same set of 7 factors. It did identify the Main Effect C and the Main Effect D as potential signals. This pair of 8-Run Plackett-Burman Studies identifies as potential signals the Main Effect C, the Main Effect D, the Main Effect F, and one of three possible two-factor interactions. Therefore, with the same amount of work, one can obtain more insight into the relationships between a set of factors and a response variable using Fractional Factorial Designs than with the naive One-At-A-Time approaches.

It should be apparent at this point that the power of Fractional Factorials lies in the orthogonal contrasts. With the proper choice of fractions for the sequence of experiments, these orthogonal contrasts make the maximum use of the data available, and they make the results interpretable. For this reason certain Fractional Factorial Designs are more useful than others. It is these designs that are the topic of the next chapter.

EXERCISE 6.2: Another Fractional Factorial Study for Data Set Eleven:

A partially reflected version of the design in Example 6.13 gives the following results when used with Data Set Eleven:

Run	A	B	C	D	E	F	G	Response
	-BC	-AC	-AB	AE	AD	-AG	-AF	
	DE	DF	DG	BF	-BG	BD	-BE	
	-FG	-EG	-EF	CG	-CF	-CE	CD	
								Y_1
120	-	-	-	+	-	-	-	46.67
121	-	-	-	-	+	+	+	47.09
76	-	+	+	-	+	-	-	47.08
69	-	+	+	+	-	+	+	57.98
18	+	+	-	+	+	+	-	49.37
31	+	+	-	-	-	-	+	45.02
46	+	-	+	-	-	+	-	50.30
35	+	-	+	+	+	-	+	55.56

Find the estimated contrasts and the SS(C) values for these data:

Contrast	$\hat{C}$	SS(C)	$\hat{l}$
A			
B			
C			
D			
E			
F			
G			

<u>EXERCISE 6.3: Combining Results Of Fractional Factorials for Data Set Eleven:</u>
Using the results of Exercise 6.2 and Examples 6.12 and 6.13:

Identify the interaction effect represented by Contrast G, and

Plot the response plot for the active factors.

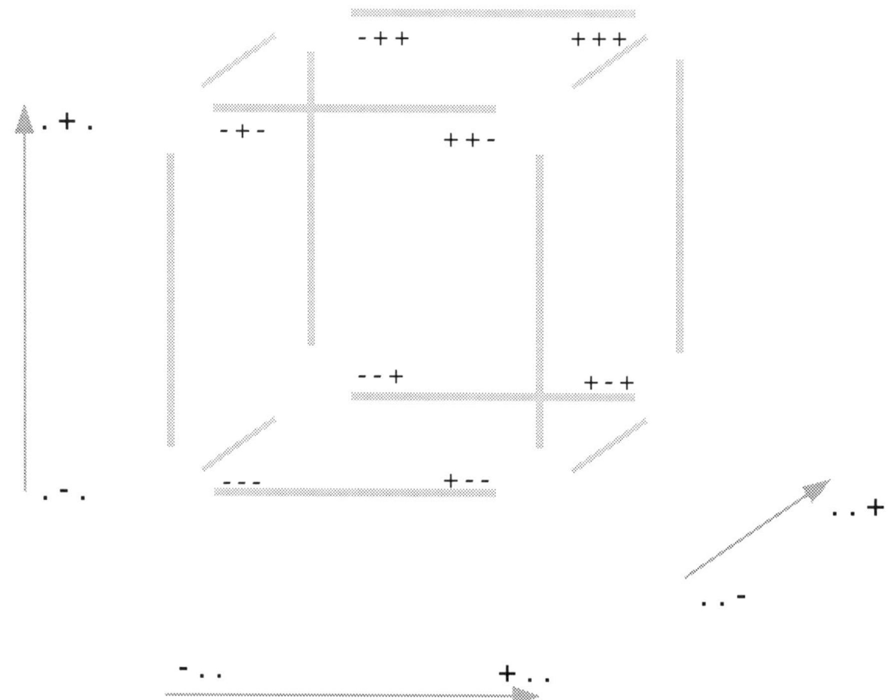

6.4 Working With 2^{k-p} Fractional Factorials

The way in which a Fractional Factorial Design is defined may be summarized by the ***Defining Relations*** for that design. In the previous examples, the first One-Half Replicate of the Flex Rate Data, shown in Example 6.5, p.189, was generated by using the plus signs of the ABC contrast. Thus, the Defining Relation for this One-Half Replicate is **I = +ABC**. The positive portion of the ABC contrast became the identity vector in the One-Half Replicate (see Example 6.4, p.188). This is usually expressed by saying that the +ABC contrast was confounded with the Grand Average.

The One-Quarter Replicate shown in Example 6.6, p.189, has Defining Relations of

I = +ABC and **I = +ABD**.

The positive portions of both the ABC and the ABD contrasts were used to form the two splits for the 2^{4-2} One-Quarter Replicate.

The third One-Quarter Replicate of the Flex Rate Data, shown in Example 6.8, p.191, has Defining Relations of **I = -ABC** and **I = +ABD**, while the fourth One-Quarter Replicate has Defining Relations of **I = -ABC** and **I = -ABD**.

The Defining Relations for any fractional design based on factors with two levels each will identify the portion of the contrasts used to split the fully crossed design. Since the contrasts used to define the Fractional Factorial cannot be estimated, it is customary to use high-order interactions when defining the fractions.

The ***Generating Relation*** for a fractional factorial is based upon the Defining Relations. It will provide a key for finding the aliases for each contrast in a Fractional Factorial Design. In order to obtain the Generating Relation one must multiply "words" in the Defining Relations. A word will consist of a group of letters that identify a contrast in the fully crossed design. Words are multiplied according to the following rules.

Letters may be rearranged in any order without affecting the word. (**ABC = CAB**)

Any word multiplied by **I** is unchanged. (**BI = B**) (**ABIC = ABC**)

Any letter multiplied by itself yields the identity **I**. (**AA = I**)

Using these rules, the Generating Relation consists of the words in the Defining Relations plus the words formed by all possible products of words in the Defining Relation.

For the first One-Quarter Replicate of the Flex Rate Data, shown in Example 6.6, p.189, the Defining Relations are **I = +ABC** and **I = +ABD**. The product of **ABC** and **ABD** is **AABBCD = CD**, so the Generating Relation is **I = +ABC = +ABD = +CD**. These three effects are the three missing effects noted following Example 6.6. Out of the original 15 effects, these three cannot be estimated. They are said to be confounded with the Grand Average.

The Generating Relation can be used to generate the pattern of aliases for a Fractional Factorial by successively multiplying the relation by the name of a contrast. For example, to find out what is confounded with main effect A, multiply the Generating Relation by A,

$$\mathbf{AI = AABC = AABD = ACD}$$

which becomes

$$\mathbf{A = BC = BD = ACD}\,.$$

Likewise, the main effect for B is confounded as follows:

$$\mathbf{BI = BABC = BABD = BCD}$$

or

$$\mathbf{B = AC = AD = BCD}$$

and the AB interaction is confounded as follows:

$$\mathbf{ABI = ABABC = ABABD = ABCD}$$

or

$$\mathbf{AB = C = D = ABCD}$$

which are exactly the sets of aliases shown in Example 6.6, p.189.

It is generally much easier to use the Generating Relation than it is to list the contrasts and look for confounded pairs. Fortunately, for many of the most useful Fractional Factorial Designs, tables of confounding patterns will make it unnecessary to use either approach.

Finally, there is the Resolution of a Fractional Factorial. The ***Resolution*** of a Fractional Factorial Design is defined to be equal to the number of letters in the *shortest* word in the Generating Relation. Of course, the identity is excluded for the purposes of this definition. The One-Half Replicate of the Flex Rate Data shown on page 189 has a Generating Relation of: **I = +ABC.** Thus, this One-Half Replicate is said to be of *Resolution III.* The One-Half Replicate shown on page 191 has a Generating Relation of **I = –ABC** and is also of *Resolution III.* The One-Quarter Replicate shown in Example 6.6, p.189, with Generating Relation

$$\mathbf{I = +ABC = +ABD = +CD,}$$

is said to be of *Resolution II.*

Designs of Resolution II will have some main effects confounded with some other main effects. This is generally undesirable.

Designs of Resolution III will not have any main effects confounded with other main effects. Here main effects will be confounded with two-factor or higher-order interactions. Many of the most useful Fractional Factorial Designs are Resolution III designs.

Designs of Resolution IV will not have any main effects confounded with two-factor interactions. Resolution IV designs are very useful for separating main effects from two-factor interactions.

Thus, the Defining Relations, the Generating Relation, and the Resolution of a Fractional Factorial Design summarize much of the interesting information about the design.

EXERCISE 6.4: Characteristics of a One Quarter Replicate:

1. What are the Defining Relations for the One-Quarter replicate shown in Example 6.7, p.190?

2. What is the Generating Relation for the One-Quarter Replicate shown in Example 6.7?

3. What is the Resolution of the One-Quarter Replicate shown in Example 6.7?

EXERCISE 6.5: Finding Aliases for a Fractional Factorial:

Given Defining Relations of **I = +ABCD** and **I = +BC** for a 2^{4-2} design:

1. Find the Generating Relation for this One-Quarter Replicate.

2. What effects are confounded with main effect A?

3. What effects are confounded with the AB interaction?

CHAPTER SEVEN

PLACKETT-BURMAN SCREENING DESIGNS

In 1946, R. L. Plackett and J. P. Burman published an article that defined a class of factorial designs based on maximal mutually orthogonal sets of contrasts. While they called these designs "Optimum Multifactorial Designs," they are now more commonly called "Screening Designs" since they provide an effective way to consider a large number of factors with a minimum number of observations. By using these designs, one can quickly separate inert factors from active factors, allowing most of the research effort to be concentrated on the active factors.

As a class, the Plackett-Burman designs have several characteristics. Among these are:

(1) The factors in a particular design will all have L levels, where L must be a prime number greater than 1, e.g. L = 2, 3, 5, 7, 11, etc.

(2) The number of runs (treatment combinations) for a particular design must be a multiple of L^2. Thus, if a design involves factors at two levels, it will have a number of runs that

is equal to some multiple of 4, while a design that involves factors at three levels will have a number of runs that is equal to some multiple of 9.

(3) All main factor effects are estimated with the same precision. This means that one does not have to anticipate which factors are the most likely to be important when setting up the study.

(4) Because of the orthogonal contrasts, all main factor effects are estimated independently of each other. While interactions may contaminate the estimates of the main factor effects, at least the main effects do not contaminate one another.

It is these last two characteristics that make the Plackett-Burman designs especially well suited for exploratory studies.

In their article, Plackett and Burman provided an essentially complete set of screening designs. This set consists of 23 designs for factors at two levels each, 3 designs for factors at three levels each, 2 designs for factors at five levels each, and one design for factors with seven levels each. The 23 designs for two-level factors range from an 8-run design to a 100-run design, thus providing a way to study simultaneously as many as 99 two-level factors.

7.1 Eight-Run Plackett-Burman Designs

The array for the basic eight-run Plackett-Burman design is shown below.

	Contrast Labels						
Run	*A*	*B*	*C*	*D*	*E*	*F*	*G*
1	-	-	-	-	-	-	-
2	-	-	-	+	+	+	+
3	-	+	+	+	+	-	-
4	-	+	+	-	-	+	+
5	+	+	-	-	+	+	-
6	+	+	-	+	-	-	+
7	+	-	+	+	-	+	-
8	+	-	+	-	+	-	+

Each column in the 8-run array shown above defines a contrast when the plus and minus signs have a coefficient of 1.0 attached.

At the same time, each row of the 8-run array defines the treatment combination under which the response(s) for that row were obtained. The minus signs indicate those factors that were set at their "low" levels, while the plus signs indicate which factors were set at their "high" levels.

The convention used by Plackett and Burman was to let the minus sign represent the "current" level of a factor, while the plus sign represented a change from the current factor level. Since they generally wanted one run to represent the current operating conditions, they always included one run with all the factors at their "minus" levels. In order to distinguish these designs from the many variations, the designs containing one run with all factors at their "minus" level will be called ***basic*** designs.

Another possible convention is to let the natural orderings of the factor levels be reflected in the minus and plus signs (minus for low level, plus for high level). If such a natural ordering exists, this convention will facilitate the interpretation of the results.

A portion of the confounding pattern (up to the three-factor interactions) for the basic 8-run Plackett-Burman design is given below.

Main Effects	A	B	C	D	E	F	G
Two-Factor	-BC	-AC	-AB	-AE	-AD	-AG	-AF
Interaction	-DE	-DF	-DG	-BF	-BG	-BD	-BE
Aliases	-FG	-EG	-EF	-CG	-CF	-CE	-CD
Three-Factor	BDG	ADG	ADF	ABG	ABF	ABE	ABD
Interaction	BEF	AEF	BDE	ACF	BCD	ACD	DEF
Aliases	CDF	CDE	BFG	BCE	DFG	DEG	BCF
	CEG	CFG	AEG	EFG	ACG	BCG	ACE

When the 8-run design is used for seven factors, it can be considered to be a 2^{7-4} fractional factorial design. As such, it is a one-sixteenth (2^{-4}) replicate of the full 2^7 factorial. and each contrast will have sixteen aliases. Those aliases not shown in the table above are those that involve four or more factors, which generally are of little practical interest. (One set of Defining Relations for the basic 8-run design with seven factors are **I** = –ABC, **I** = –ADE, **I** = –AFG, and **I** = –BDF. Of course these three-factor interactions, along with BEG, CDG, and CEF, are not confounded with any contrast shown on page 206—they are part of the generating relation for the Plackett-Burman 8-run design.)

The basic 8-run Plackett-Burman design is illustrated by Example 6.13, p.197. The calculations of contrasts, sums of squares, and estimated effects are exactly as developed in Chapter Four.

EXERCISE 7.1: The Flash Thickness Data:

A Basic 8-Run Plackett-Burman Screening Design was used in a study of the molding characteristics of a rubber part. The factors studied were (A) Pellet Weight, (B) Vulcanizing Time, (C) Demold Time, (D) Platen Temperature, (E) Pressure, (F) Rubber Batch Durometer, (G) Closing Speed of the Press. The response variable shown below is the flash thickness for the molded parts. The complete study was performed twice, giving two observations per run.

Run	A	B	C	D	E	F	G	Responses		$\bar{X}$	s^2
1	-	-	-	-	-	-	-	31	28	29.5	4.5
2	-	-	-	+	+	+	+	28	26	27.0	2.0
3	-	+	+	+	+	-	-	27	27	27.0	0.0
4	-	+	+	-	-	+	+	30	28	29.0	2.0
5	+	+	-	-	+	+	-	25	26	25.5	0.5
6	+	+	-	+	-	-	+	35	30	32.5	12.5
7	+	-	+	+	-	+	-	29	29	29.0	0.0
8	+	-	+	-	+	-	+	28	28	28.0	0.0

Find the estimated contrasts and the SS(C) values for these data:

Contrast	$\hat{C}$	SS(C)
A		
B		
C		
D		
E		
F		
G		

EXERCISE 7.2: The Flash Thickness Data:

Based on Exercise 7.1:
Find the within-run (within-subgroup) Mean Square = MSW =

Write out the ANOVA table for these data and perform the appropriate F-tests.

Find the estimated contrast effects, $\hat{l}$, for significant contrasts.

List the main effect and two-factor interaction aliases for each significant contrast.

The ability to perform F-tests in Exercise 7.2 is based upon having n = 2 observations per run. However, without further effort, one cannot distinguish between main effects and two-factor interactions. So even though one can identify the contrasts that have the greatest impact upon the response, one cannot, based on these data, know if the significant contrasts represent main effects or interaction effects.

This confounding of main-effects with two-factor interactions is characteristic of Resolution III designs. When used with 5, 6, or 7 two-level factors, the 8-run Plackett-Burman designs will be Resolution III designs regardless of the number of observations per treatment combination. Generalizing to all two-level Plackett-Burman designs, any two-level Plackett-Burman design with k runs that is used with more than k/2 factors will be a Resolution III design. When a geometric two-level Plackett-Burman design with k runs is used with k/2 or fewer factors, it is possible to assign the main effects to the contrasts in such a way to obtain designs of Resolution IV or higher.

Resolution

While the Resolution III Plackett-Burman designs do not directly allow for the separation of main effects and two-factor interactions, it is possible to combine two such Resolution III designs to obtain a Resolution IV design. In particular, any two-level, Resolution III Plackett-Burman design may be combined with its *reflection* to yield a Resolution IV design. The reflection of a two-level Plackett-Burman design is obtained by exchanging ***all*** of the plus and minus signs in the array. This means that both the contrasts and the treatment combinations change in the reflected design. Since both contrasts and treatment combinations change, each contrast will still represent the same main effect, but the confounding pattern between the main effects and the two-factor interactions will change slightly. The array and confounding pattern for a reflected 8-run design are shown below.

The Reflection of the Basic 8-Run Plackett-Burman Design

	Contrast Labels and Aliases						
Run	A	B	C	D	E	F	G
1	+	+	+	+	+	+	+
2	+	+	+	-	-	-	-
3	+	-	-	-	-	+	+
4	+	-	-	+	+	-	-
5	-	-	+	+	-	-	+
6	-	-	+	-	+	+	-
7	-	+	-	-	+	-	+
8	-	+	-	+	-	+	-
	BC	AC	AB	AE	AD	AG	AF
	DE	DF	DG	BF	BG	BD	BE
	FG	EG	EF	CG	CF	CE	CD
	BDG	ADG	ADF	ABG	ABF	ABE	ABD
	BEF	AEF	BDE	ACF	BCD	ACD	DEF
	CDF	CDE	BFG	BCE	DFG	DEG	BCF
	CEG	CFG	AEG	EFG	ACG	BCG	ACE

Reflected designs are sometimes called *fold-over* designs. The sign changes in the confounding pattern are the key for separating main effects from two-factor interactions. If a significant contrast changes sign between the original study and the reflected study, then it clearly represents some interaction effect. Since the two-factor, four-factor, six-factor and all even-factor higher order interactions will change signs between a given Two-Level Plackett-Burman Design and its Reflection, a sign change for a significant contrast will indicate that some even-order interaction is significant. Of all such interaction effects, the two-factor interactions are generally considered to be the most likely to occur.

If a significant contrast does not change sign between the original study and the reflected study, then it is probably a main effect. (It might be either a three-factor interaction or a five-factor interaction, but we can usually discount these possibilities.)

Examples of how to combine two Plackett-Burman designs to isolate the main effects from the two factor interaction effects are provided in Examples 6.12 and 6.13, pp. 196–197.

Thus, given a study to be performed using a Plackett-Burman design, one may choose to replicate the whole study, as was done with the Flash Thickness study (Exercise 7.1) or one may elect to perform both a Plackett-Burman design and its reflection. When the whole study is replicated, as in Exercise 7.1, one may use F-ratios to check for possible signals, but without reflecting the design one usually cannot isolate the main effects from the two-factor interactions. Where two different studies are done, each with one observation per run, one can use Scree Plots to identify the potential signals. Moreover, with the replication of the same potential signals from one study to another, the experimenter can be reasonably sure that the signals are real. At the same time, two different studies will allow significant main effects to be isolated from significant two-factor interactions. For these reasons, it will usually be best to run Plackett-Burman designs that are reflections of each other rather than replicating a single design. This is summed up in the following outline for using screening designs.

This outline (see page 212) is merely an application of the PDSA (Deming) Cycle. Part A is the "plan" step, Parts B and C are the "do" step, Parts D and E are the "study" step, and Part F is the "act" step. This PDSA Cycle is an adaptation of Shewhart's version of the scientific process, and it has been proven in practice to be a powerful way of acquiring knowledge.

A PROCEDURE FOR USING SCREENING DESIGNS

A. Plan the Study

1. Write out the Objective for the Experimental Study.
2. Identify factors and factor levels for the study.
3. Choose appropriate Plackett-Burman screening designs.

B. The Preliminary Experiment

1. Carry out specified runs and record the response variables.
2. Calculate estimated contrasts and sums of squares.
3. Obtain Scree Plot and estimated contrast effects.

C. The Follow-Up Experiment

1. Reflect the design used in the preliminary experiment.*
2. Carry out specified runs and record the response variables.
3. Calculate estimated contrasts and sums of squares.
4. Obtain Scree Plot and estimated contrast effects.

D. Combining the Results of Both Studies

1. Which contrasts stand out from the rubble in both studies?
2. Do any of these contrasts represent two-factor interactions?
3. Identify those factors which appear as significant main effects or in significant two-factor interactions. These are the factors that are most likely to have an impact upon the response variable. Designate these as the "active factors."

E. Obtain Graphical Summaries of Results

1. Draw the Response Plot using only the active factors.
2. Plot the ANOM graph using the subgroups defined by the active factors.
3. (Optional) In order to identify equivalence classes, carry out a Tukey Post-Hoc Test on the averages of the subgroups defined by the active factors.

F. State Conclusions

1. Interpret the plots.
2. State appropriate course of action based on these results.
   a. State factors and factor levels for next study.
   b. State best apparent operating conditions for process.
   c. Recommend changes suggested by studies.
3. Implement appropriate course of action.

* Only for Resolution III designs. For Resolution IV designs use a partial reflection, p.219, 221.

EXAMPLE 7.1: *A Basic 8-Run Study for Tensile Strength:*

A basic 8-run Plackett-Burman design was used to study the effect of 7 factors upon the tensile strength of a particular rubber product. The factors and their levels were: (A) Compound (1% Catalyst vs. 2% Catalyst), (B) Vulcanizing Time (3 m vs. 6 m), (C) Demold Time (30 s vs. 60 s), (D) Platen Temp. (100° vs. 120°), (E) Pressure (2000 psi vs. 2600 psi), (F) Transfer Time (10 s vs 20 s), and (G) Pellet Weight (80 gm vs. 100 gm). The response is the tensile strength in pounds.

	A	B	C	D	E	F	G	
	-BC	-AC	-AB	-AE	-AD	-AG	-AF	
	-DE	-DF	-DG	-BF	-BG	-BD	-BE	
Run	-FG	-EG	-EF	-CG	-CF	-CE	-CD	Response
1	-	-	-	-	-	-	-	995
2	-	-	-	+	+	+	+	1165
3	-	+	+	+	+	-	-	1150
4	-	+	+	-	-	+	+	1165
5	+	+	-	-	+	+	-	1200
6	+	+	-	+	-	-	+	980
7	+	-	+	+	-	+	-	1110
8	+	-	+	-	+	-	+	1180

ANOVA Table:

Contrast	$\hat{C}$	SS(C)	$\hat{l}$
A	-5	3.125	
B	45	253.125	
C	265	8778.125	66.25
D	-135	2278.125	-33.75
E	445	24753.125	111.25
F	335	14028.125	83.75
G	35	153.125	
Total		50246.875	

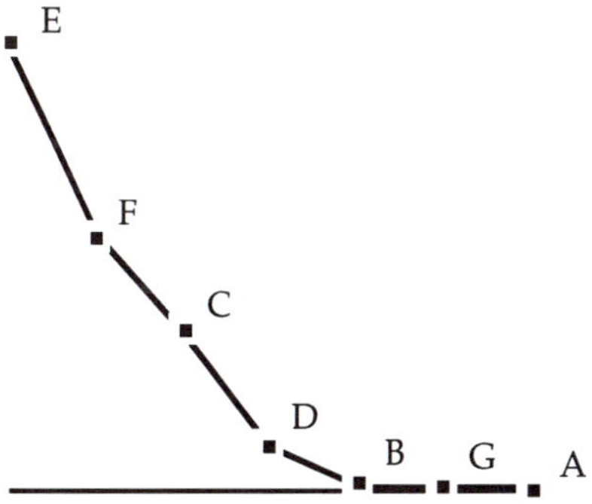

The Scree Plot suggests that factors E, F, C and possibly D have an impact upon the tensile strength. If none of these contrasts represent interaction effects, the greatest tensile strength would be obtained with factors C, E, F set at their high levels and factor D at its low level. A replication of the design array given above would allow for the computation of individual F Ratios, but it would not isolate the main effects from the two-factor interactions.

EXAMPLE 7.2: A Reflected 8-Run Study for Tensile Strength:

The reflection of the basic 8-run study in Example 7.1 was used in a follow-up study for the effect of the same seven factors upon tensile strength.

Run	A	B	C	D	E	F	G	Response
	BC	AC	AB	AE	AD	AG	AF	
	DE	DF	DG	BF	BG	BD	BE	
	FG	EG	EF	CG	CF	CE	CD	
1	+	+	+	+	+	+	+	990
2	+	+	+	-	-	-	-	1190
3	+	-	-	-	-	+	+	1180
4	+	-	-	+	+	-	-	1100
5	-	-	+	+	-	-	+	1170
6	-	-	+	-	+	+	-	1010
7	-	+	-	-	+	-	+	1150
8	-	+	-	+	-	+	-	1165

ANOVA Table:

Contrast	$\hat{C}$	SS(C)	$\hat{l}$
A	-35	153.125	
B	35	153.125	
C	-235	6903.125	-58.75
D	-105	1378.125	-26.25
E	-455	25873.125	-113.75
F	-265	8778.125	-66.25
G	25	78.125	
Total		43321.875	

Once again contrasts E, F, C and possibly D stand out from the rubble on the Scree Plot. When this study is interpreted by itself under the assumption that none of these contrasts represent interaction effects, it would appear that maximum tensile strength would result when Factors C, D, E and F were all set at their low levels.

Under the same assumption, the first study suggested that setting Factor D at its low level and setting Factors C, E and F at their high levels would maximize tensile strength!

The discrepancy between these two interpretations is resolved by combining the results of the two studies. First, observe that Contrast D does not change signs between the two studies, therefore it is likely to represent a main effect for Factor D. Next observe that contrasts C, E and F all change signs between the two studies, therefore they cannot represent a main effect. Under the assumption that they are two-factor interactions, the confounding patterns on pages 207 and 210 define the possibilities. If one is willing to ignore interactions involving C, E, or F, then Contrast E can be said to represent either the AD or the BG interaction, while Contrast F can be said to represent either the AG or the BD interaction, and Contrast C can be said to represent either the AB or the DG interaction.

Thus, rather than Factors C, E and F, it is Factors A, B, D and G that appear to actively affect the tensile strength. Factors C, E and F are essentially inert relative to tensile strength.

Notice that in the previous two examples, the use of either one of the two Resolution III designs by itself would have resulted in the identification of the wrong factors as the factors that influenced the response variable, which would have, in turn, resulted in the wrongful manipulation of the process. But by using two Resolution III designs that were reflections of each other, the results could be combined in such a way that the sequence of studies was equivalent to a Resolution IV design. This ability to properly separate main effects from two-factor interaction effects is of great import, especially when strong interactions are present. In fact, ***in the presence of interactions, attention can be properly focused on the active factors only when the interactions are clearly identified as such.***

The interactions in the Tensile Strength Data can be fully sorted out by one more study. In this third study exactly one of the four potentially active factors, A, B, D or G, should be reflected. This will change the confounding pattern in such a way that there should be no question about which interaction was represented by each of the three interaction contrasts. Of course this third study would be conducted only if the experimenter needs to fully sort out the interaction effects. In many cases such detail is not required.

EXAMPLE 7.3: *A Third 8-Run Study for Tensile Strength:*

Another follow-up study for the effect of the same seven factors upon tensile strength was carried out using a modification of the array in the first follow-up study in Example 7.2. The Contrast for Factor A was reflected in this study, changing the signs of all two-factor interactions confounded with Main Effect A or involving Factor A.

	A -BC -DE -FG	B -AC DF EG	C -AB DG EF	D -AE BF CG	E -AD BG CF	F -AG BD CE	G -AF BE CD	
Run								*Response*
1	-	+	+	+	+	+	+	1090
2	-	+	+	-	-	-	-	1125
3	-	-	-	-	-	+	+	950
4	-	-	-	+	+	-	-	1130
5	+	-	+	+	-	-	+	1055
6	+	-	+	-	+	+	-	1175
7	+	+	-	-	+	-	+	1185
8	+	+	-	+	-	+	-	940

ANOVA Table:

Contrast	$\hat{C}$	SS(C)	$\hat{l}$
A	60	450.00	
B	30	112.50	
C	240	7200.00	60
D	-220	6050.00	-55
E	510	32512.50	127.5
F	-340	14450.00	-85
G	-90	1012.50	
Total		61787.50	

Once again contrasts E, F, C and D stand out from the rubble on the Scree Plot.

Combining the results from the three studies we get:

For Contrast E:	*Example 7.1:*	*111.25 = E - AD - BG - CF*
	Example 7.2:	*-113.75 = E + AD + BG + CF*
	Example 7.3:	*127.50 = E - AD + BG + CF*

Thus, Contrast E must represent the AD interaction effect.

Likewise, for Contrast F:	*Example 7.1:*	*83.75 = F - AG - BD - CE*
	Example 7.2:	*-66.25 = F + AG + BD + CE*
	Example 7.3:	*-85.00 = F - AG + BD + CE*

Thus, Contrast F must represent the CE or the BD interaction effect.

While for Contrast C:	*Example 7.1:*	*66.25 = C - AB - DG - EF*
	Example 7.2:	*-58.75 = C + AB + DG + EF*
	Example 7.3:	*60.00 = C - AB + DG + EF*

Thus, Contrast C must represent the AB interaction effect.

Therefore, it is most likely that factors A, B, and D affect the tensile strength of the rubber part. All other factors are inert relative to tensile strength, and there are strong interactions between these three active factors. This has been discovered using 24 runs, which constitute three-sixteenths of the total number of runs in a full factorial design for seven factors at two levels each.

The response plot for Factors A, B and D is shown below using the data from all three studies.

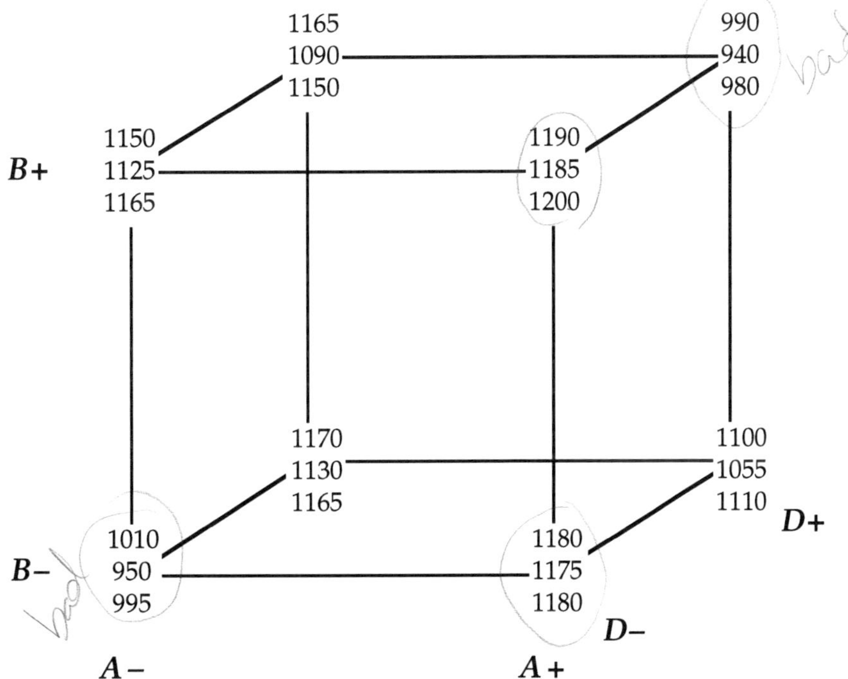

From this response plot it should be clear that $A^+B^+D^+$ and $A^-B^-D^-$ are the combinations to be avoided, and that the best tensile strengths occur when A is at its high level and D is at its low level.

In addition to the basic 8-run design, there are many other 8-run Plackett-Burman designs. Each of these designs may be obtained by reflecting one of the contrasts in either the basic design or the reflection of the basic design. This ability to get different Fractional Factorials out of the basic Plackett-Burman design with minimal effort allows one to perform a sequence of experiments without needlessly duplicating any treatment combinations.

EXERCISE 7.3: A Partially Reflected 8-Run Design:

Part I. Modify the BASIC 8-Run design by reflecting the Contrast for Main Effect C and write down the array for this partially reflected design.

Run	A	B	C	D	E	F	G
1	−	−	+	−	−	−	−
2	−	−	+	+	+	+	+
3	−	+	−	+	+	−	−
4	−	+	−	−	−	+	+
5	+	+	+	−	+	+	−
6	+	+	+	+	−	−	+
7	+	−	−	+	−	+	−
8	+	−	−	−	+	−	+

The confounding pattern for a partially reflected design is obtained from the confounding pattern of the design which is partially reflected. If you reflect one of the columns of a *Basic 8-Run P-B Design*, then you will modify the confounding pattern for the *Basic 8-Run Design*. Two sets of modifications are required when a column is reflected. Say column F is reflected, then (1) every two-factor interaction involving Factor F will have its sign changed, and (2) every two-factor interaction which is confounded with Main-Effect F will have its sign changed. (Additional sign changes will occur with the higher order interactions.) Examples 7.2 and 7.3 (pp.214, 216) show the effect of reflecting just Contrast A. Example 6.13 and Exercise 6.2 (pp.197, 199) show the effect of reflecting just Contrast D.

EXERCISE 7.3 (Continued) — A Partially Reflected 8-Run Design:

Part II. Attach the appropriate signs to the two-factor interactions in the confounding pattern below to create the confounding pattern for the partially reflected design on the previous page.

Main Effects

A	B	C	D	E	F	G

Two-Factor Interactions

BC	AC	AB	AE	AD	AG	AF
DE	DF	DG	BF	BG	BD	BE
FG	EG	EF	CG	CF	CE	CD

From these examples and exercises it should be apparent that while many two-level Plackett-Burman designs are Resolution III designs, it is possible to achieve Resolution IV or V by the judicious use of reflected and partially reflected designs. Thus, that which is unavailable from an individual Plackett-Burman design may be obtained from a *sequence* of Plackett-Burman studies. Moreover, a sequence of small studies provides an opportunity to validate our decisions regarding which contrasts are "significant." While random variation may make any contrast appear to be significant in any one study, it is highly unlikely that the same contrast would, by chance, appear to be significant in two or three successive studies. Therefore, as the same contrasts appear to be significant in successive studies, one becomes increasingly confident that they represent real effects.

When the Plackett-Burman design is already of Resolution IV or higher there is no need to run a fully reflected design. Running such a fully reflected design may merely duplicate the treatment combinations already studied. With these designs it is best to use a partially reflected design for a follow-up study. By reflecting one of the main effects, one will usually get new treatment combinations for the active factors rather than merely replicating the current combinations. In addition, the partial reflection may also facilitate the isolation of certain sets of interaction effects.

7.2 Nonsaturated 8 - Run Designs

When a two-level Plackett-Burman design with k runs is used with k-1 factors it is said to be a saturated design because every contrast is confounded with a main effect. When fewer factors are used the design is said to be nonsaturated. Saturated two-level Plackett-Burman designs are always Resolution III designs.

When an 8-run design is used with 6 factors it will be a one-eighth (2^{6-3}) replicate, and it will be of Resolution III. Any one contrast label may remain unassigned (that is, the six main effects may be assigned to any six of the seven contrasts). The confounding pattern for any such assignment may be obtained from the confounding pattern for the saturated 8-run design by simply deleting the unassigned contrast label and all interactions involving that contrast label.

For illustration, say that the six factors are assigned to labels A through F in the Basic 8-run Design, and that label G remains unassigned. The partial confounding pattern for this nonsaturated basic 8-run design would be:

Contrast Label	A	B	C	D	E	F	G
Main Effects	A	B	C	D	E	F	
Interactions	-BC -DE	-AC -DF	-AB -EF	-AE -BF	-AD -CF	-BD -CE	-AF -BE -CD
	BEF CDF	AEF CDE	ADF BDE	ACF BCE	ABF BCD	ABE ACD	ABD DEF BCF ACE

The Defining Relations for this 2^{6-3}_{III} design are: **I** = –ABC, **I** = –ADE, **I** = – BDF.

Of course complete reflections and partial reflections would affect this set of aliases in the same way that they affected the original set of aliases.

When the 8-run design is used with five factors it becomes a one-quarter (2^{5-2}) replicate of Resolution III. Any two contrast labels may remain unassigned. If, for example, the main effects are assigned to contrast labels A through E, so that labels F and G remain unassigned, the partial confounding pattern for this nonsaturated basic 8-run design is

Contrast Label	A	B	C	D	E	F	G
Main Effects	A	B	C	D	E		
Interactions	-BC -DE	-AC	-AB	-AE	-AD	-BD -CE	-BE -CD
		CDE	BDE	BCE	BCD	ABE ACD	ABD ACE

The Defining Relations for the above 2^{5-2}_{III} design are: **I** = –ABC, **I** = –ADE.

When the 8-run design is used with four factors the resolution of the design will depend upon the way the factors are assigned to the contrast labels. With some assignments the design will be a Resolution IV design, while with all other assignments it will be of Resolution III. In general, there are 35 different ways that four factors may be assigned to seven contrast labels. Of these 35 assignments, only seven result in a design of Resolution IV. These seven assignments are {A,B,D,G}, {A,B,E,F}, {A,C,D,F}, {A,C,E,G}, {B,C,D,E}, {B,C,F,G}, and {D,E,F,G}. The partial confounding pattern for the first of these Resolution IV assignments is shown for a basic 8-run design in the table below:

Contrast Labels	A	B	C	D	E	F	G
Main Effects	A	B		D			G
Interactions	BDG	ADG	-AB -DG	ABG	-AD -BG	-AG -BD	ABD

The Defining Relation for this 2^{4-1}_{IV} design is **I** = –ABDG.

Since these 8-run Resolution IV designs are, in effect, one-half replicates of a full factorial design, one should use the complementary one-half replicate for any follow-up study. This complementary one-half replicate can be obtained only from a partially reflected design. To find such a design, the experimenter should reflect *one* of the main effects (A, B, D, or G) and write out the corresponding array.

Finally, when the basic 8-run design is used with three factors it will usually be a full factorial design. Once again there are 35 different ways that three main effects can be assigned to the seven contrast labels. Of these 35 assignments, 28 will result in a fully crossed design. Seven assignments will fail to yield a fully crossed design. The seven assignments to *avoid* are {A,B,C}, {A,D,E}, {A,F,G}, {B,D,F}, {B,E,G}, {C,D,G}, and {C,E,F}.

One of the 28 assignments that yields a fully crossed design is {A,B,D}. With this assignment the confounding pattern for the basic 8-run design reduces down to the following:

Contrast Label	A	B	C	D	E	F	G
Effect	A	B	-AB	D	-AD	-BD	ABD

Except for the seven assignments noted above, any assignment of three factors to the contrasts for an 8-run design will result in one and only one effect per contrast. The assignment shown above may be easier to run than the others because it only requires that the levels of the factors be changed seven times.

7.3 16-Run Plackett-Burman Designs

The basic 16-run Plackett-Burman design for two-level factors is defined by the following array.

Run	A	B	C	D	E	F	G	H	I	J	K	L	M	N	O
1	-	-	-	-	-	-	-	-	-	-	-	-	-	-	-
2	-	-	-	-	-	-	-	+	+	+	+	+	+	+	+
3	-	-	-	+	+	+	+	+	+	+	+	-	-	-	-
4	-	-	-	+	+	+	+	-	-	-	-	+	+	+	+
5	-	+	+	+	+	-	-	-	-	+	+	+	+	-	-
6	-	+	+	+	+	-	-	+	+	-	-	-	-	+	+
7	-	+	+	-	-	+	+	+	+	-	-	+	+	-	-
8	-	+	+	-	-	+	+	-	-	+	+	-	-	+	+
9	+	+	-	-	+	+	-	-	+	+	-	-	+	+	-
10	+	+	-	-	+	+	-	+	-	-	+	+	-	-	+
11	+	+	-	+	-	-	+	+	-	-	+	-	+	+	-
12	+	+	-	+	-	-	+	-	+	+	-	+	-	-	+
13	+	-	+	+	-	+	-	-	+	-	+	+	-	+	-
14	+	-	+	+	-	+	-	+	-	+	-	-	+	-	+
15	+	-	+	-	+	-	+	+	-	+	-	+	-	+	-
16	+	-	+	-	+	-	+	-	+	-	+	-	+	-	+

The rows of this array define treatment combinations, while the columns define contrast coefficients. This design may be used for anything between a full factorial design for four factors to a 2^{-11} replicate of a design with 15 factors. As a 2^{15-11} design, each contrast will have 2048 aliases, only eight of which are main effects or two-factor interactions.

When used with 9 or more two-level factors, the 16-run Plackett-Burman designs will be of Resolution III. With the proper assignment of 6, 7 or 8 two-level factors to contrast labels, the 16-run design will be a Resolution IV design. With the proper assignment of 5 factors it may be a Resolution V design.

Just as with the 8-run Resolution III designs, a 16-run Resolution III Plackett-Burman design may be combined with its reflection to obtain a Resolution IV design. One may also reflect individual contrasts rather than the complete array in order to define different 16-run designs.

The main effect and two-factor interaction confounding pattern for the Basic 16-Run Design is shown below.

Contrast Labels and Main Effects														
A	B	C	D	E	F	G	H	I	J	K	L	M	N	O
Two-Factor Interactions														
-BC	-AC	-AB	-AE	-AD	-AG	-AF	-AI	-AH	-AK	-AJ	-AM	-AL	-AO	-AN
-DE	-DF	-DG	-BF	-BG	-BD	-BE	-BJ	-BK	-BH	-BI	-BN	-BO	-BL	-BM
-FG	-EG	-EF	-CG	-CF	-CE	-CD	-CK	-CJ	-CI	-CH	-CO	-CN	-CM	-CL
-HI	-HJ	-HK	-HL	-HM	-HN	-HO	-DL	-DM	-DN	-DO	-DH	-DI	-DJ	-DK
-JK	-IK	-IJ	-IM	-IL	-IO	-IN	-EM	-EL	-EO	-EN	-EI	-EH	-EK	-EJ
-LM	-LN	-LO	-JN	-JO	-JL	-JM	-FN	-FO	-FL	-FM	-FJ	-FK	-FH	-FI
-NO	-MO	-MN	-KO	-KN	-KM	-KL	-GO	-GN	-GM	-GL	-GK	-GJ	-GI	-GH

The reflection of the basic 16-run Plackett-Burman design would have all the minus signs in this confounding pattern changed to plus signs.

In order to obtain the main effect and two-factor interaction confounding pattern for these less than saturated 16-Run Designs one would delete certain terms from the confounding pattern above. If for example, a 16-Run Design was used for 11 factors (A, B, C, D, E, F, G, H, I, J, and K), then delete main effects for L, M, N, and O, and also delete all interactions involving L, M, N, or O. As a result, Contrast A would represent A, –BC, –DE, –FG, –HI, and –JK, while Contrast H would represent H, –AI, –BJ, and –CK.

With 8 factors at two levels each, the basic 16-run design will be of Resolution IV when the eight main effects are assigned to the contrast labels {A, B, D, G, H, K, M, and N}. The Defining Relations for this one-sixteenth replicate would be: **I** = –ABDG, **I** = –ABHK, **I** = –ABMN, and **I** = –ADHM.

With 7 factors, the basic 16-run design will be of Resolution IV when the factors are assigned to the contrast labels {A, B, D, G, H, M, and N}. The Defining Relations for this one-eighth replicate would be: **I** = –ABDG, **I** = –ABMN, and **I** = –ADHM.

With 6 factors, the basic 16-run design will be of Resolution IV when the factors are assigned to the contrast labels {A, B, D, G, H, and N}. The Defining Relations for this one-fourth replicate would be: **I** = –ABDG, and **I** = –AGHN.

With 5 factors, the basic 16-run design will be of Resolution V when the factors are assigned to the contrast labels {A, B, D, H, and O}. The Defining Relation is **I** = –ABDHO.

With 4 factors, the 16-run designs will be full factorial designs when the factors are assigned to the contrast labels {A, B, D, and H}.

The following example gives the data for an actual study. The four unassigned contrasts in this study were I, M, N, and O. There is nothing special about leaving these four unassigned, it just worked out this way when the original study was translated to conform to the orthogonal array given on page 223.

EXAMPLE 7.4: *The Carbon Foam Data:*

Surface voids with an area in excess of 0.1 square inch are considered to be defects in a certain carbon foam casting. Eleven factors were considered for their effect upon the number of voids per casting. These factors and their levels were:

(A) Type of Mold	*One unit mold (-)*	*vs*	*Two unit mold (+)*
(B) Heating of Mold	*Bottom only (-)*	*vs*	*Bottom and side (+)*
(C) Blade Height	*0.5" from bottom (-)*	*vs*	*1.5" from bottom (+)*
(D) Mixing Blade	*Single (-)*	*vs*	*Dual (+)*
(E) Pouring Speed	*Slowly (-)*	*vs*	*Quickly (+)*
(F) Mix Time	*30 sec. (-)*	*vs*	*60 sec (+)*
(G) Hitting Mold Against Floor	*No (-)*	*vs*	*Yes (+)*
(H) Blade Size/RPM	*6"/300 rpm (-)*	*vs*	*4"/800 rpm (+)*
(J) Pouring Manner	*Aim at one spot (-)*	*vs*	*Pour in circles (+)*
(K) Can Position	*Centered on turntable (-)*	*vs*	*Off-center (+)*
(L) Turntable Speed	*10 rpm (-)*	*vs*	*40 rpm (+)*

Contrast labels I, M, N, and O were unassigned. The response variable was the number of voids with areas in excess of 0.1 square inch. One carbon foam casting was made for each run and the number of voids with areas in excess of 0.1 square inch was recorded for each casting.

Run	*A*	*B*	*C*	*D*	*E*	*F*	*G*	*H*	*I*	*J*	*K*	*L*	*M*	*N*	*O*	*Voids*
1	-	-	-	-	-	-	-	-	-	-	-	-	-	-	-	0
2	-	-	-	-	-	-	-	+	+	+	+	+	+	+	+	0
3	-	-	-	+	+	+	+	+	+	+	+	-	-	-	-	4
4	-	-	-	+	+	+	+	-	-	-	-	+	+	+	+	2
5	-	+	+	+	+	-	-	-	-	+	+	+	+	-	-	0
6	-	+	+	+	+	-	-	+	+	-	-	-	-	+	+	1
7	-	+	+	-	-	+	+	+	+	-	-	+	+	-	-	6
8	-	+	+	-	-	+	+	-	-	+	+	-	-	+	+	8
9	+	+	-	-	+	+	-	-	+	+	-	-	+	+	-	5
10	+	+	-	-	+	+	-	+	-	-	+	+	-	-	+	0
11	+	+	-	+	-	-	+	+	-	-	+	-	+	+	-	0
12	+	+	-	+	-	-	+	-	+	+	-	+	-	-	+	0
13	+	-	+	+	-	+	-	-	+	-	+	+	-	+	-	5
14	+	-	+	+	-	+	-	+	-	+	-	-	+	-	+	3
15	+	-	+	-	+	-	+	+	-	+	-	+	-	+	-	0
16	+	-	+	-	+	-	+	-	+	-	+	-	+	-	+	0

EXERCISE 7.5: The Carbon Foam Data:

Find the ANOVA table for the Carbon Foam Data.

Contrast	$\hat{C}$	SS(C)
A	---------	----------
B	---------	----------
C	---------	----------
D	---------	----------
E	---------	----------
F	---------	----------
G	---------	----------
H	---------	----------
I	---------	----------
J	---------	----------
K	---------	----------
L	---------	----------
M	---------	----------
N	---------	----------
O	---------	----------

EXERCISE 7.6: Interpretation of Carbon Foam Data:

1. Pool the four unassigned contrast sums of squares to get a Mean Square Error (MSE) term, and carry out F-tests on the remaining eleven contrasts. Use a regular F critical point since this is *a priori* pooling.

2. Find the estimated contrast effect for any significant contrasts.

3. Ignoring the possibility of interaction effects, interpret any significant contrasts as significant main effects and draw a Response Plot for these main effects.

4. Summarize the results of this study in English.

Since the mean square error term was obtained by pooling the sums of squares for the unassigned contrast labels, it was independent of each of the other SS(C) values (thanks to orthogonality). Since the sums of squares that were pooled were identified prior to the collection of the data, the pooling is said to be *a priori* pooling.

When working with nonsaturated designs one may decide to pool, *a priori*, the sums of squares for the unassigned contrasts. Generally such pooling works best when the MSE term has at least 4 degrees of freedom. However, the reader is reminded that waiting to see if the unassigned contrasts are "significant" prior to pooling is still *Post Hoc* Pooling. *Post Hoc* Pooling will still require the use of the Mod F distribution critical points even when the pooled contrasts are unassigned. It is only with *A Priori* Pooling that one may use the regular F distribution critical points.

In the Carbon Foam study no formal follow up experiment was performed. With only one significant contrast it is very easy to determine if the contrast represents a main effect or some interaction between factors that otherwise appear to be inert. This can be done by collecting some data while changing the levels of the one main effect represented by the significant contrast. If this procedure fails to produce an effect like the one shown by the contrast, then the contrast must be an interaction contrast. In the Carbon Foam case, rather than perform another 16-run study, they could have held all the other factors constant (which is much easier than varying them) and make a few pieces at each of the two levels for mix time. If the pieces produced at 60 seconds of mix time show about four more voids per piece than those produced at 30 seconds of mix time, then it is definite that mix time is important. If this difference in the number of voids is not apparent, then Contrast F is likely to be an interaction effect.

7.4 Nongeometric Two-Level Plackett-Burman Designs

All Plackett-Burman designs can be classified as geometric or nongeometric designs. The two-level Plackett-Burman designs with 4, 8, 16, 32 and 64 runs will be called *geometric* designs. All other two-level Plackett-Burman designs will be said to be *nongeometric* designs. The smaller nongeometric designs are those with 12, 20, 24 and 28 runs.

The major difference between the geometric designs and the nongeometric designs is the way that the interaction effects are confounded with the main effects. With the geometric designs each interaction effect was confounded with exactly one main effect (i.e., the AB interaction might be confounded with the main effect for Factor C). This unique correspondence between an interaction and a particular contrast can be exploited to identify two-factor interaction effects as well as main effects. With the nongeometric designs this unique correspondence is missing. In these designs a two-factor interaction will be partially confounded with each of the other main effects in the study (i.e., the AB interaction will be partially confounded with C, D, E, F, G, H, I, J, K, etc.). Any particular interaction will be positively confounded with some main effects, and negatively confounded with others. At the same time, the degree of confounding for any particular interaction will vary from main effect to main effect. Thus, for nongeometric designs, the confounding pattern is so complex that it is of little use in isolating individual interaction effects. Moreover, since one significant interaction effect will impact every other main effect, interactions can appreciably cloud the interpretation of results.

The 12-run Plackett-Burman design will be illustrated using Data Set Eleven.

EXAMPLE 7.5: The Reflected 12-Run Design for Data Set Eleven:

Let contrast labels A through G represent the seven factors for Data Set Eleven. Contrasts H, I, J, and K will be unassigned.
The responses recorded are the values for Y_1 given on page 341.
The run numbers below identify the corresponding runs on page 341.

Run	A	B	C	D	E	F	G	H	I	J	K	Y_1
1	+	+	+	+	+	+	+	+	+	+	+	56.94
15	+	+	+	-	-	-	+	+	-	-	-	50.20
24	+	+	-	+	-	-	-	-	+	+	-	45.96
53	+	-	-	+	-	+	+	-	-	-	+	49.01
58	+	-	-	-	+	+	-	+	-	+	-	45.79
44	+	-	+	-	+	-	-	-	+	-	+	47.08
109	-	-	+	-	-	+	+	-	+	+	-	54.10
104	-	-	+	+	-	-	-	+	-	+	+	54.17
115	-	-	-	+	+	-	+	+	+	-	-	48.83
91	-	+	-	-	+	-	+	-	-	+	+	45.10
94	-	+	-	-	-	+	-	+	+	-	+	46.40
66	-	+	+	+	+	+	-	-	-	-	-	57.00

ANOVA Table:

Contrast	$\hat{C}$	SS(C)	F	
A	-10.62	9.399	10.5	*
B	2.62	0.572	0.6	
C	38.40	122.880	136.8	*
D	23.24	45.008	50.1	*
E	0.90	0.068	0.1	
F	17.90	26.701	29.7	*
G	7.78	5.044	5.6	*
H	4.08	1.387	pool	
I	-1.96	0.320	pool	
J	3.54	1.044	pool	
K	-3.18	0.843	pool	
Total		213.266		

C D F A G H J K B I E

A Priori Pooling of unassigned contrasts gives MSE = 0.8985 with 4 d.f.

The regular $F_{.90}(1,4) = 4.54$

Therefore, Contrasts C, D, F, A and G are found to be significant.

EXAMPLE 7.6: The Basic 12-Run Design for Data Set Eleven:

Let contrast labels A through G represent the seven factors for Data Set Eleven. Contrasts H, I, J, and K will be unassigned. The responses recorded are the values for Y_1 given in the Appendix. The run numbers below identify the corresponding runs on page 341.

Run	A	B	C	D	E	F	G	H	I	J	K	Y_1
128	-	-	-	-	-	-	-	-	-	-	-	46.31
114	-	-	-	+	+	+	-	-	+	+	+	47.33
105	-	-	+	-	+	+	+	+	-	-	+	53.51
76	-	+	+	-	+	-	-	+	+	+	-	47.08
71	-	+	+	+	-	-	+	-	+	-	+	54.77
87	-	+	-	+	-	+	+	+	-	+	-	44.99
20	+	+	-	+	+	-	-	+	-	-	+	47.41
25	+	+	-	-	+	+	+	-	+	-	-	47.01
14	+	+	+	-	-	+	-	-	-	+	+	50.27
38	+	-	+	+	-	+	-	+	+	-	-	57.91
35	+	-	+	+	+	-	+	-	-	+	-	55.56
63	+	-	-	-	-	-	+	+	+	+	+	44.71

ANOVA Table:

Contrast	$\hat{C}$	SS(C)	F
A	8.88	6.571	0.98
B	-13.80	15.870	2.37
C	41.34	142.416	21.26 *
D	19.08	30.337	4.53
E	-1.06	0.094	0.01
F	5.18	2.236	0.33
G	4.24	1.498	0.22
H	-5.64	2.651	pool
I	0.76	0.048	pool
J	-16.98	24.027	pool
K	-0.86	0.062	pool
Total		225.810	

A Priori Pooling gives MSE = 6.697 with 4. d.f.

$F_{.90}(1,4) = 4.54$

So Contrast C is significant and Contrast D may be significant

but since the MSE term contains the third largest SS(C) value we may be missing some signals.

The Scree Plot suggests that Contrasts C, D, J, B and possibly A stand out from the rubble. Since Contrasts C and D have the same sign in both studies, they are taken to be main effects. Contrast J is unassigned, so it can be only an interaction effect. Contrast A changes signs, so it can be only an interaction effect. Contrasts F and G did not show up in this study, and Contrast B did not show up on the first study.

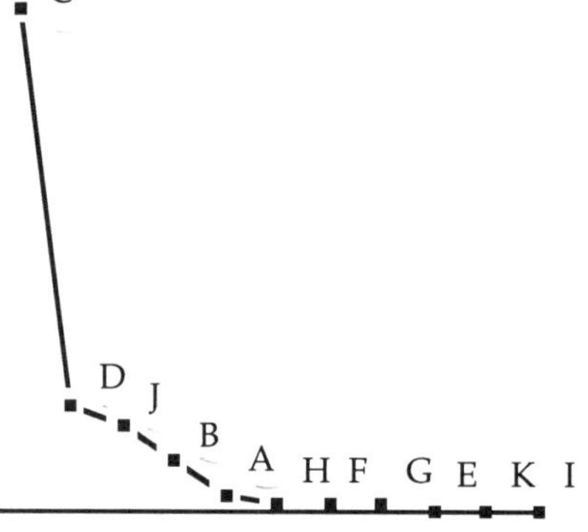

Thus, with 24 runs, these 12-run studies show significant main effects for Factor C and Factor D. In addition, there is some evidence of one or more interaction effects, but no clear indication of just which factors may be involved in the interactions.

When Data Set Eleven was studied using a sequence of 8-run studies, the results at the end of 24 runs were much clearer. Factors C, D, and F had been identified as significant main effects, and the CD interaction had been isolated as the only other appreciable source of variation.

So while both the geometric and the nongeometric designs identified the same two major factors, the complicated confounding pattern for the nongeometric designs makes them less informative in a sequential research program.

7.5 Four-Level Factors In Two-Level Designs

The orthogonality of the Plackett-Burman Designs allows for certain substitutions to be made without destroying either the analysis or the interpretation of the results. In particular, a factor with four levels may be substituted for three two-level factors. This substitution may be accomplished in the following manner.

A factor with four levels will require three degrees of freedom, and these three degrees of freedom can be represented by three orthogonal contrasts. With the proper choice, these contrasts can be made to correspond to three contrasts in a geometric two-level Plackett-Burman design. When this happens, the three contrast labels in the Plackett-Burman design will collectively represent the four-level factor, and the SS for the four-level factor will be the pooled SS for these three contrasts. For example, consider using a four-level factor with an 8-run design.

Let Factor Q have four levels, denoted by Q1, Q2, Q3, and Q4, and let Factors D, E, F, and G have two levels each. Factors Q, D, E, F and G can be studied with an 8-run design. Contrasts A, B and C (=AB) will be used for Factor Q, while contrast labels D, E, F, and G will represent the remaining factors. Two of the three contrasts for Factor Q will define the level of Q that is to be used in conjunction with each run. Say these two contrasts are A and B. The levels of Q might be assigned as follows:

Factor Q Level	A	B
Q1	-	-
Q2	-	+
Q3	+	+
Q4	+	-

Thus, using the basic 8-run design, the treatment combinations would be

Run	Q	D	E	F	G
1	Q1	-	-	-	-
2	Q1	+	+	+	+
3	Q2	+	+	-	-
4	Q2	-	-	+	+
5	Q3	-	+	+	-
6	Q3	+	-	-	+
7	Q4	+	-	+	-
8	Q4	-	+	-	+

while the contrast coefficients would remain unchanged

Run	*A*	*B*	*C*	*D*	*E*	*F*	*G*
1	-	-	-	-	-	-	-
2	-	-	-	+	+	+	+
3	-	+	+	+	+	-	-
4	-	+	+	-	-	+	+
5	+	+	-	-	+	+	-
6	+	+	-	+	-	-	+
7	+	-	+	+	-	+	-
8	+	-	+	-	+	-	+

The sums of squares for contrasts A, B and C would be pooled to obtain SS(Q). This sum of squares will have 3 degrees of freedom. If any of the individual contrasts A, B or C is significant, then the Factor Q is said to be significant. The actual comparison being made by any one contrast can be seen by inspection of the contrast coefficients and the factor levels above.

When deciding which contrasts to use to represent a four-level factor the only restriction is that the third contrast should be the one that is confounded with the interaction of the other two contrasts.

For more information on these Hybrid Designs the reader is referred to Part Four of **Tables of Screening Designs** (Wheeler, 1989). This reference contains 185 designs for combining factors having two levels with factors having from three to 16 levels. Each design is based upon a Plackett-Burman Orthogonal Array, and the confounding pattern for each design is given (up through the two-factor interactions).

7.6 The Randomization of Run Order

The randomization of run order is often recommended in connection with statistically designed experiments. The purpose of such randomization is to average out the effects of extraneous sources of variation so as to avoid two types of bias. These biases are: (1) bias in the estimated contrast effects, and (2) bias in the estimate of error used in the F-tests.

In the broad sense, an extraneous source of variation is any factor that is not included in the experiment but which influences the response variable. Since it is not reasonable to knowingly exclude such factors from a screening study, extraneous sources of variation might be more narrowly defined as those factors outside of the study that influence the response variable without our knowledge. In short, they are assignable causes of uncontrolled variation. Thus, randomization can be thought of as a technique used to overcome the effects that assignable causes have upon an experiment.

Moreover, the mechanism used by randomization is that of averaging. Randomization cannot undo the effects of assignable causes, but it can spread them out across different runs, in a random pattern, so that, hopefully, the assignable causes will not inflate any one contrast appreciably. In effect, the intent is for the effects of the assignable causes to be buried within the subgroups. When each treatment combination occurs several times during the course of an experiment, this strategy makes sense.

Finally, the rationale behind randomization suggests that the consequences of an erroneous conclusion are undesirable enough to warrant the complexity introduced by randomization.

Combining all of these observations, randomization can be seen to be of use when (1) experiments are conducted in circumstances that do not demonstrate statistical control, (2) there are multiple observations per treatment combination (subgroup), and (3) the analysis is conservative (rather than exploratory) in nature. If these conditions are not present, then randomization loses much of its usefulness.

With regard to industrial experimentation, it seems excessive to experiment with a process that is known to be out of control. A process that is out of control is trying to tell the manufacturer about the assignable causes (significant factors). It will generally be more profitable to concentrate on identifying such factors than it will be to go to the trouble and expense of conducting an experiment. The use of control charts to detect assignable causes does not require the observer to manipulate the process inputs. Instead, the experimenter waits for the active factors to voluntarily make their presence known. As assignable causes are thus identified and removed, the out of control points will become rare, and one will eventually want to shift from observing the process with control charts to actually trying to confirm suspected cause and effect relationships. This is the point at which designed experiments should be introduced, and it is also the point at which the impact of assignable causes is diminishing. Thus, establishing process stability prior to experimentation minimizes the need for randomization.

Finally, when there is only one observation per treatment combination, it is impossible for randomization to average out the effect of any assignable cause that may affect the process during the course of an experiment. With one observation per treatment combination, every degree of freedom corresponds to a contrast between the treatment combinations, and any extraneous effect will confound some contrast, even when the runs are performed in random order. Perhaps the easiest way to understand this phenomenon is to consider some examples.

EXAMPLE 7.7: The Effect of a Shift in Process Average:

A basic 8-run design was used in standard order in a particular study. The responses and ANOVA table are shown below.

Run	*Response*	*ANOVA Table*		
		Contrast	$\hat{C}$	*SS(C)*
1	15.5			
2	14.0	A	3.2	1.280
3	15.4	B	1.0	0.125
4	14.2	C	-0.8	0.080
5	16.6	D	-3.6	1.620
6	15.0	E	3.0	1.125
7	14.5	F	-2.8	0.980
8	16.2	G	-2.6	0.845
		Total		6.055

D A E F G B C

If, however, an assignable cause had happened to depress the first three responses by exactly five units, the responses and ANOVA table would be:

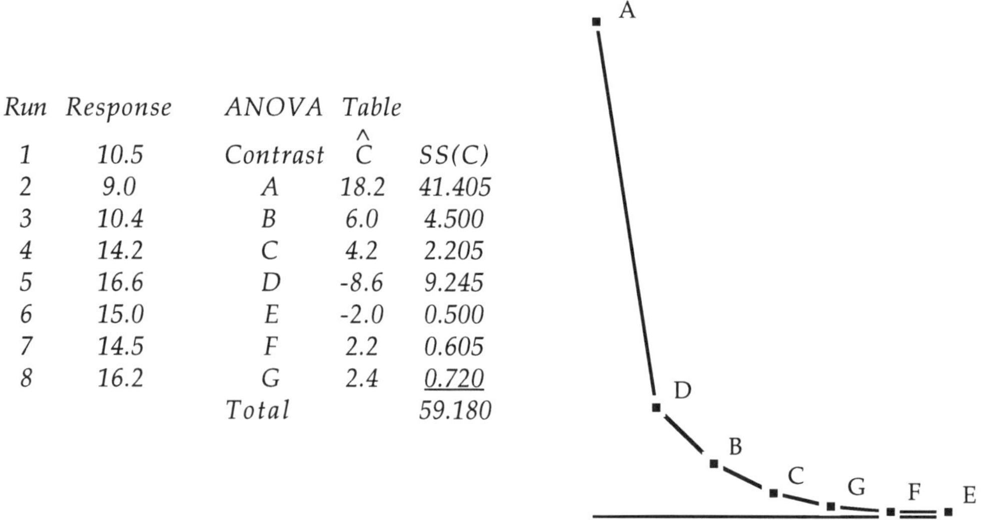

Run	*Response*	*ANOVA Table*		
		Contrast	$\hat{C}$	*SS(C)*
1	10.5			
2	9.0	A	18.2	41.405
3	10.4	B	6.0	4.500
4	14.2	C	4.2	2.205
5	16.6	D	-8.6	9.245
6	15.0	E	-2.0	0.500
7	14.5	F	2.2	0.605
8	16.2	G	2.4	0.720
		Total		59.180

Notice that **every contrast** *was affected by this shift. While most contrast estimates were shifted by 5 units, the estimate for Contrast A was shifted 15 units.*

When the runs of a Plackett-Burman design are performed in the order given in this book, a sustained shift will usually contaminate Contrast A or Contrast B most severely.

EXAMPLE 7.8: *The Effect of a Shift in Process Average (Part II):*

A basic 8-run design was performed in **random** *order. The order shown is the time order in which the runs were performed. The responses and ANOVA table are shown below.*

Run	Response	Contrast	$\hat{C}$	SS(C)
6	13.7	A	2.2	0.605
1	15.5	B	-1.0	0.125
3	14.5	C	2.6	0.845
5	16.3	D	-4.4	2.420
4	15.7	E	-1.2	0.180
2	13.9	F	3.2	1.280
8	15.4	G	-4.0	2.000
7	16.4	Total		7.455

D G F C A E B

Again, assume that an assignable cause happens to depress the first three responses by exactly five units. The responses and ANOVA table would be:

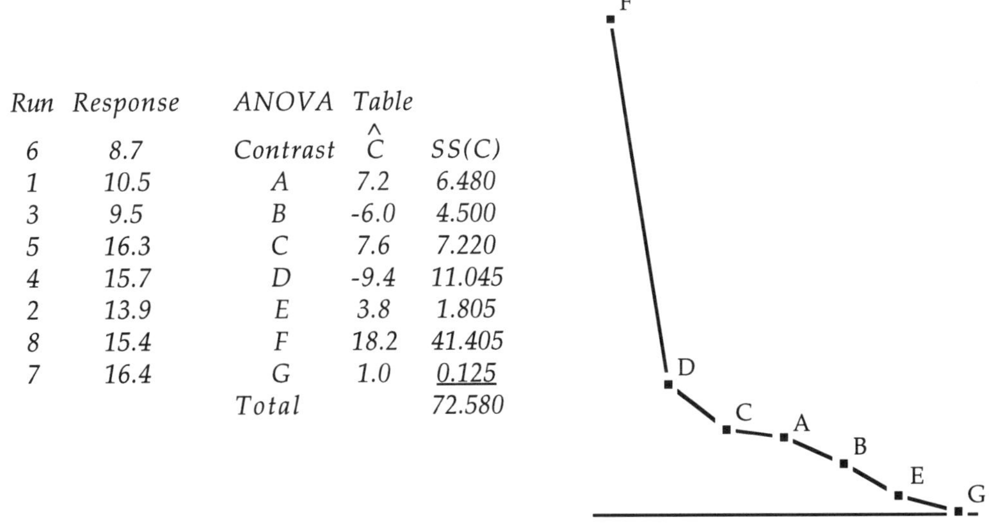

Run	Response	Contrast	$\hat{C}$	SS(C)
6	8.7	A	7.2	6.480
1	10.5	B	-6.0	4.500
3	9.5	C	7.6	7.220
5	16.3	D	-9.4	11.045
4	15.7	E	3.8	1.805
2	13.9	F	18.2	41.405
8	15.4	G	1.0	0.125
7	16.4	Total		72.580

Once again, the shift affects **every** *contrast. Most estimates are changed by 5 units. The estimate for Contrast F is changed 15 units. The same shift produces essentially the same set of effects.*

Randomizing the run order will ***not*** *change the effect of an assignable cause, it will merely randomize which contrast is most severely affected.*

Consideration of the preceding examples will show that randomization cannot cover up the effect of any assignable cause that happens to affect the response variable during the course of a study. Any false signal will, of necessity, show up in one of the contrasts, and will therefore be confounded with one of the main effects in a saturated design. Thus, ***when there is only one observation per run, the randomization of the run order cannot average out the effects of extraneous sources of variation.*** Such randomization of the run order can only add needless complexity to the experiment. For this reason, randomization of run order is of little use in screening designs.

If randomization will not protect against the effects of assignable causes when performing screening designs with one observation per run, then what will? Virtually the only tool available is the replication of results.

One way of seeking protection against the effects of assignable causes is the replication of responses at a single design point. By collecting several observations at one treatment combination during the course of an experiment the experimenter can verify the stability of the process being studied. If an assignable cause has affected the experiment then it is likely that the responses at the duplicated point will be inconsistent. This approach is described in the next section.

A stronger approach is to replicate the whole pattern of responses by a subsequent experiment. By attempting to duplicate the same cause and effect relationships in a separate experiment one will both verify that an assignable cause has not contaminated the results of either experiment, and support the claim that certain factors affect the response variable.

Therefore, the replication of results from study to study will itself serve to validate the results of each study. For while, in any one study, an assignable cause may make some contrasts appear to be significant when they are not, it is highly unlikely that assignable causes will show up in exactly the same way in successive studies. If a sequence of reflected designs is used, or even if the same design is used in succession, finding the same contrasts to be significant two or three times provides much stronger confirmation than any probabilistic procedure based on the results of just one study. For these reasons, it is much better to plan a sequence of small experiments than one large experiment.

7.7 Two-Level Plackett-Burman Designs With Added Points

As suggested in the preceding section, the two-level Plackett-Burman designs may be augmented by extra runs. Such runs will be used for special types of analyses in addition to the regular analysis already outlined for these designs.

When several runs are performed using the same treatment combination, the repeated responses will provide a way to estimate the background noise. With five extra runs at the same treatment combination, the s^2 value for these five runs may be used as a Mean Square Within value having 4 degrees of freedom. Thus, an 8-Run P-B design with Extra Runs might consist of eight subgroups of size one, and one subgroup of size five. The design layout might look like the following array:

Run	A	B	C	D	E	F	G	Response
ER1	o	o	o	o	o	o	o	y_{e1}
1	–	–	–	–	–	–	–	y_1
2	–	–	–	+	+	+	+	y_2
ER2	o	o	o	o	o	o	o	y_{e2}
3	–	+	+	+	+	–	–	y_3
4	–	+	+	–	–	+	+	y_4
ER3	o	o	o	o	o	o	o	y_{e3}
5	+	+	–	–	+	+	–	y_5
6	+	+	–	+	–	–	+	y_6
ER4	o	o	o	o	o	o	o	y_{e4}
7	+	–	+	+	–	+	–	y_7
8	+	–	+	–	+	–	+	y_8
ER5	o	o	o	o	o	o	o	y_{e5}

The "o" symbols in the design matrix may denote any specific level for each of the factors. They could be central values, mid-way between the + and – levels, or they could be either of these two levels themselves. The choice for the "o" level will be made for each factor separately. Since the responses for these Extra Runs will not be used in the direct analysis, these levels may be whatever the experimenter wants them to be in any one study. The contrasts for the Basic 8-Run design will remain the same as they were without the Extra Runs.

By spreading the 5 Extra Runs throughout the course of the experiment one will obtain a powerful check on the consistency of the experimental apparatus and environment. In fact, the five response values obtained for the Extra Runs might be plotted on a control chart for individual values and a moving range (an XmR chart). Admittedly it is only five values, and there will be only four moving ranges, but if this simple XmR chart is out of control, then it is likely that some assignable cause of uncontrolled variation has undermined your experiment. If this XmR chart displays control, then it is unlikely that any severe disturbance has occurred during the course of your study.

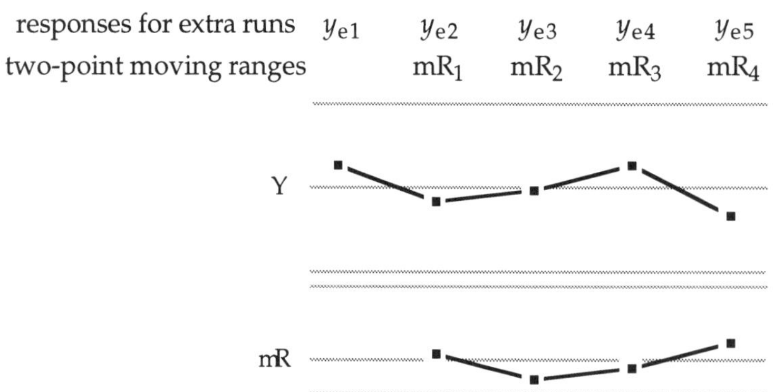

Figure 7.1 XmR Chart for Extra Run Responses

(Two conventions clash in Figure 7.1. The response variable in an experiment is typically denoted by the symbol Y, while individual values on a control chart are symbolized by X. In this case, they are one and the same, i.e. Y = X.)

If the Extra Runs are not spread throughout the 8-Run Design, then this ability to check on the stability of the underlying process is lost. The Extra Runs may still be used to compute an s^2 value, and this s^2 value may still be used to form honest F-Ratios for the between-subgroup contrasts, but an XmR chart will not have the same utility as it would otherwise.

Finally, if the Extra Runs are true centroids, that is, if the "o" levels are indeed midpoints between the + and – levels for each factor, then one may define a "curvature contrast." This curvature contrast will compare the 8 design points with the 5 Extra Run points. A significant curvature contrast will indicate curved response surfaces, while a nonsignificant curvature contrast will indicate planar response surfaces. The contrast coefficients for this and the other seven contrasts are shown in the following table:

Contrast Coefficients:

Run	A	B	C	D	E	F	G	Curvature Contrast	Response
ER1	0	0	0	0	0	0	0	8	y_{e1}
1	–1	–1	–1	–1	–1	–1	–1	–5	y_1
2	–1	–1	–1	+1	+1	+1	+1	–5	y_2
ER2	0	0	0	0	0	0	0	8	y_{e2}
3	–1	+1	+1	+1	+1	–1	–1	–5	y_3
4	–1	+1	+1	–1	–1	+1	+1	–5	y_4
ER3	0	0	0	0	0	0	0	8	y_{e3}
5	+1	+1	–1	–1	+1	+1	–1	–5	y_5
6	+1	+1	–1	+1	–1	–1	+1	–5	y_6
ER4	0	0	0	0	0	0	0	8	y_{e4}
7	+1	–1	+1	+1	–1	+1	–1	–5	y_7
8	+1	–1	+1	–1	+1	–1	+1	–5	y_8
ER5	0	0	0	0	0	0	0	8	y_{e5}

Finally, the ANOVA table for this design would have a structure like the following table. The s^2 used for MSW would be computed using the 5 Extra Run response values. The Total Sum of Squares would be obtained from an s^2 value computed using all 13 response values. (The reader is reminded that the first part of the ANOVA Worksheet will still be applicable for finding the Total Sum of Squares.)

Contrast	SS	DF	MS	F-Ratio
A	SS(A)	1		SS(A)/MSW
B	SS(B)	1		SS(B)/MSW
C	SS(C)	1		SS(C)/MSW
D	SS(D)	1		SS(D)/MSW
E	SS(E)	1		SS(E)/MSW
F	SS(F)	1		SS(F)/MSW
G	SS(G)	1		SS(G)/MSW
CURVATURE	SS(CC)	1		SS(CC)/MSW
WITHIN	$4s^2$	4	s^2 = MSW	
TOTAL	SS Total	12		

Thus, Extra Runs may be added to a two-level Plackett-Burman design in order to (1) check on consistency of the basic process during the course of an experimental study, (2) obtain several degrees of freedom for a Within-Subgroup estimate of the background noise (so that honest F-Ratios may be computed), and (3) obtain a check on the curvature of the response surface. The third objective will require that the Extra Runs are true centroid runs, with each factor set at a point half-way between its other two levels.

The first two objectives of Extra Runs are better met by performing a sequence of reflected experimental studies. Any inconsistency in the underlying process will tend to cause successive studies to have totally different results, and the persistence of results from one study to another is the strongest proof that real effects exist.

Therefore, the only unique feature of the Extra Runs is their ability to check for curvature of the response surface. If this is of great enough interest to justify the expense of the Extra Runs, and if the Extra Runs are true centroids, then this approach may be used. However, one might come out ahead by running the screening designs first in order to eliminate any inert factors, and then running one or more centroid points using only the active factors.

7.8 Three-Level Plackett-Burman Designs

Screening designs for factors at three levels each are considerably more complex than those for factors at two levels. Before discussing the properties of three-level screening designs, some discussion of the basic properties of three-level factors is appropriate.

EXAMPLE 7.9: *An Artificial Data Set:*

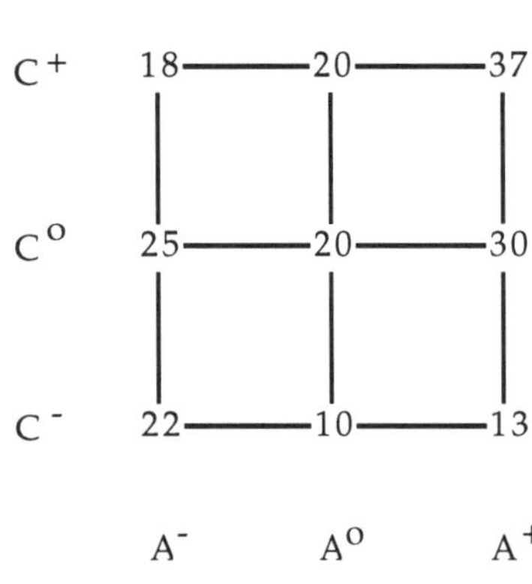

An artificial data set involving two factors at three levels each is shown in the Response Plot on the left. These two factors are fully crossed, and each factor combination defines a "subgroup." With nine subgroups of size one, there will be 8 degrees of freedom. Of these eight degrees of freedom, two will be required to explain the main effect for Factor A, two will be required to explain the main effect for Factor C, and four will be needed to explain the AC interaction effect. These eight orthogonal contrasts are shown, along with the factor levels and responses, in the table below.

			Contrasts							
			Factor A		Factor C		Interactions			
Run	Factor Levels	Response	A	A^2	C	C^2	AC	AC^2	A^2C	A^2C^2
1	$A^- C^-$	22	-1	-1	-1	-1	1	1	1	1
2	$A^- C^o$	25	-1	-1	0	2	0	-2	0	-2
3	$A^- C^+$	18	-1	-1	1	-1	-1	1	-1	1
4	$A^o C^-$	10	0	2	-1	-1	0	0	-2	-2
5	$A^o C^o$	20	0	2	0	2	0	0	0	4
6	$A^o C^+$	20	0	2	1	-1	0	0	2	-2
7	$A^+ C^-$	13	1	-1	-1	-1	-1	-1	1	1
8	$A^+ C^o$	30	1	-1	0	2	0	2	0	-2
9	$A^+ C^+$	37	1	-1	1	-1	1	-1	-1	1

The Single Degree of Freedom ANOVA Table for these values is:

Contrast	$\hat{C}$	SS(C)	$\hat{l}$
A	15	37.5	5
A^2	-45	112.5	-7.5
C	30	150.0	10
C^2	30	50.0	5
AC	28	196.0	14
AC^2	0	0	
A^2C	0	0	
A^2C^2	0	0	
Total		546.0	

The effects represented by Contrast A and Contrast C are termed "linear effects." The effects represented by Contrasts A^2 and C^2 are termed "quadratic effects." As usual, each of the interaction contrasts is the product of one Factor A contrast and one Factor C contrast. Using this terminology, the ANOVA table above shows that the variation in the data can be completely explained in terms of linear and quadratic effects for A and C and a simple (linear by linear) interaction between Factors A and C.

The terms "linear effect" and "quadratic effect" are generally used for convenience regardless of the nature of the factor. If the factor is quantitative, then the contrast for the linear effect will check for a linear relationship between the factor and the response variable. If the factor is quantitative and if the factor levels are evenly spaced, then the contrast for the quadratic effect will check for the presence of a quadratic relationship between the factor and the response variable. If the factor is qualitative, the terms "linear effect" and "quadratic effect" are names only, with no underlying physical meaning.

In addition to this new terminology, there are some other conventions used in presenting the three-level Plackett-Burman designs. Comparing the factor levels and the contrast coefficients in Example 7.9, it should be noted that, for each main effect, the coefficients for the linear contrasts (-1, 0, 1) coincide with the symbols for the factor levels (-, o, +). Because of this, the linear contrasts and the treatment combinations can be combined into one array. The contrast coefficients for the quadratic contrasts will then be given in a second array.

Just as a two-level design may be reflected, the three-level Plackett-Burman designs may be rotated. If we define the basic three-level design to be the design in which one run has all factors occur at their middle level, then one rotation will yield a design in which one run has all factors at their high level, while a second rotation will yield a design in which one run has all factors at their low level. Thus, the three rotations associated with any one three-level design may be specified as being mid-level, high-level, or low-level.

The simplest three-level Plackett-Burman designs will have 9 runs, and can handle up to four factors at three levels each. A mid-level 9-run design and its two rotations are given in the following examples. In order to show how these three-level designs are affected by an interaction between two factors the artificial data of Example 7.9 will be used with each design. In each example the response for each run will be determined by the levels of Factors A and C (which makes Factors B and D inert factors). The simple AC interaction effect present in the data of Example 7.9 will be seen to be confounded in different ways in the different designs.

EXAMPLE 7.10: A Mid-Level 9-Run Design:

Treatment Combinations

	Linear Contrasts				*Quadratic Contrasts*				
Run	A	B	C	D	A^2	B^2	C^2	D^2	*Response*
1	0	0	0	0	2	2	2	2	20
2	0	+	+	+	2	-1	-1	-1	20
3	0	-	-	-	2	-1	-1	-1	10
4	+	-	0	+	-1	-1	2	-1	30
5	+	0	+	-	-1	2	-1	-1	37
6	+	+	-	0	-1	-1	-1	2	13
7	-	+	0	-	-1	-1	2	-1	25
8	-	-	+	0	-1	-1	-1	2	18
9	-	0	-	+	-1	2	-1	-1	22

ANOVA Table

Contrast	$\hat{C}$	*SS(C)*
A	15	37.5
A^2	-45	112.5
B	0	0
B^2	42	98.0
C	30	150.0
C^2	30	50.0
D	0	0
D^2	-42	98.0
Total		546.0

The effects for Factors A and C are the same as in Example 7.9.

The AC interaction SS of 196.0 is divided between the contrasts for B^2 *and* D^2*, which means that this interaction is partially confounded with these two contrasts.*

EXAMPLE 7.11: A High-Level 9-Run Design:

	Treatment Combinations Linear Contrasts				Quadratic Contrasts				
Run	A	B	C	D	A^2	B^2	C^2	D^2	Response
1	+	+	+	+	-1	-1	-1	-1	37
2	+	-	-	-	-1	-1	-1	-1	13
3	+	0	0	0	-1	2	2	2	30
4	-	0	+	-	-1	2	-1	-1	18
5	-	+	-	0	-1	-1	-1	2	22
6	-	-	0	+	-1	-1	2	-1	25
7	0	-	+	0	2	-1	-1	2	20
8	0	0	-	+	2	2	-1	-1	10
9	0	+	0	-	2	-1	2	-1	20

ANOVA Table

Contrast	$\hat{C}$	SS(C)
A	15	37.5
A^2	-45	112.5
B	21	73.5
B^2	-21	24.5
C	30	150.0
C^2	30	50.0
D	21	73.5
D^2	21	24.5
Total		546.0

The effects for Factors A and C are the same as in Example 7.9.

The AC interaction is partially confounded with, B, B^2, D and D^2.

EXAMPLE 7.12: A Low-Level 9-Run Design:

Run	A	B	C	D	A^2	B^2	C^2	D^2	Response
	Treatment Combinations *Linear Contrasts*				*Quadratic Contrasts*				
1	-	-	-	-	-1	-1	-1	-1	22
2	-	0	0	0	-1	2	2	2	25
3	-	+	+	+	-1	-1	-1	-1	18
4	0	+	-	0	2	-1	-1	2	10
5	0	-	0	+	2	-1	2	-1	20
6	0	0	+	-	2	2	-1	-1	20
7	+	0	-	+	-1	2	-1	-1	13
8	+	+	0	-	-1	-1	2	-1	30
9	+	-	+	0	-1	-1	-1	2	37

ANOVA Table

Contrast	$\hat{C}$	SS(C)
A	15	37.5
A^2	-45	112.5
B	-21	73.5
B^2	-21	24.5
C	30	150.0
C^2	30	50.0
D	-21	73.5
D^2	21	24.5
Total		546.0

The effects for Factors A and C are the same as in Example 7.9.

The AC interaction is partially confounded with B, B^2, D and D^2.

Combining the results of the three 9-run studies we get:

Contrast	Low-Level $\hat{C}$	Mid-Level $\hat{C}$	High-Level $\hat{C}$
A	15	15	15
A^2	-45	-45	-45
B	-21	0	21
B^2	-21	42	-21
C	30	30	30
C^2	30	30	30
D	-21	0	21
D^2	21	-42	21

While the estimated contrast effects for Factors A and C are constant across all three studies, the contrasts for Factors B and D show sign changes from one study to another. These sign changes suggest that the contrasts for Factor B and Factor D represent interaction effects. (The numerical constancy of the estimates is due to the use of the same set of response values for the same levels of Factors A and C in all three examples.)

Thus, by using the three rotations of a 9-run design, it is possible to correctly identify the two active factors (A and C) and to know that the effects for the other factors (B and D) are interaction effects. Any one study by itself will not provide this insight. In each separate study, the four largest SS involve all four factors. Thus, because of the contamination introduced by the one interaction effect, no single 9-run study will "screen out" either of the two inert factors, B and D.

Finally, while performing three 9-run studies may seem excessive, a fully crossed study for 4 factors at three levels each would require 81 runs. Thus, the combined total of 27 runs is still a one-third replicate of the Fully Crossed Design. A 27-run design can be used for 13 factors at three levels. The three rotations of a 27-run design will require a total of 81 runs. A full crossing of 13 factors at three levels each would require 1.6 million runs. So even though the number of runs used by these three level designs may seem large, they are not so excessive when put in perspective.

The 9-run, 27-run and 81-run Plackett-Burman designs are geometric three-level designs. An 18-run design that may be used with seven factors at three levels each was discovered by J. P. Burman. It is a nongeometric three-level design.

7.9 Critique of Three-Level Plackett-Burman Designs

The major justification given for the use of Three-Level Plackett-Burman Designs is their ability to detect nonlinear relationships between the factors and the response variable. However, as was shown in the preceding section, this ability comes at the cost of not being able to identify any two-factor interactions which might be present. This trade-off between nonlinear main effects and two-factor interactions is the major problem with the Three-Level Plackett-Burman Designs.

Both two-factor interactions and nonlinear (quadratic) main effects are second order relationships between the factors and the response variable. Thus, there is a logical inconsistency behind the Three-Level Plackett-Burman designs: while greatly complicating the experiment in order to discover information about one type of second-order relationship (quadratic main effects), they ignore another type of second-order relationship (two-factor interactions) which is usually of equal importance.

While some experimenters are very satisfied with Three-Level Plackett-Burman Designs, their use is risky when interaction effects are likely. Since nonlinear main effects are, in essence, an interaction of a factor with itself, it is rather presumptuous to argue that nonlinear main effects are likely but that two-factor interactions are unlikely. Thus, on a purely logical level, the Three-Level Plackett-Burman Designs are not appealing.

Moreover, it will often be possible to do more with fewer runs.

For example, a 27-run Plackett-Burman design can evaluate 13 factors at three levels each. With three rotations, a total of 81 runs will be required. Only after all 81 runs have been performed will it be possible to begin to sort out the main effects from the interaction effects. It will be possible to identify which main effects are important, and the three 27-run designs will have provided an opportunity to have replicated results from one study to another, but, if interactions are present, it will be impossible to identify just which interactions are important.

On the other hand, a 16-run study can evaluate 13 (or 14 or 15) factors at two levels each. With one follow-up study (a total of 32 runs) the experimenter can isolate and identify all significant main effects and detect the presence of two-factor interactions. If necessary, another 16-run study (a total of 48 runs) will allow most of the two-factor interactions to be specifically identified. Say that only three of the 13 factors actively affect the response: these three factors could then be studied at three levels each in a fully crossed study with another 27 runs. Thus, with a total of 75 runs, one would have obtained a clear and unconfounded analysis of the three active factors, including all nonlinear main effects and all possible interactions between the three factors. (There are ways to reduce this number even more, but this will suffice for the comparison made here.)

Generally, in comparison with the Three-Level Plackett-Burman Designs, it will require fewer runs to fully study the active factors, and one will ultimately know more, if one will screen out the inert factors using Two-Level Designs, and then follow-up with multi-level full crossings or half replicate designs.

This phenomenon tends to limit the usefulness of the Three-Level Plackett-Burman Designs.

7.10 Choice of Factor Levels and Other Cautions

One advantage of the Two-Level Plackett-Burman Designs is the symmetry with which they treat each of the factors. Every contrast effect is estimated with the same precision. Thus, in terms of the design, every factor has an equal chance to display any effect it has upon the response variable. However, in practice, some factors may be handicapped relative to the others in a study.

Assume that both Factor A and Factor B have the same effect upon the Response Y. As Factor A increases by 10 units, the Response increases, on the average, by 5 units. Likewise, as Factor B increases by 10 units, the response increases, on the average, by 5 units. However, say that the levels for Factor A are set at (Current ± 5 units) while those for Factor B are set at (Current ± 2 units). Then, the design would show a stronger effect for Factor A (+5 units) than it would for Factor B (+2 units). Therefore, it is important for the experimenter to give some careful thought to the levels for the factors in any study.

When a study is purely experimental, and the product produced is not intended for sale or further use, then one may choose the factor levels boldly. By pushing the factors to reasonable extremes, one is more likely to detect the effects which are present. If one suspects that one or more factors are already near optimum, then these factors may be incremented in just one direction to avoid "straddling an optimum."

In contrast to this "bold strategy" is the concept of Evolutionary Operation (EVOP). Here one is running a production process, and all of the product is intended for sale or further use. In this case one can hardly hope to use a "bold strategy:" large changes in factors might well result in scrap or even a process shutdown. Therefore, one must be content to study the effects of small changes upon the product.

Since small changes may lead to small effects, it is easier for noise to obscure the relationships between the factors and the response variables. Thus, one will have to collect more data to offset this obscuring effect of noise.

Traditional EVOP is set up with two or three factors at two levels each. While the details of the analysis may be changed, the concept is as follows. The current operating conditions will be the centroid of the fully-crossed two-level design. The centroid and each of the vertices of the design will be run in sequence, obtaining one observation per treatment combination. After one complete round, the data will be analyzed for possible signals.

If signals are found, the centroid is moved to the best combinations of active factors, and the process is repeated. Inert factors may be dropped at this point, replaced by other factors, or retained for further study (they might have weak effects upon the responses).

If no signals are found, then the complete round of treatment combinations is repeated again, added to the previous data, and analyzed as if the subgroup size is n = 2.

If signals are found then the centroid is moved and the whole process repeated around the new centroid. If no signals are found the compete round is repeated once more.

If the subgroup size reaches n = 5 and no signals have been found, then either new factors are needed, or else bolder levels are needed for the old factors.

In performing the analyses with n = 1 one should use contrasts and Scree Plots. If repeated runs at the centroid are available, they can be used to estimate MSW and test the contrasts, or construct an ANOM Plot. When n exceeds 1, use ANOM Plots.

The Two-Level Plackett-Burman Designs provide a perfect tool for performing EVOP with more than two variables. With three factors, one could use a 4-Run Design with an added center point. The 4-Run Plackett-Burman Designs are shown below.

Basic 4-Run Plackett-Burman Design

	A	B	C
Run	-BC	-AC	-AB
1	–	–	–
2	–	+	+
3	+	+	–
4	+	–	+

Reflected 4-Run Plackett-Burman Design

	A	B	C
Run	BC	AC	AB
1	+	+	+
2	+	–	–
3	–	–	+
4	–	+	–

With four to seven factors, one could use an 8-Run Design with an added center point. Since the traditional EVOP for three factors involves eight runs plus a centroid, these designs allow for the simultaneous study of four to seven factors with no added complexity on the operational floor.

Finally, the operating principle behind any highly-fractionated design is that of effect sparsity. The experimenter is counting on no more than one effect being significant on each contrast, and that some of the contrasts will be nonsignificant. Under these conditions, highly-fractionated designs are very efficient.* However, when the number of significant main effects and interaction effects is equal to or greater than the number of runs, the confounding will tend to make it much more difficult to discover just what is happening. Therefore, one should al-

* There are the so-called Desperado Designs which will study 2^k factors with k+1 runs. This can be done only by making the strong assumption that **one and only one** of the 2^k main effects is significant. While this is highly efficient, it is also exceedingly risky.

ways keep the number of runs larger than the number of effects which are likely to be significant.

A consequence of the implicit assumption of effect sparsity is that Screening Designs are appropriate when searching through a *large* number of factors to identify the *few* which actively affect the response variable. When *all* the factors in a study are known to actively affect the response, Screening Designs are not appropriate (there is no screening to be done).

Moreover, in any highly interactive environment, where three-factor or higher order interactions are on a par with, or dominate the main effects, the highly-fractionated designs will be useless. Here one is limited to half replicates or fully-crossed factors.

Some of the basic Plackett-Burman designs have been given in this chapter. Other designs are provided in the appendix. The limited designs given there are intended to introduce and complement the more extensive tabulation of these designs in **Tables of Screening Designs** by this author.

CHAPTER EIGHT

THE PROBLEM OF PRODUCT VARIATION

The preceding chapters have outlined some techniques for quickly and efficiently sifting through a large number of factors to identify those that actively affect a response variable. In many cases this is all that is needed. The problems may be so well-defined, both in terms of the possible factors and the interesting responses, that the experimenter will know how to proceed. At other times this is not the case. The experimenter may be uncertain about what to look for, or he may not know how to use the techniques to greatest advantage. In short, for various reasons, he may lack a coherent strategy for achieving his desired goals.

It is the purpose of this chapter to present a strategy for using experimental techniques to increase product consistency while reducing product costs.

8.1 The Costs of Variation

Variation always creates costs.

This fact has been obscured by our accounting systems and ignored by manufacturers for years, yet it has always been true. As product characteristics deviate from their optimum or target values it generally becomes more difficult to use the product. These difficulties are overcome in practice by actions of one sort or another, and these actions, inevitably, cost something. Therefore, variation always creates costs.

First, there are the costs associated with inconsistent material usage. As the yield varies, so does the cost per unit. Moreover, with varying yield it is very hard to coordinate production processes. Many overruns can be rendered useless by one underrun. (Combine 12,000 pairs of shoes with 11,000 pairs of shoelaces and you get 11,000 completed pairs of shoes. Just what is the cost of the underrun of 1000 pairs of shoelaces?) The cost of the one underrun may be many times greater than the cost on that one component. Traditional cost accounting may hide this cost, but it is no less real just because it has been rendered invisible. And sometimes even the accountants cannot hide the problem: this author once saw over 20 jumbo jets parked outside the assembly plant awaiting missing parts.

Next there are the costs of getting components to assemble. Two components, A and B, may both be within their specifications, and yet they will not assemble. This problem has two typical solutions. First, the components may be sorted into subcategories within the specifications. If these subcategories are defined carefully enough, one can usually assemble the A components from one particular subcategory with the B components from a corresponding subcategory. When this approach is used one may be hopeful that the final assemblies will function more or less as intended. The second approach to the "will not assemble" problem is to redesign the components with greater clearance. Then, instead of sorting to match, one may simply use more sealant to join the components.

The chemical version of this problem is blending to achieve certain properties such as average particle size. The average particle size may be in spec, but if the shipment is blended from two batches with differing particle size distributions, there may be no particles of "average size" in the shipment!

Next there are the costs of designing jigs, fixtures, processes, and matching components to allow for variation. If a jig will not hold all of the pieces produced, many good pieces may have to be set aside as unusable, even though their functionality is satisfactory. Likewise, if a component must pass through a feed tube to reach the point of assembly, it is important that the feed tube be designed to allow all of the components to pass freely through. If this is not done, then some of the components will jam in the tube, and the assembly line will be shut down until the jam is removed. If this happens the manufacturer is faced with the choice of either (a) redesigning the feed tube, or (b) screening out product that will not fit the feed tube. Both of these options cost money, and neither adds value to the final product. They, along with the initial cost of designing the feed tube to allow for variation, are Costs of Variation.

As automation increases, Costs of Variation such as these will be greatly magnified, for people adapt to variation better than robots. Therefore, it is axiomatic that to get the full benefits of automation one should first work on reducing the variation in the product streams.

Next come the warranty costs. Variations from nominal will generally result in reduced product performance and may lead to early product failures. Some of these costs of variation come back to the manufacturer in terms of warranty costs, but the majority are borne by the customer.

Once the customer gets involved, the costs due to variation become essentially unknown and unknowable. There is just no way to quantify customer ill-will. For each dissatisfied customer that the manufacturer knows about, how many more simply resolve to take their business elsewhere? (Even a satisfied customer may decide to buy elsewhere next time.) So while customer dissatisfaction defies quantification, it still represents a real cost for the manufacturer even though it may not be apparent until some future date.

All of these costs are created or exaggerated by variation. Actions taken to deal with variation after the fact will inevitably increase the costs, while actions that reduce variation at the source will reduce subsequent costs while increasing product quality. The further upstream one works to reduce variation, the lower the Costs of Variation.

8.2 The Specification Approach to Variation

Say, for example, that Y denotes a performance characteristic for a particular product, and let the target value for Y be denoted by τ. If the target value has been properly defined, and if a particular unit of the product has a value of $Y = \tau$, then there should be no problem with that unit. The problem comes when the product performance characteristic is not equal to τ. Assume, for the purposes of this discussion, that any deviation of Y from τ will result in a degradation in the performance of the product. Generally, small differences will result in small amounts of degradation, while greater deviations will result in greater degradations.

The traditional approach to the problem of product variation has been that of specifications. By using specification limits to define some neighborhood of τ, say $\tau \pm \Delta y$, manufacturers have hoped to place acceptable bounds upon the degradation in performance for the product. As long as the quality characteristic Y falls within these specification limits, the product is said to be satisfactory. When the value for Y falls outside these limits, the product is suddenly deemed to be unsatisfactory and certain actions are invoked to remedy the situation.

To understand the natural consequences of the Specification Approach consider a stream of units produced and evaluated according to specifications of $\tau = 100$ and $\Delta y = 10$. The first unit has a value of 108 and is therefore deemed to be satisfactory. The next unit has a value of 102, and it also is passed. The third unit has a value of 96, so it is passed. Say the next unit has a value of 92. It is passed. The fifth unit has a value of 90, which is still within specifications, so it is passed and everything is still deemed to be satisfactory for the production process. However, when the sixth unit has a value of 89 the whole department is thrown into an uproar to find out why they are making nonconforming product! Inspectors are sent to inspect all incoming products. Engineers are assigned to project teams to work on the process. Managers consider if a recall is needed. And the workers adjust the process to increase the value for Y. This sudden cascade of actions will of course greatly increase the costs associated with the production of this product. (One should note that the difference between the fifth unit (90) and the sixth unit (89) was less than any of the differences between earlier successive units, yet the fifth unit was deemed to be satisfactory and the sixth unit was deemed to be unsatisfactory!)

Thus, specification limits are actually *artificial* boundaries used to make *arbitrary* decisions about what product to use. They are a naive attempt to deal with the problems created by the variation of product characteristics. All product is considered to be either good or bad, and the dividing line between good stuff and bad stuff is seen to be a sharp cliff.

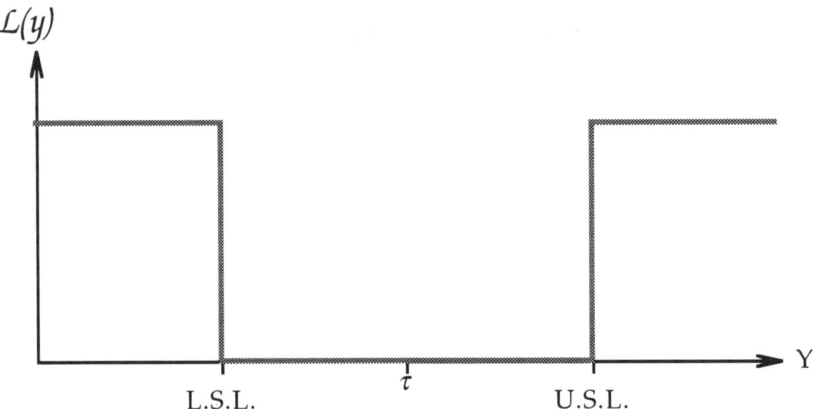

Figure 8.1: The Loss Function for the Specification Approach

The graph in Figure 8.1 shows the Loss Function for the Specification Approach to product variation. The loss associated with using a given item is viewed to be a function of the value of the product characteristic, Y. In Figure 8.1 the loss is shown on the vertical scale and is labeled as $\mathcal{L}(y)$, while values for Y are shown on the horizontal scale. With the Specification Approach to product variation the cost of using an item is defined to be zero whenever the value for Y is within the specification limits [$\mathcal{L}(y)$ is zero when Y is within the specifications]. The loss is defined to be some definite (nonzero) value whenever the value for Y is outside the specification limits. And the step function of Figure 8.1 is the result.

The world view shown in Figure 8.1 is reinforced by the cascade of actions which are initiated whenever the value for Y falls outside the specification limits. Sorting, blending, rework, scrapping of product and adjusting the process will all contribute to the loss associated with nonconforming product, and this sudden shift in the way the product is treated will essentially create a step-function in any cost curve. However, one should notice that this step-function is created by a reaction on the part of management, rather than by any sudden and dramatic change in the product characteristic, Y. The changes from 108 to 102, from 102 to 96, from 96 to 92, and from 92 to 90 were all larger than the change from 90 to 89 which triggered the responses. According to the specification world view, 108, 102, 96, 92 and 90 are equivalent, but 89 is different from 90. The shift from "operating okay" to operating "in trouble" is always seen as sudden and unexpected.

Therefore, the very nature of the Specification Approach fosters periods of neglect of the process broken by periods of intense process scrutiny. During the periods of neglect any insights into the process are usually lost, process improvements come unraveled, and the product quality begins to drift away from the target value once again. This is why Conformance to Specifications is no longer enough to remain competitive in today's world. The Conformance to Specifications Approach does not engender the constancy of purpose required to learn from, and to continually improve the production process. As long as the conformance to specifications is regarded as the main objective for any operation, it will be impossible to sustain any real process improvements. (This is why designed experiments are always more successful in an environment which has demonstrated the constancy of purpose needed to get at least some of its processes into a reasonable state of statistical control. If a plant has not demonstrated the ability to use Shewhart's Control Charts in a sustained program of process improvement, what hope is there that they can use more sophisticated and complicated techniques, and on what basis can one expect that the knowledge gained from the experiments will continue to be utilized in the weeks and months to come?)

Therefore, a different approach to the problem of product variation is needed.

8.3 The Taguchi Loss Function

In 1960, Dr. Genichi Taguchi introduced an elegant way of approaching the problem of product variation. His approach used a well known mathematical construction in a new setting, and resulted in a powerful new perspective on the problem of variation in production.

Again let Y denote some performance characteristic for a particular product, and let τ represent the target value for Y. As Y varies about τ there will be some loss associated with each particular value of Y. When Y = y, denote this loss as $\mathcal{L}(y)$. The loss function, $\mathcal{L}(y)$, is generally assumed to be;

(1) nonnegative for all values of y,

(2) equal to zero when $y = \tau$, and
(3) piecewise smooth near τ.

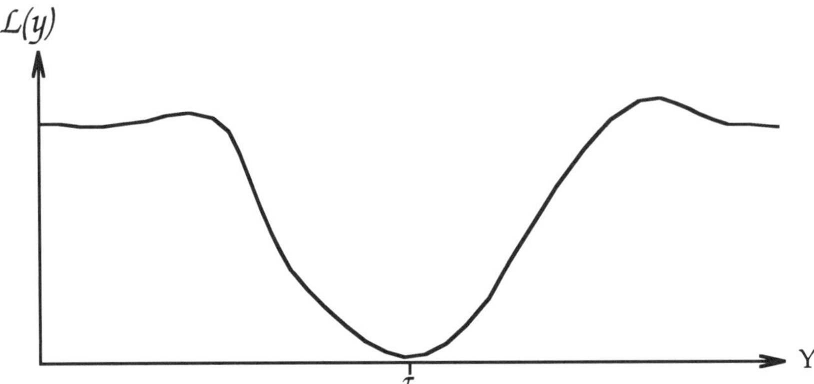

Figure 8.2: A Piecewise Smooth Loss Function

Under these rather general conditions, one may use a Taylor series expansion to approximate $\mathcal{L}(y)$ within some region close to τ. When this is done, the first three terms of the approximation are:

$$\mathcal{L}(y) \approx \mathcal{L}(\tau) + \mathcal{L}'(\tau)\,[(y-\tau)] + (1/2)\,\mathcal{L}''(\tau)\,[(y-\tau)^2]$$

Assumption (2) results in the first term on the right, $\mathcal{L}(\tau)$, being zero. Assumptions (1) and (2) imply that the loss is minimum at $y = \tau$, and since the first derivative disappears at a minimum, $\mathcal{L}'(\tau) = 0$ and the second term vanishes. Thus, the first nonzero term in the Taylor series expansion for $\mathcal{L}(y)$ in the neighborhood of τ is the term involving $(y-\tau)^2$. From this it follows that the simplest form for $\mathcal{L}(y)$ is the function

$$\mathcal{L}(y) = K(y-\tau)^2$$

for some constant K.

Since the units for $(y-\tau)^2$ will be [measurement units squared], the constant K will be expressed in terms of [dollars per measurement unit squared]. By the appropriate consideration of the costs associated with given deviations of Y from τ, one may actually define the value for K for a specific application.

It should be emphasized that the loss function $\mathcal{L}(y)$ is not likely to be represented by the quadratic function over all values for Y. It is logical to assume that there is some maximum loss that will eventually place an upper bound upon $\mathcal{L}(y)$. However, even though $\mathcal{L}(y)$ may well be more complex than the simple quadratic function defined above, this approximation will suffice in some region close to τ.

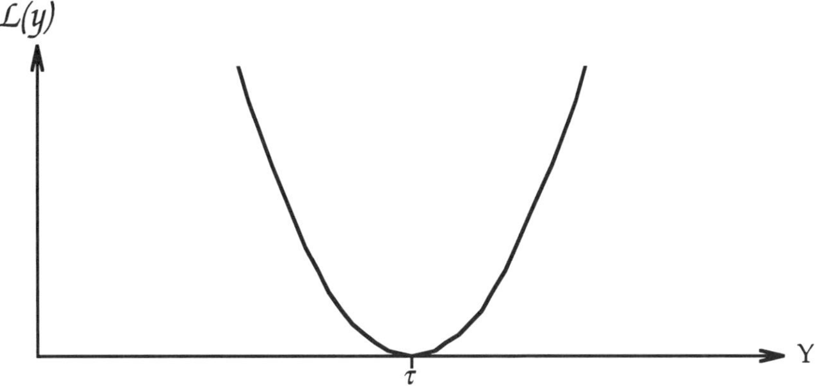

Figure 8.3: The Taguchi Loss Function: $K(y-\tau)^2$

8.4 The Average Loss Per Unit of Production

While $\mathcal{L}(y)$ defines the loss, in dollars, associated with a particular value of Y, it does not take into account the likelihood that a particular value of Y will occur. To do this one will have to consider the notion of a distribution of values for Y.

To begin with, consider a production run of 100 pieces. Each piece would have some value for the performance characteristic, $Y = y$. For each value of Y there would be some loss, say $\mathcal{L}(y)$. The 100 loss values associated with these 100 pieces could be summed up to obtain the total loss for this production run. If this total loss was divided by 100, one would then obtain an "average loss per piece" for this production run.

This same approach can be generalized to yield an average loss value for any production process which displays a reasonable degree of statistical control.

To see how to do this in practice it is helpful to first consider an idealized process which displays exact stability. Imagine that the probability density function, $f(x)$, for this process is known. Then the probability density function will define the likelihood that Y will take on each value along the horizontal axis, while the Loss Function, $\mathcal{L}(y)$, will define the loss for each value of Y. Therefore, the Average Loss per Unit of Production will be found by integrating the product of $\mathcal{L}(y)$ and $f(y)$. This integral is called the Expected Value of $\mathcal{L}(y)$:

$$E[\mathcal{L}(y)] = \int_{all\ y} \mathcal{L}(y)\, f(y)\, dy$$

When the quadratic Taguchi Loss Function, $\mathcal{L}(y) = K(y-\tau)^2$, is placed in this expression the Average Loss may be rewritten as

$$E[\mathcal{L}(y)] = \int_{all\ y} K(y-\tau)^2 f(y)\, dy = K[\sigma^2 + (\mu-\tau)^2]$$

where σ^2 is the **square of the standard deviation** (or **variance**) of the distribution of Y, and $(\mu-\tau)^2$ is the **square of the bias** of the distribution of Y. This result holds regardless of the form for the distribution of Y.

Thus, if the notion of process parameters is well defined (see Chapter One, pages 13-17), the Average Loss Per Unit of Production may be expressed in terms of parameters of the distribution of Y. In particular, the Average Loss due to product variation is seen to be proportional to the square of the process standard deviation and the square of the amount by which the process average deviates from the target value. *This means that the Average Loss will always be minimized by operating on target with minimum variance.*

Since the Average Loss is an actual cost of production, its reduction will be a desirable objective, and the equations above suggest a particular strategy for doing this. Instead of merely trying to select treatment combinations which will get the process on target, one will try to select treatment combinations which will get the process on target *while minimizing the process variation.*

While this approach can be stated very simply, it becomes slightly more complex in application. First of all, the direct evaluation of σ^2 and $(\mu-\tau)^2$ requires a knowledge of $f(y)$. Since such knowledge will generally be unavailable, we must find a way to estimate the Average Loss, $E[\mathcal{L}(y)]$.

Given a subgroup of n values, $\{y_1, y_2, y_3, \ldots, y_n\}$, all obtained under the same combination of factor levels, and under the assumption that the process displays a reasonable degree of statistical control, the Average Loss

$$E[\mathcal{L}(y)] = \int_{all\ y} K(y-\tau)^2 f(y)\, dy$$

will have a method-of-moments estimator of

$$K\left\{ \frac{1}{n} \sum_{i=1}^{n} (y_i-\tau)^2 \right\}.$$

This estimator can be expressed as K times the **Mean Square Deviation About the Target,** $K\{MSD_n(\tau)\}$. Ignoring the multiplicative constant K, the Average Loss for a particular treatment combination will be estimated by $MSD_n(\tau)$. Moreover, the $MSD_n(\tau)$ statistic can be rewritten in the following manner.

$$\text{MSD}_n(\tau) \;=\; \left\{ \frac{1}{n} \sum_{i=1}^{n} (y_i - \tau)^2 \right\} \;=\; \left[\, s_n^2 \;+\; (\bar{y} - \tau)^2 \,\right].$$

where s_n^2 and $\bar{y}$ are calculated from the n values $\{ y_1, y_2, y_3, \ldots, y_n \}$.

When an $\text{MSD}_n(\tau)$ value is calculated in this way (using the n values obtained at one treatment combination) it estimates the Average Loss associated with that treatment combination. As the $\text{MSD}_n(\tau)$ values are computed at different treatment combinations they provide a way to evaluate each combination.

Since the sample standard deviation, s, is more widely used than the root mean square deviation, s_n, the Mean Square Deviation About Target statistic is more commonly written as:

$$\text{MSD}(\tau) \;=\; \left[\, s^2 \;+\; (\bar{y} - \tau)^2 \,\right].$$

For a given set of n values for Y,

$$\text{MSD}_n(\tau) \;=\; \text{MSD}(\tau) \;-\; \frac{s^2}{n}.$$

Because of this, one may work with either form. Minimizing one of these expressions will also minimize the other expression.

8.5 Minimizing The Average Loss

Given a product performance characteristic, Y, how can the manufacturer minimize the Average Loss associated with this characteristic? By the proper choice of various parameters for (a) the product design, and (b) the production process. Since the number of such parameters will always be quite large, an obvious tool for making these choices will be screening designs. By setting up some experiments using screening designs the manufacturer can investigate how various factors (process parameters and/or product parameters) will affect the Average Loss.

Notice that working with the Average Loss involves both the average response at a given treatment combination and the variation in the responses at that treatment combination. This is a very different approach from that discussed in the preceding chapters. Up until now, the estimates of variation within the subgroups have merely been used to estimate the amount of background noise. (The implicit assumption was that the amount of background noise was relatively constant across all the subgroups.) Now the noise is itself of interest. Do some treatment combinations result in more background noise? Do some treatment combinations have less background noise than others? Given this shift in emphasis, it is now more important to think about how the variation within each subgroup is structured than it was before.

Since the estimate of the Average Loss depends upon both s^2 and $\bar{y}$, multiple observations of the performance characteristic will be needed at each point in the design. Moreover, in order for the s^2 portion of $\text{MSD}(\tau)$ to be meaningful, the n observations *must* be independent of each other. This will usually require independent set-ups of the experimental conditions. Simply collecting n observations at a given treatment combination will not generally provide a useful estimate of s^2. The whole point in using s^2 as a variable for analysis is to estimate how each treatment combination affects the variation in Y. In many cases a set of n sequential observations at one treatment combination will not display the full variation effect. For this reason, it is usually much better to replicate the whole screening design n times, with one observation per treatment combination in each replicate. Then the n observations that correspond to a given treatment combination will be independent and will fully reflect the manner in which the various factors affect the variation in Y. It should also be noted that randomization might be of interest in studies of this type.

Under these conditions, values for $\bar{y}$, s^2, and MSD(τ) may be computed for each point in the screening design.

Given these values, the analysis proceeds as follows:

(1) A regular ANOVA is performed using $\bar{y}$ as the variable for analysis. This analysis seeks to identify those factors that actively affect the average response, $\bar{y}$.

(2) An ANOVA is performed using s as the variable for analysis. This analysis seeks to identify those factors that actively affect the variation of the responses. (The use of *log s* is not recommended unless each s value is based on 10 or more observations. Bartlett and Kendall, 1946.)

(3) Once the active factors for both $\bar{y}$ and s are identified, these factors are used to construct a Response Plot or an ANOM Plot for the MSD(τ) values.

EXAMPLE 8.1: *Minimizing the Average Loss for Data Set Sixteen:*

A basic 8-run design is used to study the effects of 7 seven factors upon a product performance characteristic, Y. The 8-run design was replicated five times. The contrasts, observations, and statistics are given below. The target value is $\tau = 90$.

STUDY ONE:

Run	*A*	*B*	*C*	*D*	*E*	*F*	*G*	y_1	y_2	y_3	y_4	y_5	$\bar{y}$	s	s^2	*MSD(90)*
1	-	-	-	-	-	-	-	81	74	72	77	79	**76.6**	**3.65**	13.3	192.86
2	-	-	-	+	+	+	+	91	87	87	87	85	**87.4**	**2.19**	4.8	11.56
3	-	+	+	+	+	-	-	100	99	98	101	100	**99.6**	**1.14**	1.3	93.46
4	-	+	+	-	-	+	+	91	93	82	99	91	**91.2**	**6.10**	37.2	38.64
5	+	+	-	-	+	+	-	79	86	83	92	83	**84.6**	**4.83**	23.3	52.46
6	+	+	-	+	-	-	+	111	109	108	108	105	**108.2**	**2.17**	4.7	335.94
7	+	-	+	+	-	+	-	78	79	81	80	81	**79.8**	**1.30**	1.7	105.74
8	+	-	+	-	+	-	+	83	89	85	83	78	**83.6**	**3.97**	15.8	56.76

(1) *Using* $\bar{y}$ *as the variable for analysis gives the following ANOVA Table:*

ANOVA for $\bar{y}$

Contrast	$\hat{C}$	*SS(C)*
A	*1.4*	*1.225*
B	*56.2*	*1974.025*
C	*-2.6*	*4.225*
D	*39.0*	*950.625*
E	*-0.6*	*0.225*
F	*-25.0*	*390.625*
G	*29.8*	*555.025*
Within	*(32 df)*	*408.400*
Total		*4284.375*

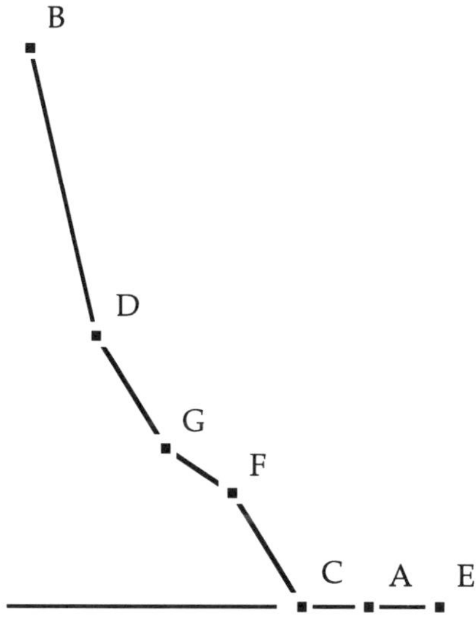

Mod F (7,3,7) = 1043.5 > Mod $F_{.95}$

Although there are 32 d.f. Within Subgroups, the regular F-Ratios were not computed. This is because the use of the F-Ratios assumes that the variation is constant in all subgroups, while the following analysis will assume that the variation is changing from subgroup to subgroup. The Scree Plot suggests that Contrasts B, D, G, and F might represent signals.

(2) *Using* s *as the variable of analysis gives the following ANOVA Table:*

ANOVA for s

Contrast	$\hat{C}$	*SS(C)*
A	*-0.81*	*0.082*
B	*3.13*	*1.225*
C	*-0.33*	*0.014*
D	*-11.75*	*17.258*
E	*-1.09*	*0.149*
F	*3.49*	*1.523*
G	*3.51*	*1.540*
Total		*21.789*

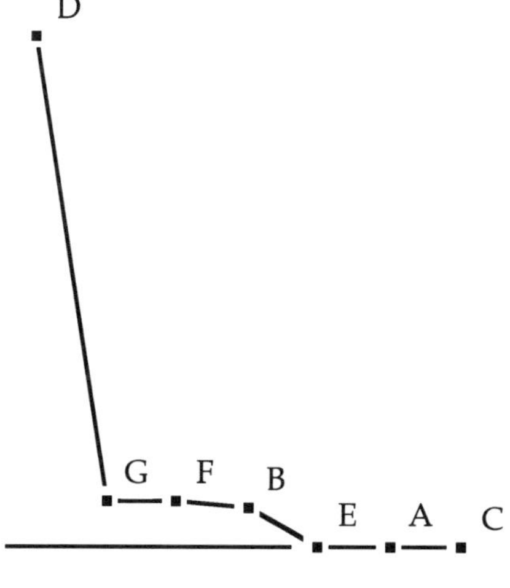

Mod F (7,3,7) = 211.3 ~ Mod $F_{.95}$

Notice that in performing this ANOVA the effective subgroup size has become $n = 1$. *(There is only one value of s for each treatment combination in the study.) Thus, the Adjustment Term for finding the SS(C) values is:*

$$\sum_{j=1}^{k} c_j^2 = 8.$$

The Scree Plot for these data suggests that Contrast D might represent a signal: that is, Contrast D appears to represent a factor which has an impact upon the variation.

(3) The analyses above identify several contrasts as potential signals, but there is no resolution as to whether these are main effects or interaction effects. Before one can draw a proper Response Plot, one must separate the main effects from the interactions. This will require a follow up study.

EXAMPLE 8.2: Minimizing the Average Loss for Data Set Sixteen (Study Two):

A reflected 8-run design is used to study the effects of 7 seven factors upon a product performance characteristic, Y. The 8-run design was replicated five times. The contrasts, observations, and statistics are given below.
The target value is $\tau = 90$.

STUDY TWO:

Run	*A*	*B*	*C*	*D*	*E*	*F*	*G*	y_1	y_2	y_3	y_4	y_5	$\bar{y}$	*s*	s^2	*MSD(90)*
1	+	+	+	+	+	+	+	109	110	110	107	110	**109.2**	**1.30**	1.7	370.34
2	+	+	+	-	-	-	-	88	87	92	85	84	**87.2**	**3.11**	9.7	17.54
3	+	-	-	-	-	+	+	84	84	75	82	90	**83.0**	**5.39**	29.0	78.00
4	+	-	-	+	+	-	-	81	81	82	81	81	**81.2**	**0.45**	0.20	77.64
5	-	-	+	+	-	-	+	88	86	86	84	87	**86.2**	**1.48**	2.2	16.64
6	-	-	+	-	+	+	-	69	80	67	80	75	**74.2**	**6.06**	36.7	286.34
7	-	+	-	-	+	-	+	84	93	89	90	90	**89.2**	**3.27**	10.7	11.34
8	-	+	-	+	-	+	-	104	104	102	105	100	**103.0**	**2.00**	4.0	173.00

(1) Interpreting ANOVA for $\bar{y}$.

Although there are 32 d.f. Within Subgroups, the regular F-Ratios were not computed. This is because the use of the F-Ratios assumes that the variation is constant in all subgroups, while the analysis of s suggests otherwise. The Scree Plot suggests that Contrasts B, D, F, and G might represent signals. Since these are the same contrasts found earlier, one can be reasonably confident that they represent real signals. Contrast F changes signs between the basic and the reflected studies, therefore it must be an interaction effect. The other contrasts are main effects for Factors B, D, and G.

(2) Interpreting the ANOVA for s.

Notice that in performing this ANOVA the effective subgroup size is again n = 1. The Adjustment Term for finding the SS(C) values is:

$$\sum_{j=1}^{k} c_j^2 = 8.$$

The Scree Plot for these data suggests that Contrasts D and F might represent signals. Contrast F did not stand out in the first study, so it will be ignored here. Since Contrast D shows up with the same sign in both studies, it would appear to represent Factor D. As Factor D goes from its low level to its high level, the variation of the response values appears to drop.

(1) *Using $\bar{y}$ as the variable of analysis gives the following ANOVA:*

ANOVA for $\bar{y}$

Contrast	$\hat{C}$	*SS(C)*
A	8.	40.0
B	64.	2560.0
C	0.4	0.1
D	46.	1322.5
E	-5.6	19.6
F	25.6	409.6
G	22.0	302.5
Within	(32 df)	376.8
Total		5031.1

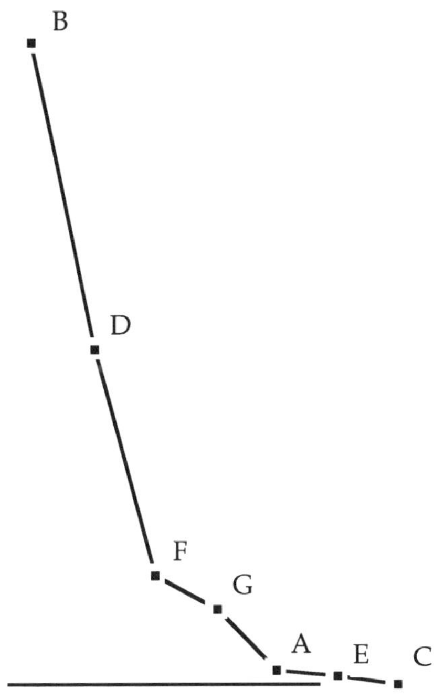

(2) *Using s as the variable of analysis gives the following ANOVA:*

ANOVA for s

Contrast	$\hat{C}$	*SS(C)*
A	-2.56	0.819
B	-3.70	1.711
C	0.84	0.088
D	-12.60	19.845
E	-0.90	0.101
F	6.44	5.184
G	-0.18	0.004
Total		27.753

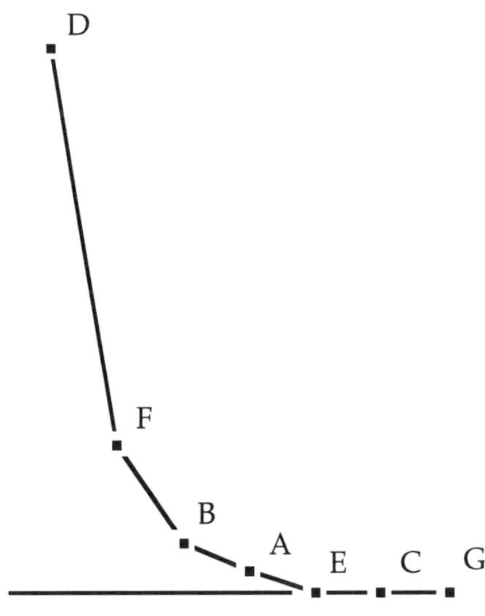

(3) Combining the results of the two studies, Factors B, D, and G affect the average response, while Factor D affects the variation of the responses. Moreover, there is an interaction effect involving the average responses. (Contrast F is likely to be either the AG interaction, the BD interaction, or the CE interaction.) Therefore, the Response Plots will be drawn using Factors B, D, and G. In this case, it is helpful to draw the Response Plots for all three statistics: MSD(90), s^2, and the subgroup averages:

The Response Plot for MSD(90) shows the best values at $B^-D^+G^+$. The next best sets of values occur at $B^+D^-G^+$ and $B^+D^-G^-$.

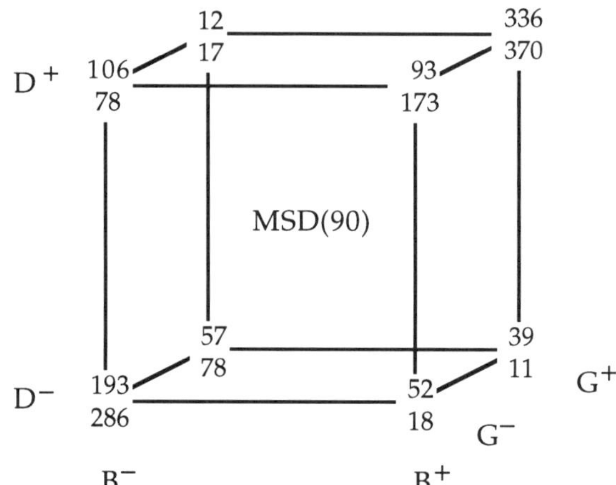

The Response Plot for s^2 shows the dramatic effect of Factor D upon the variation of the responses. The variation is minimized when Factor D is at its high level.

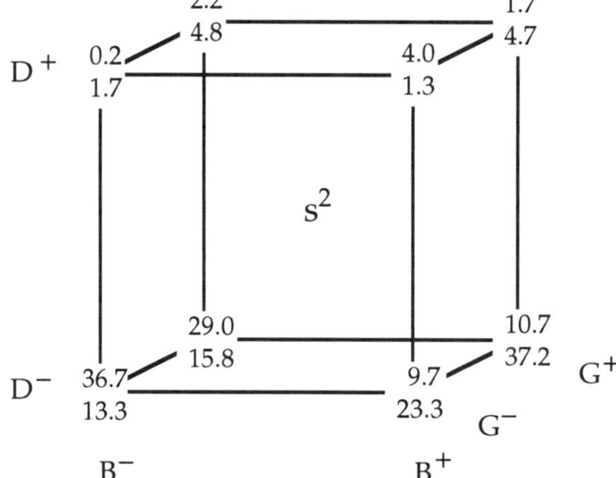

The Response Plot for $\bar{y}$ shows the effect of Factors B, D, and G upon the response variable. The shaded plane defines a region where the response is likely to be near 90. Based upon the Response Plots above, the edge of the shaded plane which intersects with the top face of the cube is the next region to investigate.

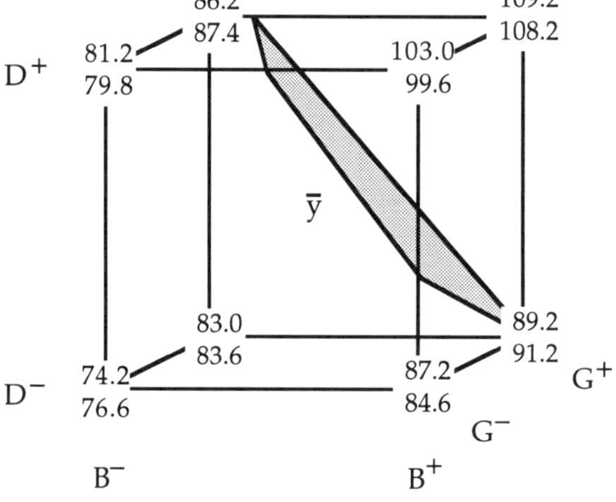

Examples 8.1 and 8.2 above illustrate an approach to minimizing the Average Loss due to product variation. By analyzing both the average of the responses and the variation in those responses at each treatment combination one can determine exactly which factors will provide the needed leverage to reduce the Average Loss. Once those factors which affect location and variation are known, it is possible to use the experimental data to construct Response Plots or ANOM Plots to examine. At this point it is suggested that the MSD(τ) be included on one of the plots. By a careful consideration of these plots it will usually be possible to identify a region of the factor space which promises to have the better values for MSD(τ).

Given that the MSD(τ) values can be obtained once the subgroup averages and subgroup variances are computed, it is natural to consider the more direct approach of using MSD(τ) as the variable for analysis. The problem with this approach is one that is always present when analyzing a composite variable: the results will usually be difficult to interpret.

EXAMPLE 8.3: Analysis of MSD(τ) for Data Set Sixteen:

Using the MSD(90) values from Example 8.1, p.260, to define the variable for analysis we get the following ANOVA Table:

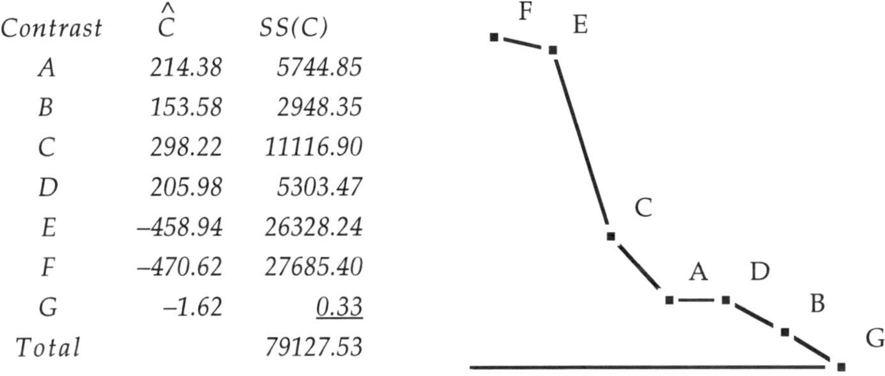

Contrast	$\hat{C}$	*SS(C)*
A	*214.38*	*5744.85*
B	*153.58*	*2948.35*
C	*298.22*	*11116.90*
D	*205.98*	*5303.47*
E	*–458.94*	*26328.24*
F	*–470.62*	*27685.40*
G	*–1.62*	*0.33*
Total		*79127.53*

As with the analysis for s, the Adjustment Term for finding SS(C) values is 8.

The Scree Plot for these values shows Contrasts F and E at the top. In Examples 8.1 and 8.2 Contrast F was found to be an interaction effect for the average responses, and Contrast E was never identified as having an impact upon either the averages or the variances. Furthermore, the slope of the "rubble" in this Scree Plot suggests that either everything is a signal, or nothing is a signal. These ambiguous results should be compared with the clear-cut results in Examples 8.1 and 8.2. Since the objective of this analysis is the identification of factors which affect both the average and the variance, the ambiguity of these results argue against the direct analysis of MSD(τ).

The insensitivity of MSD(τ) as a response variable should now be apparent. In terms of identifying active factors, the separate analyses for $\bar{y}$ and s do a far better job.

This insensitivity is an inherent feature of any analysis which uses a composite variable. When the dependent variable in an analysis is a composite of two or more variables one will always have to consider the possibility of the two variables changing in a contrary manner. In a screening design, such contrary changes can cause the experimenter to delete a factor which should not be deleted. This is why the MSD(τ) values were not used in Examples 8.1 and 8.2 until after the unified set of active factors had been identified.

8.6 A Strategy For Experimentation

The examples in the preceding section outline a strategy for minimizing the average loss due to the variation of a product performance characteristic about some target value, τ. The steps in this strategy are as follows:

First screening designs are used to identify the factors that have an impact upon either the average performance characteristic or the variation of the performance characteristic. (By identifying the active factors for these two variables separately one is less likely to miss an important factor.) These two sets of factors are then combined into a unified set of factors that can be said to influence the Average Loss due to that performance characteristic. Unlike simple screening designs which often have one observation per subgroup, it may be worthwhile to consider the randomization of these experimental runs as long as the complexity introduced by randomization does not make the experiment impossible to perform.

Next, the mean square deviation about the target value, MSD(τ), is analyzed using the unified set of factors. This analysis will usually consist of Response Plots or ANOM plots. Those treatment combinations that have the smaller MSD(τ) values will define the regions of greatest interest for the manufacturer. These plots may be combined with Response Plots for both the average performance characteristic and the variance of performance characteristics.

Based on the results of the analyses of $\bar{y}$, s^2, and MSD(τ), further investigations may be made, or specific sets of parameter values may be selected.

This simple strategy will allow the experimenter to consistently pick the better treatment combinations among those considered. Moreover, since the MSD(τ) values are proportional to the Average Loss due to product variation, they will permit a relative comparison of the benefits associated with any particular change in parameters.

Finally, two special situations need to be covered in order to make this strategy for experimentation complete. They are the modifications needed to accommodate special target values of $\tau = 0$ and $\tau = \infty$.

When the target is $\tau = 0.0$ the loss function becomes

$$\mathcal{L}(y) = K\, y^2.$$

The average loss then becomes

$$E[\mathcal{L}(y)] = K\{\sigma^2 + \mu^2\}$$

and the Mean Squared Deviation About Target becomes

$$MSD(0) = s^2 + \bar{y}^2 .$$

Thus, when $\tau = 0$ the MSD statistic is still a function of both the sample variance and the sample average, and the strategy outlined earlier is still appropriate.

When the target is $\tau = \infty$, the calculations are somewhat different. Assume that the product performance characteristic, Y, is nonnegative, and that the loss function is approximated by

$$\mathcal{L}(y) = K\, y^{-2} .$$

The average loss then becomes

$$E[\mathcal{L}(y)] = K \int_{\text{all } y} y^{-2} f(y)\, dy .$$

This expression cannot be simplified in the same manner that the earlier versions were. Thus, ignoring the constant K, the method of moments estimator of the average loss is

$$MSD(\infty) = \frac{1}{n} \sum_{i=1}^{n} (y_i)^{-2}$$

For lack of a better name, call this quantity the Mean Square Deviation About Infinity. This statistic will not reduce down as the others did. Moreover, it is very difficult to interpret. Just as with the other MSD(τ) statistics, this statistic makes a poor variable for analysis for a screening design. Some of the ambiguity inherent in this statistic can be seen in the following example.

EXAMPLE 8.9: *MSD(∞) for Five Subgroups:*

Subgroup	*Values*					$\bar{y}$	s^2	$\frac{1}{n}\sum_{i=1}^{n}(y_i)^{-2}$
1	10	10	10	10	10	10	0	0.0100
2	8.75	10	10	10	12	10.15	1.3625	0.0100
3	6.42	12	12	13	13	11.284	7.6433	0.0100
4	7.39	7.39	15	15	15	11.956	17.374	0.0100
5	6.86	6.86	20	20	20	14.744	51.798	0.0100

So, just as with the other MSD(τ) functions, it is possible to have offsetting changes in the average and the variance. So while MSD(∞) provides an overall estimate of the average loss, it is still a composite function, and should be interpreted only in context.

Therefore, when working with $\tau = \infty$, one should once more identify the active factors for $\bar{y}$ and s^2 separately, and then combine these active factors for an analysis of MSD(∞). This analysis should then be interpreted in the context of minimizing MSD(∞) while making a reasonable trade-off between the variation and the average.

CHAPTER NINE

THE TAGUCHI APPROACH TO INDUSTRIAL EXPERIMENTATION

In addition to the application of the quadratic loss function to the problem of product variation cited earlier, Dr. Genichi Taguchi has made other unique contributions to the area of industrial experimentation. Two contributions that are commonly associated with Dr. Taguchi are the use of Noise Matrices and the use of Signal-to-Noise Ratios. It is the purpose of this chapter to illustrate these two aspects of the Taguchi Approach.

Since some of the nomenclature used in the translations of Dr. Taguchi's work is rather unusual, a short glossary will be helpful. The Noise Matrix is sometimes called the "outer array." The experimental design (Design Matrix) is sometimes referred to as the "inner array." The variation in a product performance characteristic is referred to as "functional variation," and the factors being studied in the experiment (design parameters for product or process) are referred to as "control factors."

9.1 The Noise Matrix

As was seen in the preceding chapter, the use of the quadratic loss function introduces a new element into the analysis of experimental data. Instead of simply analyzing the subgroup averages, one must also analyze the subgroup variances as response values. This new role for the s^2 values, in turn, places more importance upon the way the observations are obtained. In particular, when one is analyzing only averages some dependency between the responses inside the subgroups is not a serious problem. It may inflate the F-ratios slightly, but Scree Plots and Normal Probability Plots can get around this problem. Therefore, if all the observations in a subgroup come from product produced during one production run, it is of little consequence as long as we are interested only in analyzing the averages. In effect, when the average is the only variable used in the analysis, the only question pertaining to the variances is "Are the variances homogeneous?"

On the other hand, if we wish to analyze both the average and the variance, then correlated values within the subgroups present a real problem. Such correlations will usually reduce the value for s^2. Such reductions will affect the analysis, and if they do not happen to be uniform, they can even skew the analysis. In other words, the homogeneity of the variances is no longer the key issue. The analysis will decide if they are homogeneous or not. The key questions now become "Are the variances realistic estimates of the variation? Do they reflect the variation that will actually occur in practice?" Of course the reason for this change in emphasis should be clear; by analyzing the variance, we hope to predict where the variation will be minimum, and if our statistics are skewed and distorted, our predictions will be erroneous.

In Chapter 8, this problem of the quality of the subgroup variances was addressed in the protocol for performing the experiments. In particular, it was recommended that the screening designs be performed in replicates (sometimes called blocks). That is, each replicate is a complete screening design with one observation per run. By performing a sequence of these replicates, the observations for corresponding runs can be collected into subgroups. This procedure forces one to set up the experimental conditions for each observation independently of the others. While this adds complexity to the experiment, it also guarantees the maximum independence of observations within the subgroups.

The approach outlined in Chapter 8 and described in the preceding paragraph essentially trusts in the variation from one replicate to another to provide reasonable estimates of the variation associated with each particular treatment combination. While this may be perfectly appropriate in some circumstances, it will be unsatisfactory in others. In order to remedy this situation, Dr. Taguchi introduced the concept of a Noise Matrix. Instead of trusting to natural variation, Dr. Taguchi suggests performing the experiment in such a way that the variation within the subgroups is introduced systematically.

In order to understand the role of a Noise Matrix it is helpful to divide the factors that affect a particular performance characteristic into three categories. The first category will consist of those factors that are under the direct control of the manufacturer. An obvious example of such factors would be the product parameters. The second category will consist of those factors that are only partially under the control of the manufacturer. Examples of such factors would be the materials used, the production conditions, the operator techniques, and other variable factors associated with the production process. The third category will consist of those factors that are completely beyond the control of the manufacturer. Examples of these factors might be the environmental conditions under which the product is used and different modes of operation for the product. For convenience, denote these three sets of variables as (1) product parameters, (2) process parameters, and (3) environmental factors.

If one wishes to optimize the performance characteristic over the product life, and from one product to another, one must consider all three sets of variables described above. The product parameter variables and the process parameter variables can be combined into a screening design. But what about the third category, the environmental factors or other variables beyond the control of the manufacturer? In an experimental context it is usually possible to select certain sets of conditions to represent this third set of variables. This is the role of the Noise Matrix. Typically, noise matrices consist of screening designs defined upon the environmental factors.

Thus, a typical Taguchi style experiment would begin with a screening design for the product and process parameters. This design will be called the **Design Matrix**. Then, at each run in this Design Matrix, a second screening design is applied to vary the environmental factors. This second screening design is called the **Noise Matrix**. The general layout of such a study is shown on the next page.

EXAMPLE 9.1: General Layout for a Taguchi Product Design Study:

Let {A, B, C, D, E, F, G} denote the design factors (product or process parameters). An 8-run Plackett-Burman design will be used for the Design Matrix. Let {M, N, O} denote the noise factors (environmental factors). A 4-run Plackett-Burman design will be used for the Noise Matrix.

								Noise Matrix:	Run i	Run ii	Run iii	Run iv		
								M	+	-	+	-		
								N	+	+	-	-		
	Design Matrix							O	-	+	+	-		
Run	A	B	C	D	E	F	G						$\bar{y}$	s^2
1	-	-	-	-	-	-	-		y_1	y_9	y_{17}	y_{25}	$\bar{y}_1$	s_1^2
2	-	-	-	+	+	+	+		y_2	y_{10}	y_{18}	y_{26}	$\bar{y}_2$	s_2^2
3	-	+	+	+	+	-	-		y_3	y_{11}	y_{19}	y_{27}	$\bar{y}_3$	s_3^2
4	-	+	+	-	-	+	+		y_4	y_{12}	y_{20}	y_{28}	$\bar{y}_4$	s_4^2
5	+	+	-	-	+	+	-		y_5	y_{13}	y_{21}	y_{29}	$\bar{y}_5$	s_5^2
6	+	+	-	+	-	-	+		y_6	y_{14}	y_{22}	y_{30}	$\bar{y}_6$	s_6^2
7	+	-	+	+	-	+	-		y_7	y_{15}	y_{23}	y_{31}	$\bar{y}_7$	s_7^2
8	+	-	+	-	+	-	+		y_8	y_{16}	y_{24}	y_{32}	$\bar{y}_8$	s_8^2

The Design Matrix is defined in the usual manner. Each row is a treatment combination, while each column defines a contrast.

The Noise Matrix is flipped on its side. Each column defines a combination of the levels of the Noise Factors.

Thus, responses y_1 through y_8 are all collected under the same *combination of environmental factors, $M^+N^+O^-$.*

Responses y_9 through y_{16} are collected with the environmental factors set at $M^-N^+O^+$.

Responses y_{17} through y_{24} are collected with the environmental factors set at $M^+N^-O^+$.

And responses y_{25} through y_{32} are collected with the environmental factors set at $M^-N^-O^-$.

Moreover, since the Noise Factors represent variables which, in practice, are beyond the control of the manufacturer, there is no explicit attempt to evaluate the impact of the Noise Factors upon the response variable. The only way that the

Noise Factors enter into the analysis is through their impact upon the variation within each subgroup.

Each run in the Design Matrix has four observations, and each of these groups of four responses will define a subgroup. The average and the standard deviation would be calculated for each subgroup and these values are then used to find the $MSD(\tau)$ values (or the Signal to Noise Ratios).

Moreover, while undue complexity is to be avoided, the order of the runs within each column might be randomized if it was felt that the run order could have an impact upon the results.

By varying the environmental factors in the same way in connection with each run in the Design Matrix, the Taguchi approach attempts to force a certain and consistent amount of environmental variation into each subgroup. If some subgroups respond to this environmental variation with a smaller variation from response to response within the subgroup than do other subgroups, then this is useful information. In short, the use of a Noise Matrix is intended to see if certain treatment combinations result in *less* variation being transmitted from the environmental factors to the product characteristic, Y.

Of course, before an environmental factor may be used in conjunction with the Noise Matrix, it has to be subject to control during the course of the experiment. That is, only those factors for which the experimenter can change the levels at will during the experiment can be used as noise factors. Environmental factors that cannot be changed by the experimenter cannot be used as noise factors.

This brings about an important distinction. In the list on page 271 the possible variables were divided into (1) design parameters, (2) process parameters, and (3) environmental factors. There it was noted that the environmental factors category included all those variables that are completely beyond the control of the manufacturer. These variables are distinct from those variables that *could* be under the control of the manufacturer, but which currently, for whatever reason, are not being controlled. Variables in this latter group properly belong in either the product parameter category or the process parameter category. Given these three categories, the major function of the Noise Matrix is the inclusion in the experiment of variables from the environmental factors category.

An example of a Taguchi-style experiment that has been reported in the literature is summarized in the following example.

EXAMPLE 9.2: *The Spring Height Data:*

Four design factors and one noise factor were studied for their effect upon the free height of a leaf spring. The design factors were:

(A) High Heat Temp, 1840°(-) vs 1880°(+);
(B) High Heat Time, 25 sec.(-) vs 23 sec.(+);
(D) Transfer Time, 12 sec.(-) vs 10 sec.(+);
(G) Hold Down Time, 2 sec(-) vs 3 sec(+).

A Reflected 8-Run Placket-Burman Design was used as the Design Matrix. With the assignment above it was a One-Half Replicate, Resolution IV design (see p.221 for details and the confounding pattern for this design).
The noise factor was :

(O) Quench Oil Temperature, 140°(-) vs 160°(+).

In order to control the quench oil temperature a portable heat exchanger was installed for this experiment. Prior to this the quench oil temperature had not been controlled. Thus, the Noise Matrix consists of one factor at two levels.
This two-run design is fully crossed with the Design Matrix to yield 16 subgroups.
Three pieces were produced and measured at each of the 16 sets of conditions.
The response variable is the free height of a leaf spring.
The target value is $\tau = 8.0$ in.

Run	A	B	AB	D	AD	AG	G	O	responses	$\bar{y}$	s^2	MSD(8)
1a	+	+	+	+	+	+	+	-	7.56 7.81 7.69			
1b	+	+	+	+	+	+	+	+	7.81 7.50 7.59	7.6600	.01728	.1329
2a	+	+	+	-	-	-	-	-	7.59 7.56 7.75			
2b	+	+	+	-	-	-	-	+	7.63 7.75 7.56	7.6400	.00792	.1375
3a	+	-	-	-	-	+	+	-	8.15 8.18 7.88			
3b	+	-	-	-	-	+	+	+	7.88 7.88 7.44	7.9017	.07074	.0807
4a	+	-	-	+	+	-	-	-	7.69 8.09 8.06			
4b	+	-	-	+	+	-	-	+	7.56 7.69 7.62	7.7850	.05291	.0970
5a	-	-	+	+	-	-	+	-	7.94 8.00 7.88			
5b	-	-	+	+	-	-	+	+	7.32 7.44 7.44	7.6700	.09084	.1997
6a	-	-	+	-	+	+	-	-	7.78 7.78 7.81			
6b	-	-	+	-	+	+	-	+	7.50 7.25 7.12	7.5400	.09004	.3016
7a	-	+	-	-	+	-	+	-	7.50 7.56 7.50			
7b	-	+	-	-	+	-	+	+	7.50 7.50 7.56	7.5200	.00096	.2314
8a	-	+	-	+	-	+	-	-	7.56 7.62 7.44			
8b	-	+	-	+	-	+	-	+	7.18 7.18 7.25	7.3717	.03802	.4349

ANOVA for $\bar{y}$*:*

Contrast	$\hat{C}$	SS(C)
A, BDG	0.885	0.5874
B, ADG	-0.705	0.3728
AB, DG	-0.068	0.0035
D, ABG	-0.115	0.0099
AD, BG	-0.078	0.0046
AG, BD	-0.142	0.0151
G, ABD	0.415	0.1292
Within	(40 d.f.)	1.8435
Total		2.9660

A B G AG D AD AB

Ignoring three-factor interactions, Factors A, B and G appear to affect $\bar{y}$*.*

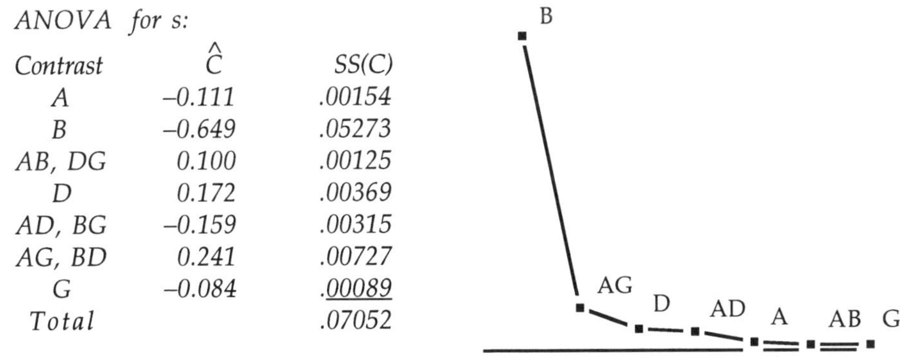

ANOVA for s:

Contrast	$\hat{C}$	SS(C)
A	–0.111	.00154
B	–0.649	.05273
AB, DG	0.100	.00125
D	0.172	.00369
AD, BG	–0.159	.00315
AG, BD	0.241	.00727
G	–0.084	.00089
Total		.07052

Ignoring three-factor interactions, Factor B might affect s^2

Using either $\bar{y}$ *or MSD(8), the optimum combination of those conditions studied is* A$^+$B$^-$G$^+$ *(Runs 3a and 3b) 1880°, 25 sec. at high heat, and 3 sec. hold down. In addition, note that the effect of Quench Oil Temperature has not been explicitly considered in this analysis. Factor O does not appear in either ANOVA above.*

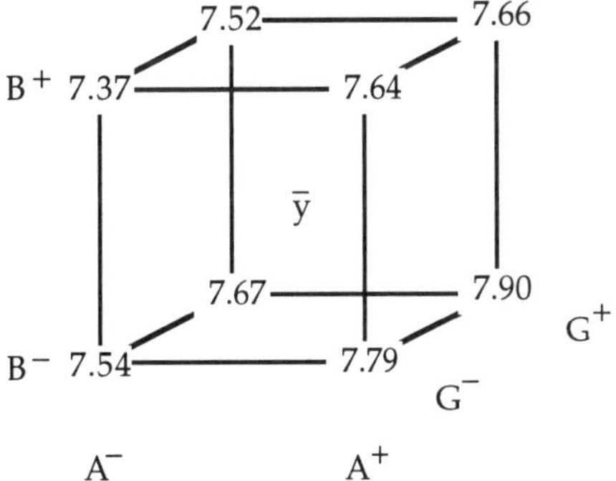

This study does not use the Noise Matrix correctly. It has introduced a variable (Quench Oil Temperature) from the Process Parameter category into the Noise Matrix. As can be seen from this example, the factors included in the Noise Matrix do not directly enter into the analysis. They show up in the analysis only through their influence upon the subgroup variances, and no explicit effect is ever considered for the Noise Factors. Thus, in this example, there is no analysis of the effect of Oil Temperature upon the response variable. The following analysis corrects this situation.

EXAMPLE 9.3: *The Spring Height Data (Second Analysis):*

The previous analysis arranged these data into 8 subgroups of size 6. The Quench Oil Temperature went from high to low within each subgroup. Therefore, the effect of Quench Oil Temperature was buried within each subgroup. Quench Oil Temperature can be included in the analysis by arranging the data into 16 subgroups of size 3.

Run	A	B	AB	D	AD	AG	G	O	responses			$\bar{y}$	s^2	MSD(8)
1a	+	+	+	+	+	+	+	-	7.56	7.81	7.69	7.6867	.01563	.1138
1b	+	+	+	+	+	+	+	+	7.81	7.50	7.59	7.6333	.02543	.1623
2a	+	+	+	-	-	-	-	-	7.59	7.56	7.75	7.6333	.01043	.1473
2b	+	+	+	-	-	-	-	+	7.63	7.75	7.56	7.6467	.00923	.1317
3a	+	-	-	-	-	+	+	-	8.15	8.18	7.88	8.0700	.02730	.0322
3b	+	-	-	-	-	+	+	+	7.88	7.88	7.44	7.7333	.06453	.1374
4a	+	-	-	+	+	-	-	-	7.69	8.09	8.06	7.9467	.04963	.0521
4b	+	-	-	+	+	-	-	+	7.56	7.69	7.62	7.6233	.00423	.1486
5a	-	-	+	+	-	-	+	-	7.94	8.00	7.88	7.9400	.00360	.0072
5b	-	-	+	+	-	-	+	+	7.32	7.44	7.44	7.4000	.00480	.3648
6a	-	-	+	-	+	+	-	-	7.78	7.78	7.81	7.7900	.00030	.0444
6b	-	-	+	-	+	+	-	+	7.50	7.25	7.12	7.2900	.03730	.5414
7a	-	+	-	-	+	-	+	-	7.50	7.56	7.50	7.5200	.00120	.2316
7b	-	+	-	-	+	-	+	+	7.50	7.50	7.56	7.5200	.00120	.2316
8a	-	+	-	+	-	+	-	-	7.56	7.62	7.44	7.5400	.00840	.2200
8b	-	+	-	+	-	+	-	+	7.18	7.18	7.25	7.2033	.00163	.6416

Of course, with 16 subgroups, there are 15 degrees of freedom between subgroups. Eight contrasts are given above. The other 7 contrasts are obtained by multiplying each of the first 7 contrasts by Contrast O. These are listed in the table below, along with the summary statistics for the 16 subgroups of size 3 shown above.

Run	AO	BO	ABO	DO	ADO	AGO	GO	$\bar{y}$	s^2	MSD(8)
1a	–	–	–	–	–	–	–	7.6867	.01563	.1138
1b	+	+	+	+	+	+	+	7.6333	.02543	.1623
2a	–	–	–	+	+	+	+	7.6333	.01043	.1473
2b	+	+	+	–	–	–	–	7.6467	.00923	.1317
3a	–	+	+	+	+	–	–	8.0700	.02730	.0322
3b	+	–	–	–	–	+	+	7.7333	.06453	.1374
4a	–	+	+	–	–	+	+	7.9467	.04963	.0521
4b	+	–	–	+	+	–	–	7.6233	.00423	.1486
5a	+	+	–	–	+	+	–	7.9400	.00360	.0072
5b	–	–	+	+	–	–	+	7.4000	.00480	.3648
6a	+	+	–	+	–	–	+	7.7900	.00030	.0444
6b	–	–	+	–	+	+	–	7.2900	.03730	.5414
7a	+	–	+	+	–	+	–	7.5200	.00120	.2316
7b	–	+	–	–	+	–	+	7.5200	.00120	.2316
8a	+	–	+	–	+	–	+	7.5400	.00840	.2200
8b	–	+	–	+	–	+	–	7.2033	.00163	.6416

The 15 contrasts given in columns above will be used below to analyze both the average responses, and the variance of the responses.

ANOVA for $\bar{y}$:

Contrast	$\hat{C}$	$SS(C)x10^4$
A, BDG	1.7700	5874
B, ADG	-1.4100	3728
AB, DG	-0.1367	35
D, ABG	-0.2300	99
AD, BG	-0.1567	46
AG, BD	-0.2833	151
G, ABD	0.8300	1292
O	-2.0767	8086
AO	0.6767	859
BO	1.3233	3284
ABO, DGO	-0.0833	13
DO	-0.4300	347
ADO, BGO	0.3233	196
AGO, BDO	-0.3767	266
GO	0.2167	88
Within	(32 d.f.)	5298
Total		29660

Mod F (15,9,15) = 58.6 ≈ Mod $F_{.95}$

Ignoring three-factor interactions, Factors O, A, B and possibly G appear to affect $\bar{y}$. The previous analysis simply missed the most powerful signal present in these data!

ANOVA for s:

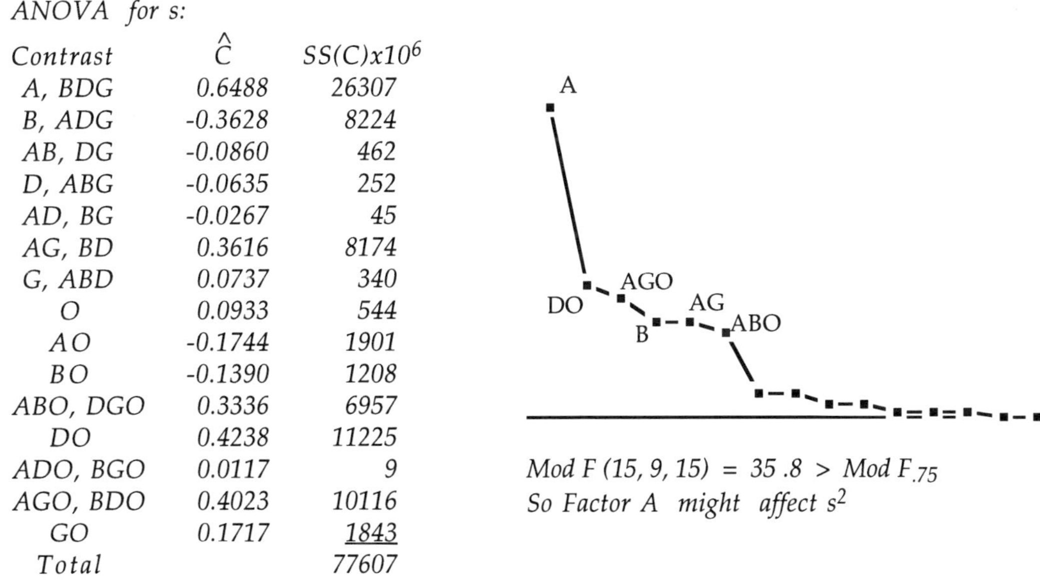

Contrast	$\hat{C}$	$SS(C)x10^6$
A, BDG	0.6488	26307
B, ADG	-0.3628	8224
AB, DG	-0.0860	462
D, ABG	-0.0635	252
AD, BG	-0.0267	45
AG, BD	0.3616	8174
G, ABD	0.0737	340
O	0.0933	544
AO	-0.1744	1901
BO	-0.1390	1208
ABO, DGO	0.3336	6957
DO	0.4238	11225
ADO, BGO	0.0117	9
AGO, BDO	0.4023	10116
GO	0.1717	1843
Total		77607

Mod F (15, 9, 15) = 35 .8 > Mod $F_{.75}$
So Factor A might affect s^2

Therefore, of the five factors (A, B, D, G, and O) in the study, all but D show up as potential signals on the Scree Plots. However, given the strength of the indications for Factors O, A, and B, and given the relative weakness of the indication for Factor G, it was decided to use only Factors A, B, and O in constructing the Response Plots.

The Response Plot for 100 x MSD(8) shows consistently small values along the edge defined by B^-O^-.

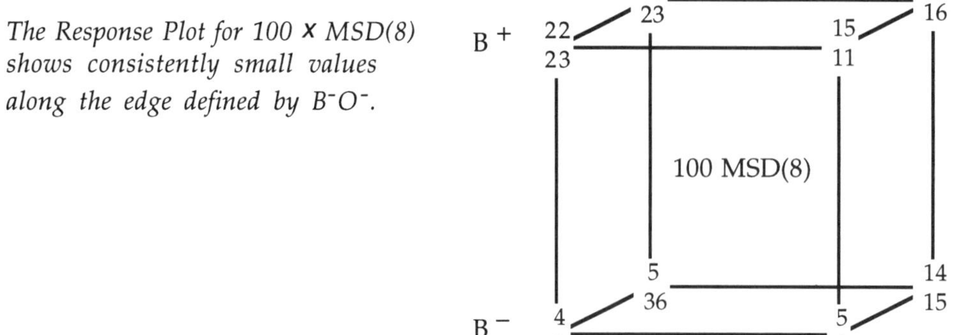

The Response Plot for $\bar{y}$ shows the best Spring Heights at $A^+B^-O^-$, with a close second place at $A^-B^-O^-$

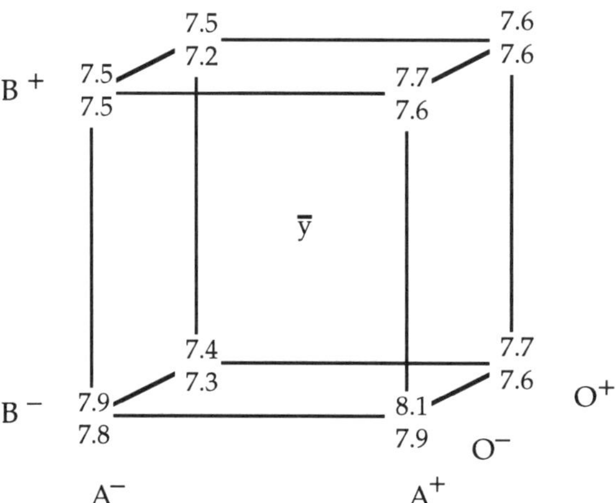

The Response Plot for 1000 x s^2 shows the smallest product variances are located on the A^- side of the cube. In particular, the smallest variances are found at $A^-B^-O^-$and at $A^-B^+O^+$.

Thus, $A^-B^-O^-$and $A^+B^-O^-$appear to be the best points out of those considered for minimizing the Average Loss.

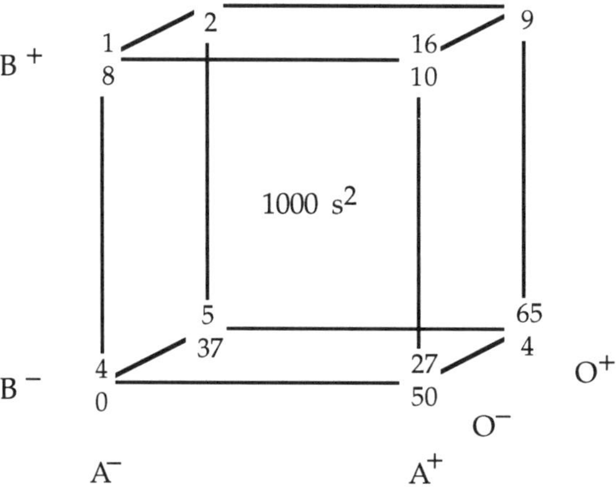

Thus, the best compromises involve either level for High Heat Temp (1840° or 1880°), but have High Heat Time at 25 sec. and Quench Oil Temp at 140°. These conditions should do the best job of producing springs with ***consistent*** *free heights near 8.0 inches. In fact, they did.*

It is a serious mistake for a factor that represents a product or process parameter to be placed in the Noise Matrix when it is not also explicitly included in the Design Matrix at the same time.

When a factor occurs in both the Design Matrix and the Noise Matrix, the Noise Matrix is being used to perform a sensitivity analysis with respect to that factor. To understand just how such a sensitivity analysis might work, consider the following scenario. Say that Mold Temperature is difficult to control exactly. While the actual temperature of the mold varies cyclically as the heating elements go on and off, say that the smallest incremental adjustment that can be made results in a 5° change in the average temperature. In the Design Matrix, the change in Mold Temperature between the low level and the high level might be 20°, 30° or more, while the change between high level and low level in the Noise Matrix might be 5° or 10°. In this manner the Noise Matrix functions to examine the sensitivity of the response to the small variations in Mold Temperature, while the Design Matrix examines the effect upon the response of large variations in Mold Temperature. In this role, as in the other, the purpose of the Noise Matrix is the introduction of variation into the subgroups in a systematic manner.

The Noise Matrix should be reserved for those factors (or those fluctuations in factor levels) that will remain outside the control of the manufacturer.

9.2 Signal to Noise Ratios

Another of Dr. Taguchi's unique contributions to industrial experimentation is the use of Signal-to-Noise Ratios as the variable of analysis for screening designs. These ratios are generally some logarithmic transformation of the summary statistics for a given subgroup. While many such ratios may be defined, three are commonly being used in this country today.

For a target of $\tau = 0$ the customary Signal-to-Noise Ratio is

$$Z = -10 \log \text{MSD}(0) = -10 \log [\, s^2 + \bar{y}^2 \,].$$

The negative sign creates a variable for analysis which is optimum when *maximized*. Multiplying by 10 changes the Signal-to-Noise Ratio into decibel units. The use of the logarithm will be discussed below.

For a target of $\tau = \infty$ the customary Signal-to-Noise Ratio is

$$Z = -10 \log \text{MSD}(\infty) = -10 \log \left[\frac{1}{n} \sum_{i=1}^{n} (y_i)^{-2} \right]$$

The negative sign and the multiplication by 10 serve the same function as before. In each of these cases the assumptions and conditions outlined relative to MSD(0) and MSD(∞) still apply.

For a general target of τ, the customary Signal-to-Noise Ratio is

$$Z = 10 \log [\bar{y}^2 / s^2]$$

In each case, the use of the logarithmic transformation is an attempt to minimize interactions between the factors. The rationale used in support of this practice is that interactions behave like the products of factors. Unfortunately, this is only partially true. So while the use of logarithms will increase the additivity of the factors, it will not always remove the contamination introduced by interaction effects. Moreover, all of these Signal-to-Noise Ratios are composite functions, and therefore they will suffer the same problems discussed relative to MSD(τ) in the previous chapter. (See Example 8.3, p.265.) Using a Signal-to-Noise Ratio as the variable for analysis with a screening design will not allow the experimenter to reliably identify the active factors. If one or more active factors are missed the experimenter is denied access to some of the leverage needed to change the response variable in a favorable manner.

Thus, the use of Signal-to-Noise Ratios with screening designs entails the risk of an inefficient and flawed analysis.

EXAMPLE 9.4: Analysis for Spring Height Data S/N ratios:

The Signal-to-Noise ratios for Example 9.3 are listed below. The form used is:

$$Z = 10 \log [\bar{y}^2 / s^2]$$

Run	A	B	AB	D	AD	AG	G	O	responses			$\bar{y}$	s^2	MSD(8)	Z
1a	+	+	+	+	+	+	+	-	7.56	7.81	7.69	7.6867	.01563	.1117	35.78
1b	+	+	+	+	+	+	+	+	7.81	7.50	7.59	7.6333	.02543	.1623	33.60
2a	+	+	+	-	-	-	-	-	7.59	7.56	7.75	7.6333	.01043	.1473	37.47
2b	+	+	+	-	-	-	-	+	7.63	7.75	7.56	7.6467	.00923	.1317	38.02
3a	+	-	-	-	-	+	+	-	8.15	8.18	7.88	8.0700	.02730	.0322	33.78
3b	+	-	-	-	-	+	+	+	7.88	7.88	7.44	7.7333	.06453	.1374	29.67
4a	+	-	-	+	+	-	-	-	7.69	8.09	8.06	7.9467	.04963	.0521	31.05
4b	+	-	-	+	+	-	-	+	7.56	7.69	7.62	7.6233	.00423	.1486	41.37
5a	-	-	+	+	-	-	+	-	7.94	8.00	7.88	7.9400	.00360	.0072	42.43
5b	-	-	+	+	-	-	+	+	7.32	7.44	7.44	7.4000	.00480	.3648	40.57
6a	-	-	+	-	+	+	-	-	7.78	7.78	7.81	7.7900	.00030	.0444	53.06
6b	-	-	+	-	+	+	-	+	7.50	7.25	7.12	7.2900	.03730	.5414	31.54
7a	-	+	-	-	+	-	+	-	7.50	7.56	7.50	7.5200	.00120	.2316	46.73
7b	-	+	-	-	+	-	+	+	7.50	7.50	7.56	7.5200	.00120	.2316	46.73
8a	-	+	-	+	-	+	-	-	7.56	7.62	7.44	7.5400	.00840	.2200	38.30
8b	-	+	-	+	-	+	-	+	7.18	7.18	7.25	7.2033	.00163	.6416	45.02

Run	AO	BO	ABO	DO	ADO	AGO	GO	$\bar{y}$	s^2	MSD(8)	Z
1a	–	–	–	–	–	–	–	7.6867	.01563	.1117	35.78
1b	+	+	+	+	+	+	+	7.6333	.02543	.1623	33.60
2a	–	–	–	+	+	+	+	7.6333	.01043	.1473	37.47
2b	+	+	+	–	–	–	–	7.6467	.00923	.1317	38.02
3a	–	+	+	+	+	–	–	8.0700	.02730	.0322	33.78
3b	+	–	–	–	–	+	+	7.7333	.06453	.1374	29.67
4a	–	+	+	–	–	+	+	7.9467	.04963	.0521	31.05
4b	+	–	–	+	+	–	–	7.6233	.00423	.1486	41.37
5a	+	+	–	–	+	+	–	7.9400	.00360	.0072	42.43
5b	–	–	+	+	–	–	+	7.4000	.00480	.3648	40.57
6a	+	+	–	+	–	–	+	7.7900	.00030	.0444	53.06
6b	–	–	+	–	+	+	–	7.2900	.03730	.5414	31.54
7a	+	–	+	+	–	+	–	7.5200	.00120	.2316	46.73
7b	–	+	–	–	+	–	+	7.5200	.00120	.2316	46.73
8a	+	–	+	–	+	–	+	7.5400	.00840	.2200	38.30
8b	–	+	–	+	–	+	–	7.2033	.00163	.6416	45.02

The first part of Taguchi's Approach is to use Z as the variable for analysis.

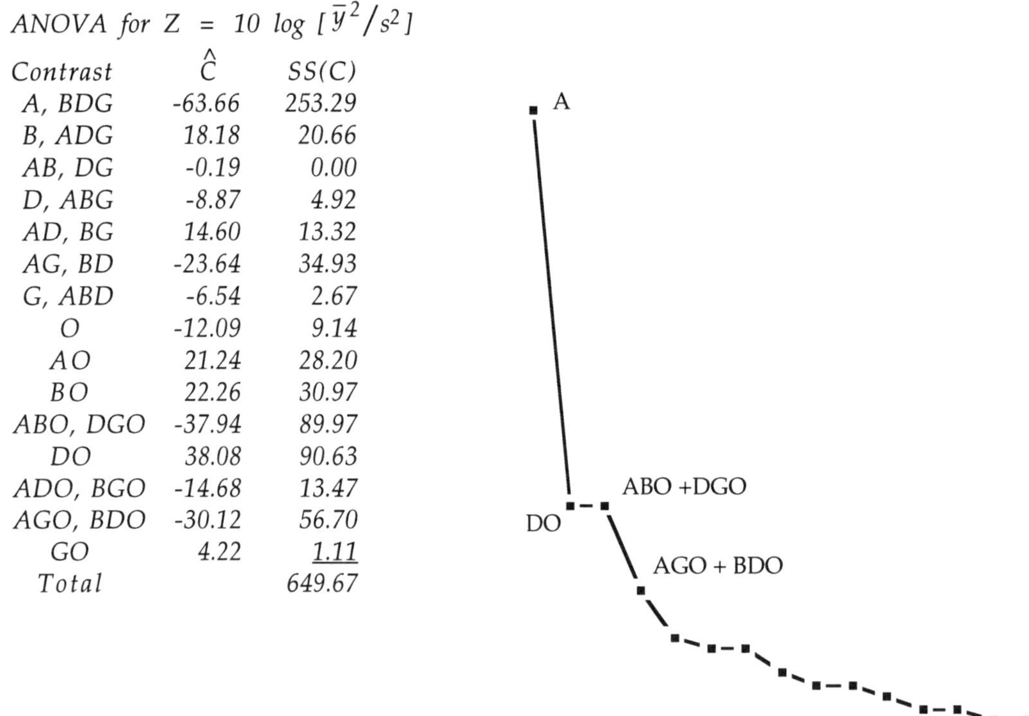

ANOVA for $Z = 10 \log [\bar{y}^2/s^2]$

Contrast	$\hat{C}$	*SS(C)*
A, BDG	*-63.66*	*253.29*
B, ADG	*18.18*	*20.66*
AB, DG	*-0.19*	*0.00*
D, ABG	*-8.87*	*4.92*
AD, BG	*14.60*	*13.32*
AG, BD	*-23.64*	*34.93*
G, ABD	*-6.54*	*2.67*
O	*-12.09*	*9.14*
AO	*21.24*	*28.20*
BO	*22.26*	*30.97*
ABO, DGO	*-37.94*	*89.97*
DO	*38.08*	*90.63*
ADO, BGO	*-14.68*	*13.47*
AGO, BDO	*-30.12*	*56.70*
GO	*4.22*	*1.11*
Total		*649.67*

Mod F(15,7,15) = 38.9 > Mod $F_{.50}$
So Factor A may affect the Signal-to-Noise Ratio.
(The cliff is not very high relative to the rubble.)

If Factor A affects the Signal-to-Noise Ratio, then the Signal-to-Noise Ratios are maximum when Factor A is at its low level. (This may be seen from the table on page 281.)

After the analysis of the Signal-to-Noise Ratios, the next step in the Taguchi approach is the use of Adjustment Factors. (Adjustment Factors are factors that affect the average response but which do not influence the Signal-to-Noise Ratio.) Thus, the Taguchi approach is a two step approach: first identify those factors (and their levels) which will maximize the Signal-to-Noise Ratio, then identify additional (Adjustment) factors which may be used to minimize the bias, $(\bar{y} - \tau)$.

In order for this approach to work there must be at least one Adjustment Factor.

One way to search for such Adjustment Factors is to obtain the ANOVA table using $\bar{y}$ as the variable of analysis. This was done in Example 9.3, and is repeated below.

Thus, the basic argument of Taguchi's Approach is the following: if one can identify some factors which have an effect upon $\bar{y}$ but which do not have an effect upon Z, then one may be able to adjust out the bias while maximizing the Signal-to-Noise Ratio.

EXAMPLE 9.5: *Adjustment Factors for Spring Height Data:*

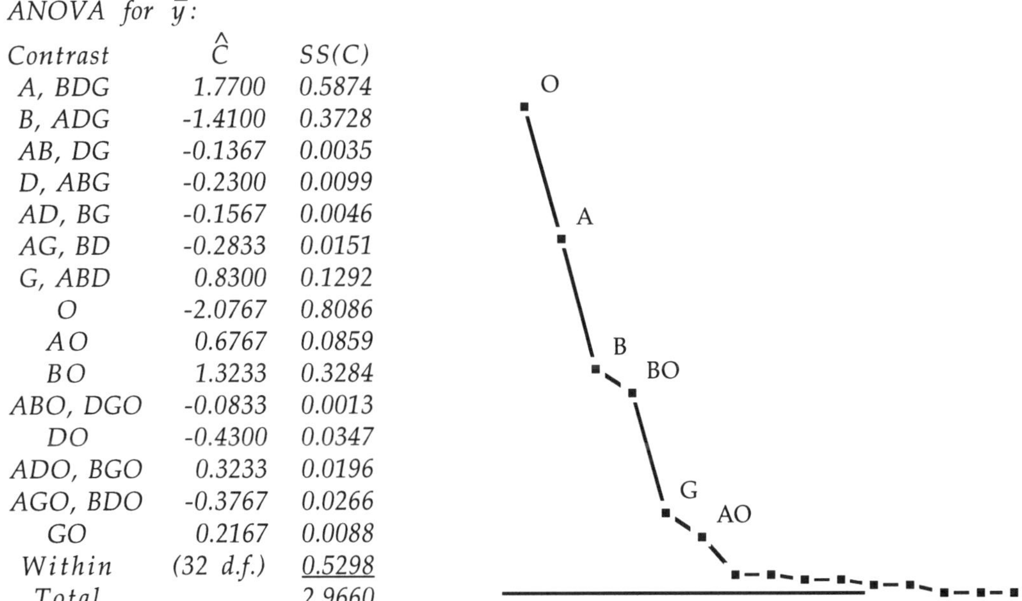

ANOVA for $\bar{y}$:

Contrast	$\hat{C}$	SS(C)
A, BDG	1.7700	0.5874
B, ADG	-1.4100	0.3728
AB, DG	-0.1367	0.0035
D, ABG	-0.2300	0.0099
AD, BG	-0.1567	0.0046
AG, BD	-0.2833	0.0151
G, ABD	0.8300	0.1292
O	-2.0767	0.8086
AO	0.6767	0.0859
BO	1.3233	0.3284
ABO, DGO	-0.0833	0.0013
DO	-0.4300	0.0347
ADO, BGO	0.3233	0.0196
AGO, BDO	-0.3767	0.0266
GO	0.2167	0.0088
Within	(32 d.f.)	0.5298
Total		2.9660

This analysis is the same as that on p.277.

Factors O, A, B and possibly G appear to affect $\bar{y}$.

Combining the analysis for Z with the analysis for $\bar{y}$, we get a Response Plot for $\bar{y}$ in order to determine how to adjust the aim toward the target while maximizing the Signal-to-Noise Ratio.

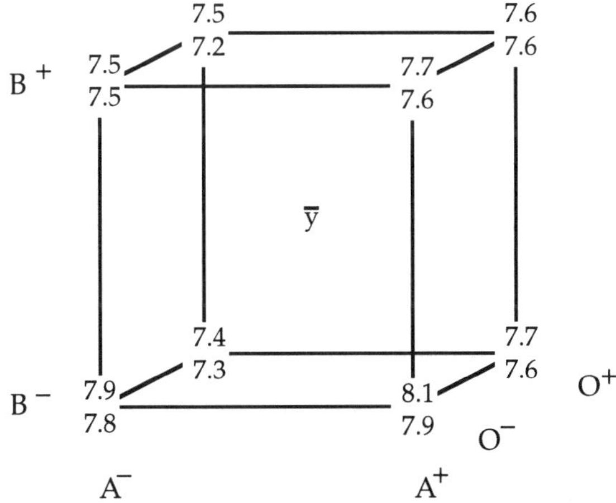

The Response Plot for $\bar{y}$ as a function of Factors A, B, and O is shown above. Holding Factor A at its low level to maximize the Signal-to-Noise Ratio, the best value for the average free height of the springs is found at B^-O^-.

A weakness of the Taguchi Approach outlined above is its attempt to manipulate both Z and $\bar{y}$ separately. Since Z is a function of $\bar{y}$, this can be quite difficult. Given strong signals in the analysis for $\bar{y}$, and weak signals in the analysis for Z, it is possible that any potential Adjustment Factors may not be able to compensate for the effect of trying to maximize Z. That is, the choice of factor levels required to maximize Z may result in values for the response variable that are so far off target that the Adjustment Factors cannot get the process back on target.

For this reason, the Taguchi Approach and the approach in Chapter 8 will not always identify the same combination of factor levels as "best." With the Chapter 8 Approach one will have the advantage of examining the MSD(τ) values directly (remember that they are proportional to the Average Loss per unit of production). The Taguchi Approach transforms the data, creates a dependency between the two quantities being manipulated, and then singles out one treatment combination as "best." By encouraging the examination of the MSD(τ) values, the Chapter 8 Approach is better suited than the Taguchi Approach for making those trade-offs which are always required in practice.

This difference between these two approaches is illustrated in the following example. Even though the results from the two approaches are the same, the Chapter 8 Approach gives the user more options and more latitude in making the final decision about how to modify the process.

EXAMPLE 9.6: *The Cable Casing Data:*

A 16-run Plackett-Burman design was used to study the effect of fifteen factors upon the shrinkage of a speedometer cable casing. The cable casing consists of an extruded tubular polyethylene liner, surrounded with a wire braid, and covered with an extruded polyethylene cover. With age these casings would shrink to the extent that they would interfere with the operation of the steel speedometer cable. Thus, the problem is how to minimize the shrinkage of the casing. The fifteen factors considered were:

	Factor	*Low*	*High*
A.	*Outside Diameter of Liner*	*current*	*new*
B.	*Die for Extruding Liner*	*current*	*new*
C.	*Liner Material*	*current*	*new*
D.	*Speed of Liner Extrusion*	*current*	*slower*
E.	*Type of Wire Braid*	*current*	*new*
F.	*Braiding Tension*	*current*	*new*
G.	*Wire Diameter*	*current*	*smaller*
H.	*Liner Tension While Extruding Cover*	*current*	*greater*
I.	*Liner Temperature While Extruding Cover*	*ambient*	*heated*
J.	*Cover Material*	*current*	*new*
K.	*Die for Extruding Cover*	*current*	*new*
L.	*Melt Temperature for Cover Material*	*current*	*cooler*
M.	*Screen Pack*	*existing*	*denser*
N.	*Cooling Method*	*current*	*new*
O.	*Speed of Cover Extrusion*	*current*	*slower*

Each run in the 16-run design defines a different treatment combination for the study. Given the difficulty of changing some of these factors, each treatment combination was set-up once and a minimum amount of product was produced at each setting. Four sample pieces were cut from each of these minimum runs, giving a total of 64 pieces for use in the study.

One set of 16 pieces (one piece from each run) was heat soaked on Day 1. Another complete set of 16 pieces was heat soaked on Day 2. Likewise for Days 3 and 4. Therefore, the four different short-term heat-soak tests could be thought of as a Noise Matrix. (The 16 runs are crossed with the Days of Heat Soak.) Since each group of four pieces were produced in the same (minimum) production run, the variation within each subgroup represents the variation due to the heat soak test rather than the variation due to different set-ups of the manufacturing operation.

The response recorded is the percent shrinkage for each sample piece. The target value is $\tau = 0$. The shrinkages, averages, variances, MSD(0) and Signal-to-Noise ratios for each subgroup of size 4 are shown in the following tables:

Run	*A*	*B*	*C*	*D*	*E*	*F*	*G*	*H*	*I*	*J*	*K*	*L*	*M*	*N*	*O*	*Percent Shrinkage* *Day 1*	*Day 2*	*Day 3*	*Day 4*
1	-	-	-	-	-	-	+	-	-	-	-	-	-	-	-	0.49	0.54	0.46	0.45
2	-	-	-	-	-	-	+	+	+	+	+	+	+	+	+	0.55	0.60	0.57	0.58
3	-	-	-	+	+	+	-	+	+	+	+	-	-	-	-	0.16	0.16	0.19	0.19
4	-	-	-	+	+	+	-	-	-	-	-	+	+	+	+	0.07	0.09	0.11	0.08
5	-	+	+	+	+	-	+	-	-	+	+	+	+	-	-	0.24	0.22	0.19	0.25
6	-	+	+	+	+	-	+	+	+	-	-	-	-	+	+	0.13	0.19	0.19	0.19
7	-	+	+	-	-	+	-	+	+	-	-	+	+	-	-	0.16	0.17	0.13	0.12
8	-	+	+	-	-	+	-	-	-	+	+	-	-	+	+	0.13	0.22	0.20	0.23
9	+	+	-	-	+	+	+	-	+	+	-	-	+	+	-	0.13	0.17	0.21	0.17
10	+	+	-	-	+	+	+	+	-	-	+	+	-	-	+	0.28	0.26	0.26	0.30
11	+	+	-	+	-	-	-	+	-	-	+	-	+	+	-	0.58	0.62	0.59	0.54
12	+	+	-	+	-	-	-	-	+	+	-	+	-	-	+	0.34	0.32	0.30	0.41
13	+	-	+	+	-	+	+	-	+	-	+	+	-	+	-	0.48	0.49	0.44	0.41
14	+	-	+	+	-	+	+	+	-	+	-	-	+	-	+	0.54	0.53	0.53	0.54
15	+	-	+	-	+	-	-	+	-	+	-	+	-	+	-	0.07	0.04	0.19	0.18
16	+	-	+	-	+	-	-	-	+	-	+	-	+	-	+	0.08	0.10	0.14	0.18

Run	*A*	*B*	*C*	*D*	*E*	*F*	*G*	*H*	*I*	*J*	*K*	*L*	*M*	*N*	*O*	$\bar{y}$	s^2	*MSD(0)*	*Z*
1	-	-	-	-	-	-	+	-	-	-	-	-	-	-	-	0.485	.001633	0.2369	6.255
2	-	-	-	-	-	-	+	+	+	+	+	+	+	+	+	0.575	.000433	0.3311	4.801
3	-	-	-	+	+	+	-	+	+	+	+	-	-	-	-	0.175	.000300	0.0309	15.097
4	-	-	-	+	+	+	-	-	-	-	-	+	+	+	+	0.0875	.000292	0.0079	20.997
5	-	+	+	+	+	-	+	-	-	+	+	+	+	-	-	0.225	.000700	0.0513	12.897
6	-	+	+	+	+	-	+	+	+	-	-	-	-	+	+	0.175	.000900	0.0315	15.013
7	-	+	+	-	-	+	-	+	+	-	-	+	+	-	-	0.145	.000567	0.0216	16.657
8	-	+	+	-	-	+	-	-	-	+	+	-	-	+	+	0.195	.002033	0.0401	13.973
9	+	+	-	-	+	+	+	-	+	+	-	-	+	+	-	0.170	.001067	0.0300	15.234
10	+	+	-	-	+	+	+	+	-	-	+	+	-	-	+	0.275	.000367	0.0760	11.192
11	+	+	-	+	-	-	-	+	-	-	+	-	+	+	-	0.5825	.001092	0.3404	4.680
12	+	+	-	+	-	-	-	-	+	+	-	+	-	-	+	0.3425	.002292	0.1196	9.223
13	+	-	+	+	-	+	+	-	+	-	+	+	-	+	-	0.455	.001367	0.2084	6.811
14	+	-	+	+	-	+	+	+	-	+	-	-	+	-	+	0.535	.000033	0.2863	5.432
15	+	-	+	-	+	-	-	+	-	+	-	+	-	+	-	0.120	.005800	0.0202	16.946
16	+	-	+	-	+	-	-	-	+	-	+	-	+	-	+	0.125	.001967	0.0176	17.547

Example 9.7: Taguchi Approach to Analyzing the Cable Casing Data:

The Taguchi Approach is a two-step approach:
Maximize the Signal-to-Noise Ratio and find Adjustment Factors.

ANOVA for $Z = -10 \log [\, s^2 + \bar{y}^2 \,]$

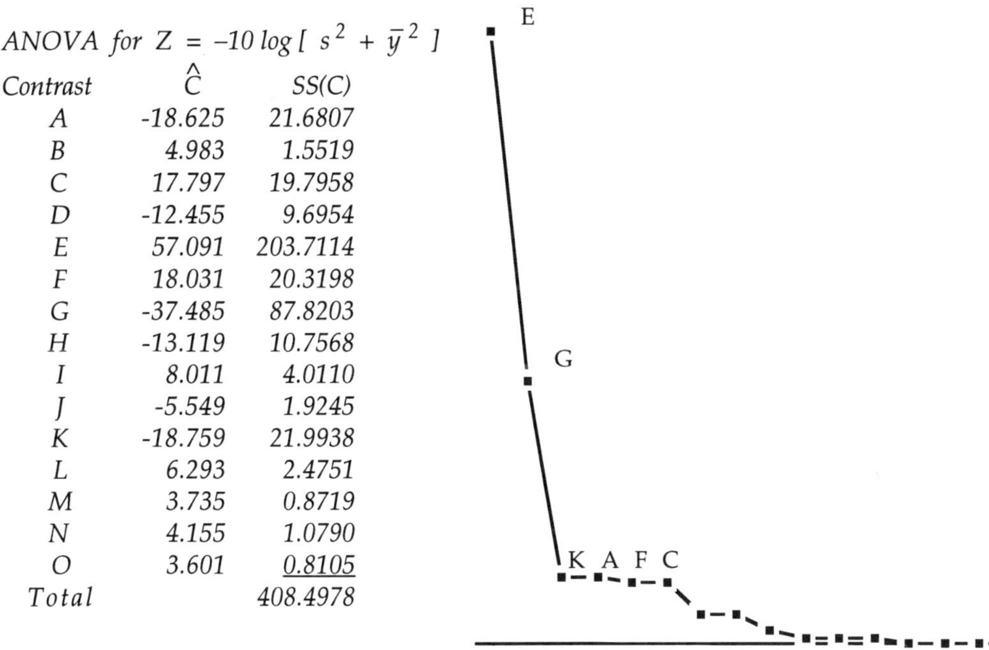

Contrast	$\hat{C}$	SS(C)
A	-18.625	21.6807
B	4.983	1.5519
C	17.797	19.7958
D	-12.455	9.6954
E	57.091	203.7114
F	18.031	20.3198
G	-37.485	87.8203
H	-13.119	10.7568
I	8.011	4.0110
J	-5.549	1.9245
K	-18.759	21.9938
L	6.293	2.4751
M	3.735	0.8719
N	4.155	1.0790
O	3.601	0.8105
Total		408.4978

Contrasts E and G stand out from the rubble.
Interpreting these as main effects, use Factors E and G to maximize the Signal-to-Noise Ratio.

A Response Plot of the Z values versus these two factors gives

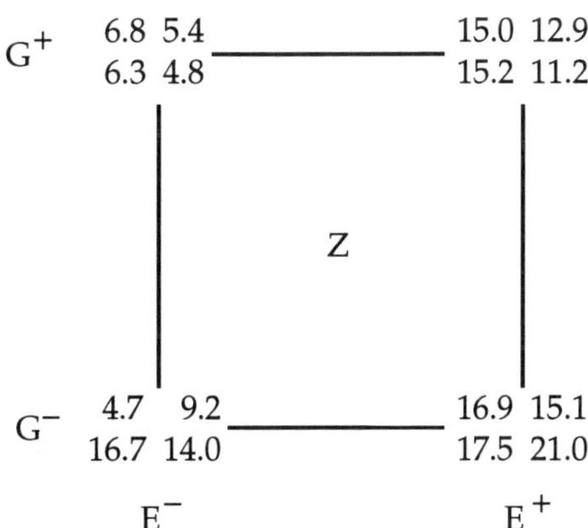

Maximum Z values coincide with E^+ *and* G^-*. Since the low level of each factor represents the current value, this combination translates into {New Type of Braid, Current Wire Diameter}*

In order to find some Adjustment Factors the ANOVA for $\bar{y}$ is needed:

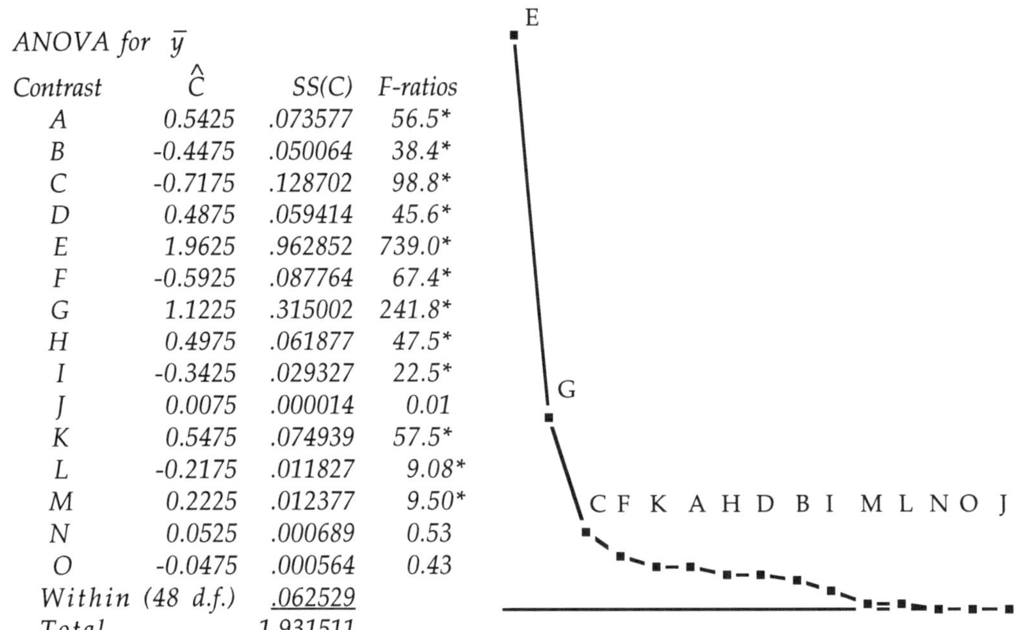

ANOVA for $\bar{y}$

Contrast	$\hat{C}$	*SS(C)*	*F-ratios*
A	0.5425	.073577	56.5*
B	-0.4475	.050064	38.4*
C	-0.7175	.128702	98.8*
D	0.4875	.059414	45.6*
E	1.9625	.962852	739.0*
F	-0.5925	.087764	67.4*
G	1.1225	.315002	241.8*
H	0.4975	.061877	47.5*
I	-0.3425	.029327	22.5*
J	0.0075	.000014	0.01
K	0.5475	.074939	57.5*
L	-0.2175	.011827	9.08*
M	0.2225	.012377	9.50*
N	0.0525	.000689	0.53
O	-0.0475	.000564	0.43
Within (48 d.f.)		.062529	
Total		1.931511	

MSW = 0.001303 with 48 d.f., and $F_{.99}(1,40) = 7.31$.

Twelve of the F-ratios exceed the 99th percentile of the F-distribution. So there may be many signals in these data. However, recall that the variation within the subgroups was strictly the variation introduced by the heat soak tests. The four pieces within each subgroup were all produced during a minimum production run, and therefore may have been very much alike. In other words, there may be twelve significant F-ratios simply because there was too little variation within each subgroup.

Even if the F-tests are right, and there really are many different signals in these data, the Scree Plot shows that two signals dominate all the others: namely Contrasts E and G. Since these Contrasts also affected the Signal-to-Noise ratio, they are not useful as Adjustment Factors. Thus, the first potential Adjustment Factor is represented by Contrast C.

A Response Plot of the average shrinkage versus Factors E, G, and C is

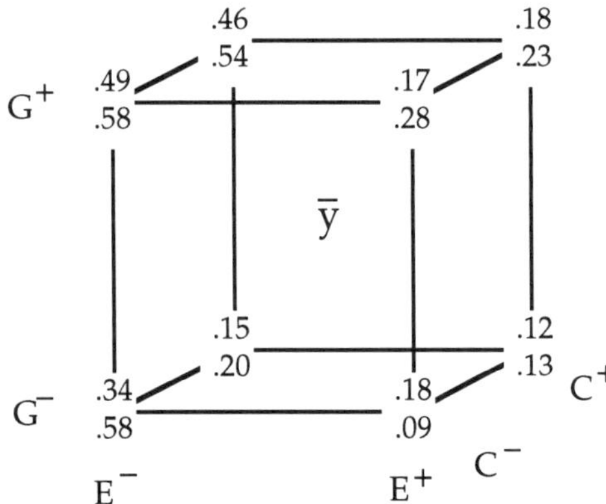

At E^+G^- the Adjustment Factor C does not appear to have much impact. The average percent shrinkage at $E^+G^-C^-$ is 0.135, while at $E^+G^-C^+$ it is 0.125. The manufacturer chose to change the Type of Braid while leaving other factors at their current levels ($E^+G^-C^-$).

EXAMPLE 9.8: Chapter 8 Approach to Analyzing the Cable Casing Data:

Since the F-ratios shown for the ANOVA for $\bar{y}$ are suspect, one could analyze the data from each day separately. This would provide a way to check for consistency of signals from one set of 16 responses to another.

ANOVA for y values from Day One of Heat Soak Tests

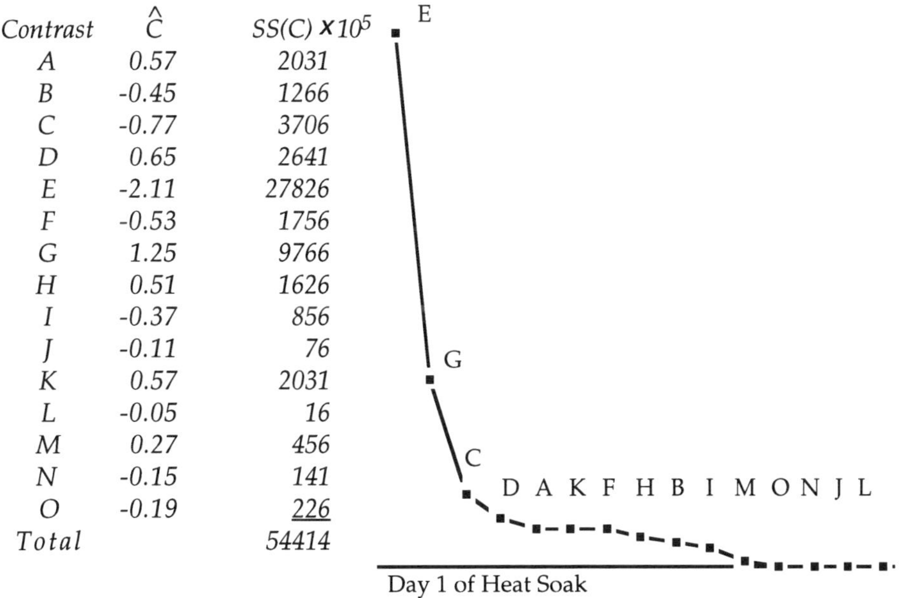

Contrast	$\hat{C}$	SS(C) x 10^5
A	0.57	2031
B	-0.45	1266
C	-0.77	3706
D	0.65	2641
E	-2.11	27826
F	-0.53	1756
G	1.25	9766
H	0.51	1626
I	-0.37	856
J	-0.11	76
K	0.57	2031
L	-0.05	16
M	0.27	456
N	-0.15	141
O	-0.19	226
Total		54414

ANOVA for y values from Day Two of Heat Soak Tests

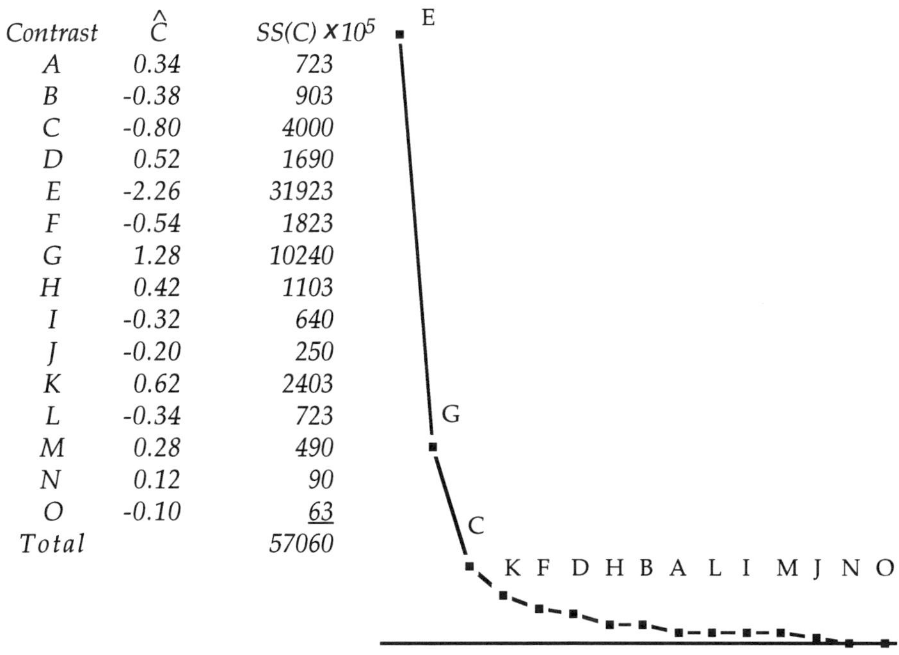

Contrast	$\hat{C}$	SS(C) x 10^5
A	0.34	723
B	-0.38	903
C	-0.80	4000
D	0.52	1690
E	-2.26	31923
F	-0.54	1823
G	1.28	10240
H	0.42	1103
I	-0.32	640
J	-0.20	250
K	0.62	2403
L	-0.34	723
M	0.28	490
N	0.12	90
O	-0.10	63
Total		57060

ANOVA for y values from Day Three of Heat Soak Tests

Contrast	$\hat{C}$	SS(C) x10^5
A	0.62	2403
B	-0.56	1960
C	-0.68	2890
D	0.38	903
E	-1.74	18923
F	-0.56	1960
G	1.00	6250
H	0.60	2250
I	-0.36	810
J	0.06	23
K	0.46	1323
L	-0.32	640
M	0.24	360
N	0.30	563
O	-0.10	63
Total		41318

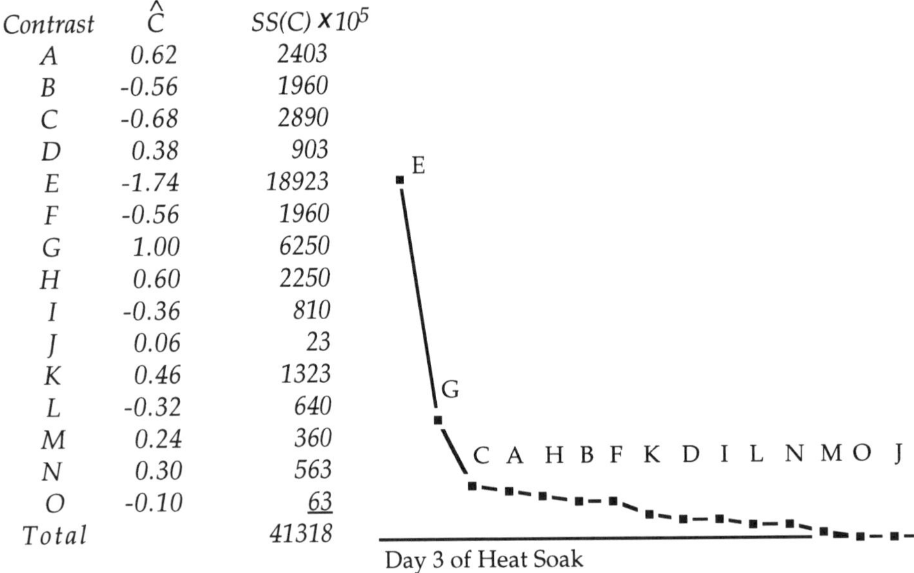

ANOVA for y values from Day Four of Heat Soak Tests

Contrast	$\hat{C}$	SS(C) x10^5
A	0.64	2560
B	-0.40	1000
C	-0.62	2403
D	0.40	1000
E	-1.74	18923
F	-0.74	3423
G	0.96	5760
H	0.46	1323
I	-0.32	640
J	0.28	490
K	0.54	1823
L	-0.16	160
M	0.10	63
N	-0.06	23
O	0.20	250
Total		39838

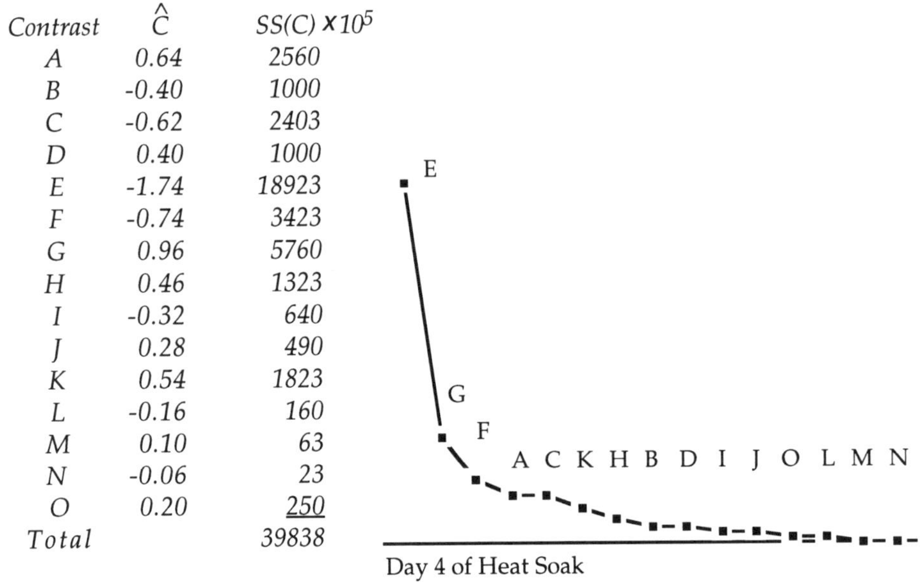

The order of the Contrasts on each day provides a clue to just what may be a signal and what is likely to be noise.

Day One	E	G	C	D	A	K	F	H	B	I	M	O	N	J	L
Day Two	E	G	C	K	F	D	H	B	A	L	I	M	J	N	O
Day Three	E	G	C	A	H	B	F	K	D	I	L	N	M	O	J
Day Four	E	G	F	A	C	K	H	B	D	I	J	O	L	M	N

Clearly Contrasts E and G must represent signals.
Contrast C is rather likely to be a signal.

The next six Contrasts, A, B, D, F, H, and K, as a group, stay in the same place from Day to Day. They do not mix with the contrasts having smaller SS(C) values. Thus, they are all likely to represent signals which are about the same size.

Finally, Contrasts I, J, L, M, N, and O are likely to represent noise.

This analysis suggests that as many as nine contrasts may represent signals, which is fewer than the 12 contrasts found by the F-ratios in Example 9.7. Since Contrasts A, B, D, F, H, and K may all represent signals with similar effects, one should include or exclude them as a group. In the interest of simplicity, the following analysis will consider only Factors E, G, and C to affect the average percent shrinkage.

Next those factors which actively affect the variation in the percent shrinkage are sought:

ANOVA for s:

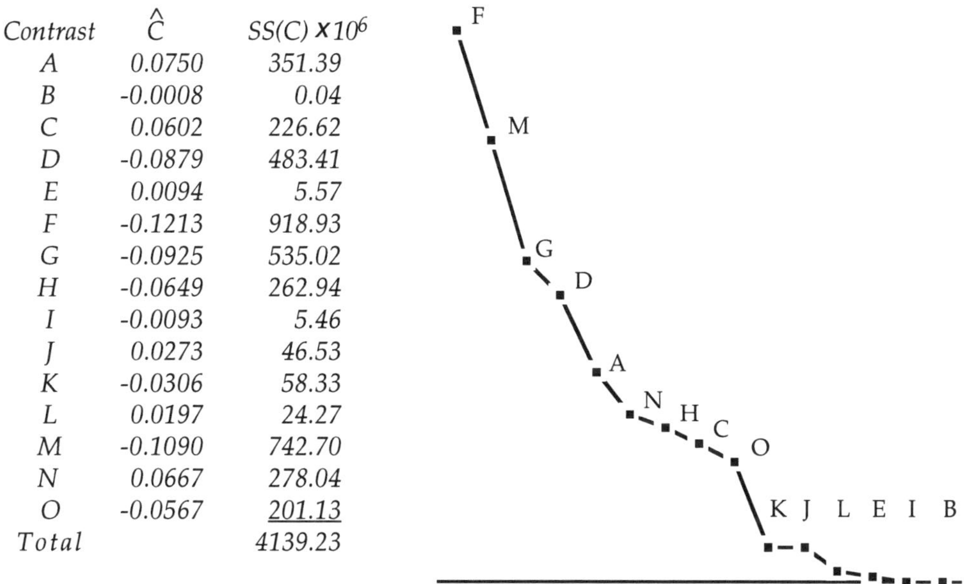

Contrast	$\hat{C}$	*SS(C)* x 10^6
A	*0.0750*	*351.39*
B	*-0.0008*	*0.04*
C	*0.0602*	*226.62*
D	*-0.0879*	*483.41*
E	*0.0094*	*5.57*
F	*-0.1213*	*918.93*
G	*-0.0925*	*535.02*
H	*-0.0649*	*262.94*
I	*-0.0093*	*5.46*
J	*0.0273*	*46.53*
K	*-0.0306*	*58.33*
L	*0.0197*	*24.27*
M	*-0.1090*	*742.70*
N	*0.0667*	*278.04*
O	*-0.0567*	*201.13*
Total		*4139.23*

Pooling the 6 smallest SS(C) values gives a Mod F ratio for the largest SS(C) of:

$$\text{Mod } F\,(15,6,15) = 39.3 < \text{Mod } F_{.50}$$

which is less than the median of the Mod F distribution. Low values for the Mod F ratio may occur in two ways; either most of the contrasts represent signals, or else the contrasts all represent noise. Since factors which affect the variation are relatively rare, it is not likely that all of these contrasts represent signals, therefore, one can conclude that none of these contrasts are likely to have an effect upon the variation in percent shrinkage.

If no factors have a detectable impact upon s^2*, then the variation in the percent shrinkage must be taken to be a constant across the factor space, and the MSD(τ) function can be characterized as:*

$$MSD(0) = \bar{y}^{\,2} + \text{constant}.$$

Thus, the problem of minimizing the Average Loss reduces to the problem of getting on target. The appropriate Response Plot using Factors E, G, and C is shown below: The values shown are the averages of the $\bar{y}^2$ *values.*

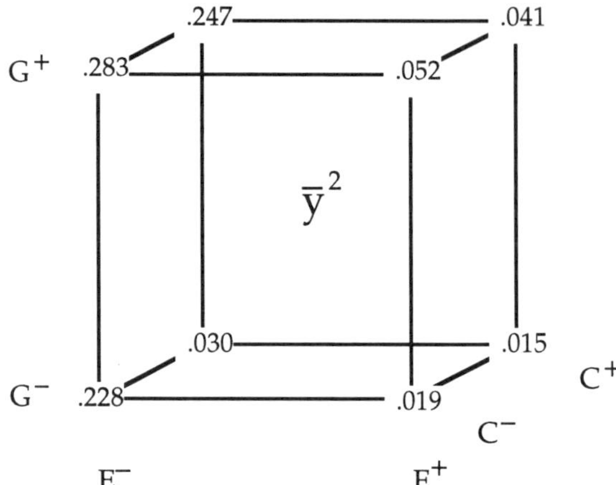

Recall that Factor E is the Type of Wire Braid, Factor G is the Wire Diameter, and Factor C is the Liner Material. The current operating conditions are $E^-G^-C^-$. *Five of the treatment combinations studied yield lower MSD(0) values than the current conditions. Thus, by changing the levels of some of these three factors, the Average Loss due to shrinkage can be dramatically reduced.*

Changing the Liner material ($E^-G^-C^+$ *) will result in an Average Loss that is estimated to be approximately one-seventh of the current value.*

Changing the Type of Braid ($E^+G^-C^-$ *) will result in an average loss that is estimated to be one-twelfth of the current value.*

Changing both the Type of Braid and the Liner Material ($E^+G^-C^+$ *) will result in an average loss that is estimated to be one-fifteenth of the current value.*

Thus, the Chapter 8 approach yields a set of options for the consideration of the manufacturer. In this case, the set of options included the action specified by the Taguchi Approach. As noted, the manufacturer did change the Type of Wire Braid, and this did result in a dramatic drop in the percent shrinkage.

9.3 Summary

Whenever a performance characteristic deviates from its target a loss is incurred. In attempting to find ways to minimize these losses the experimenter is faced with several choices. Among these will be the choices regarding how to conduct the experiment, and how to perform the analysis. This chapter has introduced certain options in both of these areas.

With regard to how to conduct the experiment, the use of a Noise Matrix provides an alternative to simply replicating the whole Design Matrix several times. With the replication approach, one is simply trusting random variation to introduce the appropriate amount of variation into each subgroup. With the Noise Matrix, one is seeking to guarantee a certain spectrum of environmental conditions within each subgroup. The robustness this offers is the advantage of the Noise Matrix. Countering this advantage is the complexity introduced by the use of a Noise Matrix. If one is trying to design a product, and wants to choose those product parameters which will result in a product which is robust to a certain spectrum of environmental conditions, then the use of a Noise Matrix makes sense. However, if one is simply searching for the best set of process parameters, the advantages of a Noise Matrix are much less clear. Moreover, one should always be careful to use to Noise Matrix only for Environmental Factors, while placing both Product and Process Parameters in the Design Matrix.

With regard to how to perform the analysis, the Signal-to-Noise Ratios provide an alternative to the analysis outlined in Chapter 8. The examples in this chapter provide side-by-side comparisons of these two approaches to analysis. The Signal-to-Noise Ratios suffer the same drawbacks as MSD(τ) when used as the variable of analysis for a screening design: they can be misleading because of their composite nature (see Example 8.3, p.265). Because of this insensitivity of the Signal-to-Noise Ratios, the Taguchi Approach will not always find the treatment combination (among those in the study) with the minimum MSD(τ) value.

Finally, Signal-to-Noise Ratios are, for many, rather exotic and difficult to interpret. This lack of an underlying interpretation for the Signal-to-Noise Ratios can make obscure that which an analysis of averages and variances make clear and understandable. Thus, the use of Signal-to-Noise Ratios adds complexity to the analysis, undermines the sensitivity of the analysis, may unnecessarily restrict one's choices, and results in values that are difficult to interpret.

The two step approach of Chapter 8, which uses those factors that affect both the bias and the variation of the product performance characteristic in a direct evaluation of MSD(τ), is at once easier, more flexible, and more understandable than the use of the Signal-to-Noise Ratios and Adjustment Factors.

Finally, with regard to replication, the Taguchi Approach emphasizes the use of a follow-up run to verify the experimental results. However, a single confirmatory run simply will not provide any real confirmation of results. The dangers of this approach have hopefully been made clear in the last four chapters of this book. Instead of a follow-up run, one should use a follow-up design. This is because nontrivial confirmation rests on the replication of both the ***presence*** and the ***absence*** of an effect. Until such confirmation is at hand, changing a production process or a product design may result in serious mistakes (especially when interaction effects are present). The scientific method has always required the nontrivial replication of results, and no transformation of the data, nor any probabilistic test of a hypothesis, can replace this tried and true methodology.

All data contain noise. Some data contain signals. Before one can detect or interpret signals, one must filter out the noise. Therefore, one must have some way to separate potential signals from background noise. Both graphic and algebraic techniques for doing this have been presented. While some of these techniques have made use of probability theory to obtain a decision rule for separating potential signals from that which is likely to be noise, this does not mean that the analysis is exact, or that the theoretical risks can be extrapolated into the real world. The theory merely serves as a guideline for practice.

Finally, the purpose of analysis is insight. The best analysis is the simplest analysis which provides and communicates the insight which can be obtained from the data. Complexity, for the sake of sophistication, is obfuscation. Unfortunately, there are people who are afraid of clarity because they fear that it may not seem profound.

APPENDICES

DATA SET INDEX

DATA SETS ONE AND TWO

DATA SET THREE

COATING WEIGHT DATA

GAUGE STUDY DATA

REFLECTIVITY DATA

(Irwin Miller and John E. Freund, *Probability and Statistics for Engineers, 2nd Ed.*, p.389.)

STEEL PARTS DATA

(Ellis R. Ott, *Journal of Quality Technology, Vol. 15*, 1983, pp.10-18.)

TENSILE STRENGTH DATA

(A. J. Duncan, *Quality Control and Industrial Statistics, 3rd Ed.*, pp.620-621.)

LAB COMPARISON DATA

(Ronald E. Walpole and Raymond H. Myers, *Probability and Statistics for Engineers and Scientists*, p.393.)

ADHESIVE STRENGTH DATA

FLEX RATE DATA

DATA SET ELEVEN

FLASH THICKNESS DATA

TENSILE STRENGTH DATA II

CARBON FOAM DATA

(W. E. Level, et.al., "Statistical Analysis of the Results of a Series of Experiments on the Carbon Foam Production Process," Union Carbide Corporation unclassified report Y-1682, August 5, 1969.)

AN ARTIFICIAL DATA SET

DATA SET SIXTEEN

SPRING HEIGHT DATA

(Joseph J. Pignatiello and John S. Ramberg, *Journal of Quality Technology, Vol.17*, 1985, pp.198-206.)

CABLE CASING DATA

(Jim Quinlan, *Third Supplier Symposium on Taguchi Methods*, 1985, American Supplier Institute, pp.367-384.)

Table A
Bias Correction Factors For Estimating Standard Deviations

Subgroup Size	d_2	c_2	c_4	d_3	Subgroup Size	d_2	c_2	c_4	d_3
2	1.128	0.5642	.7979	0.8525	21	3.778	0.9638	.9876	0.7272
3	1.693	0.7236	.8862	0.8884	22	3.819	0.9655	.9882	0.7199
4	2.059	0.7979	.9213	0.8798	23	3.858	0.9670	.9887	0.7159
5	2.326	0.8407	.9400	0.8641	24	3.895	0.9684	.9892	0.7121
6	2.534	0.8686	.9515	0.8480	25	3.931	0.9695	.9896	0.7084
7	2.704	0.8882	.9594	0.8332	30	4.086	0.9748	.9915	0.6927
8	2.847	0.9027	.9650	0.8198	35	4.213	0.9784	.9927	0.6799
9	2.970	0.9139	.9693	0.8078	40	4.322	0.9811	.9936	0.6692
10	3.078	0.9227	.9727	0.7971	45	4.415	0.9832	.9943	0.6601
11	3.173	0.9300	.9754	0.7873	50	4.498	0.9849	.9949	0.6521
12	3.258	0.9359	.9776	0.7785	60	4.639	0.9874	.9957	0.6389
13	3.336	0.9410	.9794	0.7704	70	4.755	0.9892	.9963	0.6283
14	3.407	0.9453	.9810	0.7630	80	4.854	0.9906	.9968	0.6194
15	3.472	0.9490	.9823	0.7562	90	4.939	0.9916	.9972	0.6118
16	3.532	0.9523	.9835	0.7499	100	5.015	0.9925	.9975	0.6052
17	3.588	0.9551	.9845	0.7441					
18	3.640	0.9576	.9854	0.7386					
19	3.689	0.9599	.9862	0.7335					
20	3.735	0.9619	.9869	0.7287					

"Unbiased" estimates of various Standard Deviations will be obtained from any of the following:

$$\text{An Unbiased Est. of SD(X)} = \frac{s_n}{c_2} \text{ or } \frac{\bar{s}_n}{c_2}$$

$$\text{An Unbiased Est. of SD(X)} = \frac{s}{c_4} \text{ or } \frac{\bar{s}}{c_4}$$

$$\text{An Unbiased Est. of SD(X)} = \frac{R}{d_2} \text{ or } \frac{\bar{R}}{d_2}$$

$$\text{An Unbiased Est. of SD(R)} = \frac{d_3\,\bar{R}}{d_2}$$

Table B

Bias Correction Factors and Degrees of Freedom For Average Ranges

k = number of subgroups used for $\bar{R}$ $\quad$ ν = degrees of freedom for $\bar{R}$

n = number of observations per subgroup $\quad$ d_2 = limiting value for d_2^*

$(d_2^*)^2$ = bias correction factor for $(\bar{R})^2$

$\Delta\nu$ = increment in degrees of freedom per additional subgroup

	n=2		n=3		n=4		n=5		n=6	
k	ν	d_2^*	ν	d_2^*	ν	d_2^*	ν	d_2^*	ν	d_2^*
1	*1.0*	1.414	*2.0*	1.906	*2.9*	2.237	*3.8*	2.477	*4.7*	2.669
2	*1.9*	1.276	*3.8*	1.806	*5.7*	2.149	*7.5*	2.404	*9.2*	2.603
3	*2.9*	1.227	*5.7*	1.767	*8.4*	2.120	*11.1*	2.378	*13.6*	2.580
4	*3.7*	1.206	*7.5*	1.749	*11.2*	2.105	*14.7*	2.365	*18.1*	2.569
5	*4.6*	1.189	*9.3*	1.738	*13.9*	2.096	*18.4*	2.358	*22.6*	2.562
6	*5.5*	1.179	*11.1*	1.731	*16.6*	2.090	*22.0*	2.352	*27.1*	2.557
7	*6.4*	1.172	*12.9*	1.726	*19.4*	2.086	*25.6*	2.349	*31.5*	2.554
8	*7.2*	1.167	*14.8*	1.722	*22.1*	2.082	*29.3*	2.346	*36.0*	2.552
9	*8.1*	1.163	*16.6*	1.718	*24.8*	2.080	*32.9*	2.344	*40.5*	2.550
10	*9.0*	1.159	*18.4*	1.716	*27.6*	2.078	*36.5*	2.342	*44.9*	2.548
11	*9.9*	1.157	*20.2*	1.714	*30.3*	2.076	*40.1*	2.341	*49.4*	2.547
12	*10.8*	1.154	*22.1*	1.712	*33.0*	2.075	*43.7*	2.339	*53.8*	2.546
13	*11.6*	1.152	*23.9*	1.711	*35.8*	2.073	*47.4*	2.338	*58.4*	2.545
14	*12.5*	1.151	*25.7*	1.709	*38.5*	2.072	*51.0*	2.337	*62.8*	2.544
15	*13.4*	1.149	*27.5*	1.708	*41.3*	2.071	*54.6*	2.337	*67.3*	2.543
16	*14.2*	1.148	*29.3*	1.707	*44.0*	2.071	*58.2*	2.336	*71.7*	2.543
17	*15.1*	1.147	*31.1*	1.707	*46.8*	2.070	*61.8*	2.335	*76.2*	2.542
18	*16.0*	1.145	*33.0*	1.706	*49.5*	2.069	*65.5*	2.335	*80.6*	2.542
19	*16.9*	1.145	*34.8*	1.705	*52.3*	2.069	*69.1*	2.334	*85.1*	2.541
20	*17.7*	1.144	*36.6*	1.705	*55.0*	2.068	*72.7*	2.334	*89.6*	2.541
21	*18.6*	1.143	*38.4*	1.704	*57.7*	2.068	*76.3*	2.333	*94.0*	2.541
22	*19.5*	1.143	*40.2*	1.704	*60.5*	2.068	*80.0*	2.333	*98.5*	2.540
23	*20.4*	1.142	*42.1*	1.703	*63.2*	2.067	*83.6*	2.333	*103.0*	2.540
24	*21.2*	1.141	*43.9*	1.703	*65.9*	2.067	*87.2*	2.333	*107.4*	2.540
25	*22.1*	1.141	*45.7*	1.702	*68.7*	2.066	*90.8*	2.332	*111.9*	2.540
$\Delta\nu$	*0.88*		*1.82*		*2.74*		*3.62*		*4.47*	
d_2		1.128		1.693		2.059		2.326		2.534

$$d_2^* \approx d_2 + \frac{d_2}{4\nu}$$

Table B
Bias Correction Factors and Degrees of Freedom for Average Ranges

	n = 7		n = 8		n = 9		n = 10		n = 11	
k	ν	d_2^*	ν	d_2^*	ν	d_2^*	ν	d_2^*	ν	d_2^*
1	5.5	2.827	6.3	2.961	7.0	3.076	7.7	3.178	8.4	3.268
2	10.8	2.767	12.3	2.905	13.8	3.024	15.2	3.129	16.5	3.221
3	16.1	2.746	18.3	2.886	20.5	3.006	22.6	3.112	24.6	3.205
4	21.3	2.736	24.4	2.876	27.3	2.997	30.1	3.104	32.7	3.197
5	26.6	2.729	30.4	2.870	34.0	2.992	37.5	3.098	40.9	3.192
$\Delta\nu$	5.26		6.03		6.75		7.46		8.13	
d_2		2.704		2.847		2.970		3.078		3.173

	n = 12		n = 13		n = 14		n = 15		n = 16	
k	ν	d_2^*	ν	d_2^*	ν	d_2^*	ν	d_2^*	ν	d_2^*
1	9.0	3.348	9.6	3.423	10.2	3.490	10.8	3.552	11.4	3.610
2	17.8	3.304	19.0	3.380	20.2	3.449	21.3	3.513	22.5	3.571
3	26.6	3.289	28.4	3.365	30.2	3.435	31.9	3.499	33.6	3.558
4	35.3	3.281	37.8	3.358	40.2	3.428	42.4	3.492	44.7	3.552
5	44.1	3.276	47.2	3.354	50.2	3.424	52.9	3.488	55.8	3.548
$\Delta\nu$	8.77		9.39		9.90		10.57		11.11	
d_2		3.258		3.336		3.407		3.472		3.532

	n = 17		n = 18		n = 19		n = 20		n = 21	
k	ν	d_2^*	ν	d_2^*	ν	d_2^*	ν	d_2^*	ν	d_2^*
1	11.9	3.663	12.4	3.713	12.9	3.760	13.4	3.805	13.9	3.846
2	23.6	3.626	24.6	3.677	25.6	3.725	26.5	3.770	27.5	3.812
3	35.2	3.613	36.8	3.665	38.2	3.713	39.6	3.759	41.1	3.801
4	46.9	3.607	48.9	3.659	50.9	3.707	52.7	3.753	54.7	3.795
5	58.5	3.603	61.1	3.655	63.6	3.704	65.9	3.749	68.4	3.792
$\Delta\nu$	11.66		12.17		12.66		13.12		13.62	
d_2		3.588		3.640		3.689		3.735		3.788

	n = 22		n = 23		n = 24		n = 25		n = 30	
k	ν	d_2^*	ν	d_2^*	ν	d_2^*	ν	d_2^*	ν	d_2^*
1	14.3	3.886	14.8	3.923	15.2	3.959	15.6	3.994	16.9	4.147
2	28.4	3.853	29.3	3.891	30.2	3.927	31.0	3.963	33.5	4.116
3	42.5	3.841	43.8	3.880	45.1	3.917	46.3	3.952	50.1	4.106
4	56.5	3.836	58.3	3.875	60.1	3.911	61.7	3.947	66.7	4.101
5	70.6	3.833	72.8	3.871	75.1	3.908	77.1	3.944	83.3	4.098
$\Delta\nu$	14.07		14.51		14.96		15.36		16.6	
d_2		3.819		3.858		3.895		3.931		4.086

Table C
Control Chart Factors

For any number of subgroups of size n with

Grand Average, $\overline{\overline{X}}$, and Average Range, $\overline{R}$,

or

Grand Average, $\overline{\overline{X}}$, and Average Standard Deviation, $\overline{s}$

Subgroup Size n	A_2	D_3	D_4
2	1.880	--	3.268
3	1.023	--	2.574
4	0.729	--	2.282
5	0.577	--	2.114
6	0.483	--	2.004
7	0.419	0.076	1.924
8	0.373	0.136	1.864
9	0.337	0.184	1.816
10	0.308	0.223	1.777

Subgroup Size n	A_3	B_3	B_4
2	2.659	--	3.267
3	1.954	--	2.568
4	1.628	--	2.266
5	1.427	--	2.089
6	1.287	0.030	1.970
7	1.182	0.118	1.882
8	1.099	0.185	1.815
9	1.032	0.239	1.761
10	0.975	0.284	1.716

$$UCL_{\overline{X}} = \overline{\overline{X}} + A_2 \overline{R}$$

$$CL_{\overline{X}} = \overline{\overline{X}}$$

$$LCL_{\overline{X}} = \overline{\overline{X}} - A_2 \overline{R}$$

$$UCL_{\overline{X}} = \overline{\overline{X}} + A_3 \overline{s}$$

$$CL_{\overline{X}} = \overline{\overline{X}}$$

$$LCL_{\overline{X}} = \overline{\overline{X}} - A_3 \overline{s}$$

$$UCL_R = D_4 \overline{R}$$

$$CL_R = \overline{R}$$

$$LCL_R = D_3 \overline{R}$$

$$UCL_s = B_4 \overline{s}$$

$$CL_s = \overline{s}$$

$$LCL_s = B_3 \overline{s}$$

And for $n > 10$:

$$A_2 = \frac{3}{d_2 \sqrt{n}}$$

$$A_3 = \frac{3}{c_4 \sqrt{n}}$$

$$D_3 = \left[1 - \frac{3 d_3}{d_2}\right]$$

$$D_4 = \left[1 + \frac{3 d_3}{d_2}\right]$$

$$B_3 = \left[1 - \frac{3}{c_4}\sqrt{1 - c_4^2}\right]$$

$$B_4 = \left[1 + \frac{3}{c_4}\sqrt{1 - c_4^2}\right]$$

Table D
ANOM CRITICAL VALUES

$\alpha = 0.25$

v = degrees of freedom for estimate of standard deviation
k = number of averages being compared
Tabled values = H

v	k=2	k=3	k=4	k=5	k=6	k=7	k=8	k=9	k=10	k=12	k=15	k=20
2	1.13	1.91	2.31	2.59	2.80	2.97	3.11	3.23	3.34	3.51	3.72	3.96
3	1.00	1.67	2.00	2.23	2.41	2.55	2.67	2.76	2.86	3.00	3.16	3.36
4	0.95	1.56	1.86	2.08	2.24	2.36	2.47	2.56	2.64	2.77	2.92	3.11
5	0.92	1.50	1.78	1.99	2.14	2.26	2.36	2.44	2.51	2.64	2.78	2.95
6	0.90	1.46	1.74	1.93	2.07	2.19	2.28	2.37	2.44	2.56	2.70	2.86
7	0.88	1.44	1.71	1.89	2.04	2.15	2.24	2.32	2.38	2.50	2.63	2.79
8	0.88	1.42	1.68	1.86	2.00	2.11	2.20	2.27	2.34	2.45	2.58	2.74
9	0.87	1.40	1.66	1.84	1.97	2.08	2.17	2.24	2.31	2.42	2.55	2.70
10	0.86	1.40	1.65	1.82	1.95	2.06	2.15	2.23	2.29	2.39	2.52	2.67
12	0.86	1.38	1.63	1.80	1.93	2.03	2.11	2.19	2.25	2.36	2.47	2.62
15	0.85	1.36	1.60	1.77	1.90	2.00	2.09	2.15	2.21	2.32	2.43	2.57
20	0.83	1.35	1.58	1.74	1.87	1.97	2.05	2.12	2.18	2.28	2.40	2.53
24	0.83	1.34	1.58	1.74	1.86	1.95	2.03	2.10	2.16	2.26	2.37	2.50
30	0.83	1.33	1.57	1.73	1.85	1.93	2.02	2.08	2.14	2.24	2.35	2.49
40	0.83	1.32	1.55	1.71	1.83	1.93	2.00	2.06	2.13	2.22	2.33	2.46
60	0.82	1.31	1.54	1.70	1.82	1.91	1.98	2.05	2.11	2.20	2.31	2.44
120	0.82	1.31	1.53	1.69	1.81	1.90	1.97	2.04	2.09	2.17	2.28	2.41
inf.	0.81	1.30	1.52	1.67	1.79	1.88	1.96	2.02	2.07	2.15	2.26	2.39

These values are exact values, rather than upper bounds.
The values in Table D were derived from values tabled by Lloyd S. Nelson and Peter R. Nelson.
The author would like to especially thank Peter Nelson for providing the expanded version of his tables while they were still in manuscript form.

Table D
ANOM CRITICAL VALUES

$\alpha = 0.10$

v = degrees of freedom for estimate of standard deviation
k = number of averages being compared
Tabled values = $H_{.}$

v	k=2	k=3	k=4	k=5	k=6	k=7	k=8	k=9	k=10
2	2.06	3.31	3.95	4.40	4.74	5.01	5.24	5.43	5.61
3	1.66	2.58	3.03	3.35	3.60	3.79	3.96	4.09	4.21
4	1.51	2.29	2.68	2.94	3.15	3.31	3.45	3.56	3.66
5	1.42	2.15	2.49	2.73	2.91	3.06	3.18	3.28	3.37
6	1.37	2.06	2.37	2.60	2.77	2.90	3.01	3.11	3.19
7	1.34	1.99	2.29	2.51	2.67	2.80	2.90	2.99	3.07
8	1.32	1.95	2.24	2.44	2.60	2.72	2.82	2.90	2.98
9	1.30	1.91	2.20	2.40	2.55	2.67	2.76	2.84	2.91
10	1.28	1.89	2.17	2.36	2.50	2.62	2.71	2.79	2.87
12	1.26	1.85	2.12	2.30	2.44	2.55	2.64	2.72	2.78
15	1.24	1.81	2.07	2.25	2.38	2.48	2.57	2.64	2.70
20	1.22	1.78	2.03	2.19	2.32	2.43	2.51	2.57	2.64
24	1.21	1.76	2.01	2.17	2.29	2.39	2.47	2.54	2.60
30	1.20	1.74	1.98	2.15	2.26	2.36	2.44	2.51	2.56
40	1.19	1.72	1.97	2.12	2.24	2.33	2.40	2.47	2.52
60	1.18	1.71	1.94	2.09	2.21	2.31	2.38	2.44	2.50
120	1.17	1.69	1.92	2.08	2.18	2.27	2.35	2.40	2.46
inf.	1.16	1.67	1.90	2.05	2.15	2.24	2.31	2.38	2.42

v	k =12	k=15	k=20	k=24	k=30	k=40	k=60
7	3.20	3.35	3.54				
8	3.10	3.25	3.43				
9	3.04	3.17	3.34				
10	2.98	3.11	3.27				
12	2.86	3.02	3.18				
15	2.81	2.94	3.09				
20	2.74	2.85	2.99				
24	2.70	2.81	2.94	3.03			
30	2.65	2.76	2.90	2.98	3.08		
40	2.61	2.72	2.86	2.93	3.02	3.14	
60	2.59	2.69	2.81	2.88	2.97	3.08	3.23
120	2.55	2.65	2.77	2.83	2.92	3.02	3.16
inf.	2.51	2.60	2.72	2.79	2.87	2.97	3.10

Table D
ANOM CRITICAL VALUES

$\alpha = 0.05$

v = degrees of freedom for estimate of standard deviation
k = number of averages being compared
Tabled values = H

v	k=2	k=3	k=4	k=5	k=6	k=7	k=8	k=9	k=10
2	3.04	4.81	5.72	6.35	6.84	7.23	7.55	7.82	8.06
3	2.25	3.41	3.98	4.39	4.70	4.94	5.14	5.33	5.47
4	1.96	2.91	3.37	3.69	3.93	4.12	4.28	4.42	4.54
5	1.82	2.65	3.06	3.33	3.54	3.70	3.84	3.97	4.07
6	1.73	2.51	2.87	3.12	3.30	3.45	3.58	3.69	3.78
7	1.67	2.40	2.75	2.98	3.15	3.30	3.40	3.51	3.60
8	1.63	2.34	2.66	2.87	3.04	3.18	3.28	3.38	3.45
9	1.60	2.28	2.59	2.80	2.96	3.08	3.19	3.28	3.36
10	1.58	2.24	2.54	2.75	2.89	3.02	3.11	3.21	3.27
12	1.54	2.18	2.47	2.66	2.80	2.92	3.01	3.09	3.16
15	1.51	2.12	2.39	2.58	2.71	2.82	2.91	2.99	3.05
20	1.48	2.07	2.32	2.50	2.63	2.73	2.82	2.88	2.95
24	1.46	2.04	2.29	2.46	2.58	2.68	2.77	2.84	2.89
30	1.44	2.02	2.26	2.42	2.55	2.64	2.72	2.79	2.85
40	1.43	1.98	2.23	2.39	2.51	2.60	2.68	2.74	2.80
60	1.41	1.96	2.20	2.35	2.46	2.56	2.63	2.70	2.75
120	1.40	1.94	2.17	2.32	2.43	2.52	2.59	2.65	2.69
inf.	1.39	1.91	2.14	2.29	2.39	2.48	2.54	2.60	2.66

v	k=12	k=15	k=20	k=24	k=30	k=40	k=60
7	3.73	3.90	4.11				
8	3.59	3.74	3.95				
9	3.49	3.63	3.82				
10	3.40	3.55	3.72				
12	3.27	3.41	3.58				
15	3.16	3.28	3.44				
20	3.04	3.17	3.31				
24	3.00	3.11	3.25	3.33			
30	2.94	3.05	3.19	3.26	3.36		
40	2.88	2.99	3.12	3.20	3.29	3.40	
60	2.83	2.94	3.06	3.13	3.22	3.33	3.47
120	2.79	2.88	3.00	3.06	3.14	3.25	3.38
inf.	2.74	2.83	2.94	3.01	3.08	3.18	3.31

Table D
ANOM CRITICAL VALUES

$\alpha = 0.01$

v = degrees of freedom for estimate of standard deviation
k = number of averages being compared
Tabled values = H

v	k=2	k=3	k=4	k=5	k=6	k=7	k=8	k=9	k=10
2	7.02	10.9	13.0	14.4	15.5	16.4	17.1	17.7	18.3
3	4.13	6.13	7.12	7.80	8.33	8.76	9.11	9.41	9.68
4	3.26	4.69	5.38	5.85	6.22	6.51	6.75	6.96	7.13
5	2.85	4.03	4.58	4.96	5.25	5.48	5.68	5.84	5.98
6	2.62	3.66	4.13	4.45	4.71	4.91	5.07	5.20	5.33
7	2.47	3.41	3.85	4.14	4.36	4.54	4.69	4.81	4.91
8	2.37	3.25	3.65	3.92	4.12	4.29	4.42	4.53	4.63
9	2.30	3.13	3.51	3.76	3.95	4.10	4.22	4.33	4.42
10	2.24	3.05	3.39	3.64	3.82	3.96	4.08	4.18	4.26
12	2.16	2.91	3.24	3.46	3.63	3.76	3.86	3.96	4.03
15	2.08	2.79	3.10	3.30	3.46	3.57	3.67	3.75	3.82
20	2.01	2.68	2.96	3.16	3.30	3.40	3.49	3.56	3.63
24	1.98	2.62	2.90	3.09	3.21	3.31	3.40	3.48	3.54
30	1.94	2.57	2.84	3.01	3.14	3.24	3.32	3.38	3.44
40	1.91	2.52	2.78	2.94	3.07	3.17	3.24	3.30	3.36
60	1.88	2.47	2.72	2.88	3.00	3.09	3.16	3.22	3.28
120	1.85	2.42	2.66	2.82	2.93	3.02	3.09	3.15	3.20
inf.	1.82	2.38	2.61	2.75	2.87	2.94	3.01	3.07	3.12

v	k=12	k=15	k=20	k=24	k=30	k=40	k=60
7	5.09	5.30	5.57				
8	4.80	4.99	5.22				
9	4.58	4.75	4.97				
10	4.40	4.57	4.78				
12	4.16	4.32	4.50				
15	3.94	4.08	4.25				
20	3.73	3.86	4.02				
24	3.64	3.76	3.90	3.98			
30	3.54	3.65	3.79	3.87	3.97		
40	3.45	3.56	3.68	3.76	3.85	3.96	
60	3.36	3.47	3.59	3.66	3.74	3.84	3.98
120	3.27	3.37	3.49	3.53	3.61	3.70	3.83
inf.	3.20	3.28	3.39	3.45	3.53	3.62	3.73

Table D2
Critical Values for the Analysis of Mean Ranges (ANOMR)

$\alpha = 0.05$

n = subgroup size in original One-Way ANOM
L = number of Average Ranges being compared (number of levels in Main Effect)
m = number of subgroup ranges averaged for each level of the Main Effect

$$LDL_R = H_L \bar{\bar{R}} \qquad UDL_R = H_U \bar{\bar{R}}$$

n = 2

	L=2		L=3		L=4		L=5		L=6		L=7	
m	H_L	H_U	H_L	H_U	H_L	H_U	H_L	H_U	H_L	H_U	H_L	H_U
1	0.156	1.844	0.068	2.438	0.053	2.795	0.044	3.058	0.038	3.263	0.034	3.431
2	0.334	1.666	0.206	2.073	0.178	2.286	0.160	2.440	0.148	2.558	0.138	2.653
3	0.436	1.564	0.305	1.886	0.272	2.046	0.251	2.159	0.236	2.246	0.225	2.316
4	0.503	1.497	0.375	1.770	0.341	1.900	0.319	1.992	0.303	2.063	0.292	2.119
5	0.550	1.450	0.427	1.690	0.393	1.801	0.371	1.880	0.355	1.939	0.343	1.987
6	0.587	1.413	0.468	1.630	0.434	1.728	0.413	1.797	0.397	1.850	0.385	1.892
7	0.615	1.385	0.502	1.583	0.468	1.671	0.447	1.734	0.432	1.781	0.420	1.818
8	0.639	1.361	0.529	1.545	0.497	1.626	0.476	1.683	0.461	1.726	0.449	1.760
10	0.675	1.325	0.573	1.486	0.542	1.557	0.552	1.605	0.507	1.642	0.496	1.672
12	0.703	1.297	0.606	1.443	0.577	1.506	0.557	1.549	0.543	1.582	0.532	1.608
14	0.724	1.276	0.633	1.410	0.604	1.466	0.585	1.506	0.572	1.535	0.561	1.559
16	0.741	1.259	0.654	1.383	0.627	1.435	0.609	1.471	0.596	1.498	0.586	1.520
20	0.768	1.232	0.688	1.342	0.663	1.387	0.646	1.418	0.633	1.442	0.624	1.461

n = 3

	L=2		L=3		L=4		L=5		L=6		L=7	
m	H_L	H_U	H_L	H_U	H_L	H_U	H_L	H_U	H_L	H_U	H_L	H_U
1	0.343	1.657	0.215	2.055	0.186	2.263	0.168	2.412	0.155	2.527	0.146	2.620
2	0.511	1.489	0.383	1.757	0.349	1.884	0.327	1.973	0.312	2.041	0.300	2.096
3	0.594	1.406	0.476	1.618	0.443	1.714	0.421	1.782	0.405	1.833	0.393	1.874
4	0.645	1.355	0.537	1.535	0.504	1.614	0.483	1.669	0.468	1.711	0.457	1.745
5	0.681	1.319	0.580	1.477	0.549	1.546	0.529	1.594	0.514	1.630	0.503	1.659
6	0.708	1.292	0.613	1.435	0.583	1.496	0.564	1.538	0.550	1.570	0.539	1.596
7	0.729	1.271	0.639	1.402	0.611	1.458	0.592	1.496	0.579	1.525	0.568	1.548
8	0.746	1.254	0.660	1.376	0.633	1.427	0.615	1.462	0.602	1.488	0.592	1.509
10	0.772	1.228	0.694	1.335	0.668	1.380	0.651	1.410	0.639	1.433	0.630	1.452
12	0.792	1.208	0.719	1.305	0.695	1.345	0.679	1.373	0.667	1.393	0.658	1.409
14	0.807	1.193	0.738	1.282	0.716	1.319	0.701	1.343	0.690	1.362	0.681	1.377
16	0.819	1.181	0.754	1.264	0.733	1.297	0.719	1.320	0.708	1.337	0.700	1.351
20	0.838	1.162	0.779	1.235	0.759	1.265	0.746	1.285	0.736	1.300	0.729	1.312

The values in Table D2 are those given by Neil R. Ullman in his 1989 article cited in the References.

Table D2
Critical Values for the Analysis of Mean Ranges (ANOMR)

$\alpha = 0.05$

n = subgroup size in original One-Way ANOM
L = number of Average Ranges being compared (number of levels in Main Effect)
m = number of subgroup ranges averaged for each level of the Main Effect

$$LDL_R = H_L \bar{\bar{R}} \qquad UDL_R = H_U \bar{\bar{R}}$$

n = 4

	L=2		L=3		L=4		L=5		L=6		L=7	
m	H_L	H_U	H_L	H_U	H_L	H_U	H_L	H_U	H_L	H_U	H_L	H_U
1	0.446	1.554	0.315	1.869	0.282	2.203	0.261	2.133	0.246	2.217	0.235	2.285
2	0.595	1.405	0.477	1.617	0.444	1.712	0.422	1.779	0.407	1.830	0.395	1.871
3	0.665	1.335	0.561	1.502	0.529	1.576	0.509	1.626	0.494	1.665	0.483	1.696
4	0.709	1.291	0.614	1.434	0.584	1.495	0.565	1.537	0.551	1.569	0.540	1.594
5	0.739	1.261	0.651	1.387	0.623	1.440	0.605	1.477	0.592	1.504	0.582	1.526
6	0.761	1.239	0.679	1.353	0.653	1.400	0.636	1.433	0.623	1.457	0.613	1.477
7	0.778	1.222	0.701	1.326	0.676	1.369	0.660	1.399	0.648	1.421	0.638	1.439
8	0.792	1.208	0.719	1.305	0.695	1.344	0.680	1.371	0.668	1.392	0.659	1.408
10	0.814	1.186	0.747	1.272	0.725	1.306	0.711	1.330	0.700	1.348	0.692	1.362
12	0.830	1.170	0.768	1.248	0.748	1.279	0.734	1.300	0.724	1.316	0.716	1.329
14	0.842	1.158	0.785	1.229	0.765	1.257	0.752	1.277	0.743	1.291	0.736	1.303
16	0.853	1.147	0.798	1.214	0.780	1.240	0.767	1.258	0.758	1.272	0.751	1.282
20	0.868	1.132	0.819	1.191	0.802	1.214	0.791	1.230	0.782	1.242	0.776	1.251

n = 5

	L=2		L=3		L=4		L=5		L=6		L=7	
m	H_L	H_U	H_L	H_U	H_L	H_U	H_L	H_U	H_L	H_U	H_L	H_U
1	0.510	1.490	0.383	1.757	0.349	1.884	0.327	1.974	0.312	2.042	0.300	2.097
2	0.645	1.355	0.536	1.535	0.504	1.615	0.483	1.670	0.468	1.712	0.456	1.746
3	0.708	1.292	0.612	1.436	0.583	1.497	0.564	1.539	0.550	1.571	0.539	1.597
4	0.746	1.254	0.660	1.376	0.633	1.427	0.615	1.462	0.602	1.489	0.592	1.510
5	0.772	1.228	0.693	1.336	0.668	1.380	0.651	1.411	0.639	1.434	0.629	1.452
6	0.792	1.208	0.718	1.306	0.694	1.346	0.679	1.373	0.667	1.393	0.658	1.410
7	0.807	1.193	0.738	1.283	0.715	1.319	0.700	1.344	0.689	1.363	0.681	1.377
8	0.819	1.181	0.754	1.264	0.733	1.297	0.718	1.320	0.708	1.338	0.700	1.351
10	0.838	1.162	0.779	1.236	0.759	1.265	0.746	1.285	0.736	1.300	0.729	1.312
12	0.852	1.148	0.797	1.215	0.779	1.241	0.767	1.259	0.758	1.273	0.751	1.293
14	0.863	1.137	0.812	1.198	0.795	1.223	0.783	1.239	0.774	1.252	0.768	1.261
16	0.872	1.128	0.824	1.185	0.807	1.208	0.796	1.233	0.788	1.235	0.782	1.244
20	0.885	1.115	0.842	1.166	0.827	1.185	0.817	1.199	0.809	1.209	0.804	1.217

Table D2
Critical Values for the Analysis of Mean Ranges (ANOMR)

$\alpha = 0.01$

n = subgroup size in original One-Way ANOM
L = number of Average Ranges being compared (number of levels in Main Effect)
m = number of subgroup ranges averaged for each level of the Main Effect

$$LDL_R = H_L \bar{\bar{R}} \qquad UDL_R = H_U \bar{\bar{R}}$$

n = 2

	L=2		L=3		L=4		L=5		L=6		L=7	
m	H_L	H_U	H_L	H_U	H_L	H_U	H_L	H_U	H_L	H_U	H_L	H_U
1	0.061	1.939	0.027	2.650	0.207	3.125	0.017	3.474	0.015	3.745	0.014	3.965
2	0.202	1.798	0.126	2.284	0.108	2.564	0.097	2.761	0.090	2.911	0.085	3.030
3	0.305	1.695	0.214	2.076	0.190	2.280	0.176	2.423	0.166	2.530	0.158	2.614
4	0.377	1.623	0.282	1.941	0.256	2.106	0.240	2.219	0.228	2.304	0.220	2.371
5	0.432	1.568	0.336	1.846	0.309	1.985	0.292	2.058	0.279	2.151	0.270	2.208
6	0.474	1.526	0.379	1.774	0.352	1.895	0.331	1.978	0.322	2.040	0.312	2.089
7	0.509	1.491	0.415	1.717	0.387	1.826	0.370	1.900	0.357	1.955	0.347	1.998
8	0.537	1.463	0.445	1.671	0.418	1.770	0.400	1.837	0.387	1.887	0.378	1.926
10	0.582	1.418	0.494	1.560	0.467	1.684	0.449	1.742	0.437	1.784	0.427	1.817
12	0.616	1.384	0.532	1.547	0.505	1.622	0.488	1.672	0.475	1.709	0.466	1.738
14	0.642	1.358	0.562	1.506	0.536	1.573	0.519	1.618	0.507	1.652	0.498	1.678
16	0.664	1.336	0.587	1.473	0.561	1.534	0.545	1.575	0.533	1.606	0.524	1.629
20	0.698	1.302	0.626	1.422	0.602	1.475	0.586	1.510	0.575	1.536	0.567	1.557

n = 3

	L=2		L=3		L=4		L=5		L=6		L=7	
m	H_L	H_U	H_L	H_U	H_L	H_U	H_L	H_U	H_L	H_U	H_L	H_U
1	0.211	1.789	0.133	2.265	0.115	2.536	0.104	2.728	0.096	2.873	0.090	2.989
2	0.386	1.614	0.291	1.925	0.265	2.085	0.248	2.195	0.237	2.278	0.228	2.343
3	0.483	1.517	0.388	1.760	0.360	1.878	0.328	1.959	0.330	2.020	0.320	2.067
4	0.545	1.455	0.454	1.659	0.426	1.755	0.408	1.821	0.395	1.869	0.386	1.907
5	0.589	1.411	0.502	1.589	0.474	1.671	0.457	1.727	0.445	1.768	0.435	1.801
6	0.622	1.378	0.539	1.537	0.512	1.610	0.495	1.659	0.483	1.695	0.474	1.723
7	0.649	1.351	0.569	1.497	0.543	1.562	0.526	1.606	0.514	1.639	0.505	1.664
8	0.670	1.330	0.593	1.464	0.568	1.524	0.552	1.564	0.540	1.594	0.531	1.617
10	0.704	1.297	0.632	1.414	0.608	1.466	0.593	1.500	0.582	1.526	0.573	1.546
12	0.728	1.272	0.661	1.377	0.639	1.423	0.624	1.454	0.613	1.477	0.605	1.494
14	0.748	1.252	0.684	1.349	0.663	1.390	0.649	1.418	0.639	1.439	0.631	1.455
16	0.764	1.236	0.703	1.326	0.682	1.364	0.669	1.389	0.659	1.408	0.652	1.423
20	0.788	1.212	0.733	1.290	0.713	1.324	0.701	1.346	0.692	1.362	0.685	1.375

Table D2
Critical Values for the Analysis of Mean Ranges (ANOMR)

$\alpha = 0.01$

n = subgroup size in original One-Way ANOM
L = number of Average Ranges being compared (number of levels in Main Effect)
m = number of subgroup ranges averaged for each level of the Main Effect

$$LDL_R = H_L \bar{\bar{R}} \qquad UDL_R = H_U \bar{\bar{R}}$$

n = 4

	L=2		L=3		L=4		L=5		L=6		L=7	
m	H_L	H_U	H_L	H_U	H_L	H_U	H_L	H_U	H_L	H_U	H_L	H_U
1	0.315	1.685	0.223	2.056	0.200	2.254	0.185	2.391	0.175	2.495	0.167	2.576
2	0.484	1.516	0.389	1.758	0.361	1.876	0.344	1.957	0.331	2.016	0.322	2.064
3	0.569	1.431	0.480	1.619	0.453	1.708	0.435	1.768	0.423	1.812	0.413	1.847
4	0.623	1.377	0.540	1.536	0.513	1.608	0.496	1.657	0.484	1.693	0.475	1.721
5	0.661	1.339	0.583	1.478	0.557	1.540	0.541	1.582	0.529	1.613	0.520	1.637
6	0.689	1.311	0.615	1.436	0.591	1.491	0.575	1.528	0.564	1.555	0.555	1.577
7	0.711	1.289	0.641	1.403	0.617	1.453	0.602	1.486	0.591	1.511	0.583	1.530
8	0.729	1.271	0.662	1.376	0.639	1.422	0.625	1.453	0.614	1.475	0.606	1.493
10	0.757	1.243	0.695	1.336	0.674	1.375	0.660	1.402	0.650	1.421	0.643	1.437
12	0.778	1.222	0.720	1.306	0.700	1.341	0.687	1.365	0.678	1.382	0.670	1.396
14	0.794	1.206	0.739	1.283	0.720	1.315	0.708	1.336	0.699	1.352	0.692	1.364
16	0.807	1.193	0.755	1.264	0.737	1.294	0.725	1.313	0.717	1.328	0.710	1.339
20	0.827	1.173	0.780	1.235	0.763	1.261	0.752	1.279	0.744	1.291	0.738	1.301

n = 5

	L=2		L=3		L=4		L=5		L=6		L=7	
m	H_L	H_U	H_L	H_U	H_L	H_U	H_L	H_U	H_L	H_U	H_L	H_U
1	0.386	1.614	0.290	1.929	0.264	2.086	0.248	2.196	0.236	2.279	0.227	2.344
2	0.544	1.456	0.453	1.659	0.425	1.756	0.408	1.821	0.395	1.870	0.385	1.908
3	0.622	1.378	0.538	1.537	0.512	1.610	0.495	1.659	0.483	1.696	0.473	1.724
4	0.670	1.330	0.593	1.464	0.568	1.524	0.552	1.565	0.540	1.594	0.531	1.618
5	0.703	1.297	0.632	1.414	0.608	1.466	0.592	1.501	0.581	1.526	0.573	1.546
6	0.728	1.272	0.661	1.378	0.638	1.424	0.624	1.454	0.613	1.477	0.605	1.495
7	0.748	1.252	0.684	1.349	0.662	1.391	0.648	1.419	0.638	1.439	0.630	1.455
8	0.764	1.236	0.703	1.326	0.682	1.364	0.669	1.390	0.659	1.409	0.651	1.423
10	0.788	1.212	0.732	1.291	0.713	1.324	0.701	1.346	0.691	1.363	0.684	1.375
12	0.806	1.194	0.755	1.265	0.736	1.295	0.724	1.315	0.716	1.329	0.709	1.341
14	0.820	1.180	0.772	1.245	0.755	1.272	0.743	1.290	0.735	1.303	0.729	1.314
16	0.832	1.168	0.786	1.229	0.770	1.254	0.759	1.271	0.751	1.283	0.745	1.292
20	0.849	1.151	0.808	1.204	0.793	1.226	0.783	1.241	0.776	1.251	0.770	1.260

Table E
Percentiles of the Studentized Range Distribution
90th Percentiles

k = Number of Subgroup Averages Being Compared

v	2	3	4	5	6	7	8	9	10	11	12	13	14	15	16	17	18	19	20	24
1	8.93	13.4	13.4	18.5	20.2	21.5	22.6	23.6	24.5	25.2	25.9	26.5	27.1	27.6	28.1	28.5	29.0	29.3	29.7	31.0
2	4.13	5.73	6.77	7.54	8.14	8.63	9.05	9.41	9.72	10.0	10.3	10.5	10.7	10.9	11.1	11.2	11.4	11.5	11.7	12.2
3	3.33	4.47	5.20	5.74	6.16	6.51	6.81	7.06	7.29	7.49	7.67	7.83	7.98	8.12	8.25	8.37	8.48	8.58	8.68	9.03
4	3.01	3.98	4.59	5.03	5.39	5.68	5.93	6.14	6.33	6.49	6.65	6.78	6.91	7.02	7.13	7.23	7.33	7.41	7.50	7.79
5	2.85	3.72	4.26	4.66	4.98	5.24	5.46	5.65	5.82	5.97	6.10	6.22	6.34	6.44	6.54	6.63	6.71	6.79	6.86	7.14
6	2.75	3.56	4.07	4.44	4.73	4.97	5.17	5.34	5.50	5.64	5.76	5.87	5.98	6.07	6.16	6.25	6.32	6.40	6.47	6.71
7	2.68	3.45	3.93	4.28	4.55	4.78	4.97	5.14	5.28	5.41	5.53	5.64	5.74	5.83	5.91	5.99	6.06	6.13	6.19	6.42
8	2.63	3.37	3.83	4.17	4.43	4.65	4.83	4.99	5.13	5.25	5.36	5.46	5.56	5.64	5.72	5.80	5.87	5.93	6.00	6.21
9	2.59	3.32	3.76	4.08	4.34	4.54	4.72	4.87	5.01	5.13	5.23	5.33	5.42	5.51	5.58	5.66	5.72	5.79	5.85	6.05
10	2.56	3.27	3.70	4.02	4.26	4.47	4.64	4.78	4.91	5.03	5.13	5.23	5.32	5.40	5.47	5.54	5.61	5.67	5.73	5.93
11	2.54	3.23	3.66	3.96	4.20	4.40	4.57	4.71	4.84	4.95	5.05	5.15	5.23	5.31	5.38	5.45	5.51	5.57	5.63	5.83
12	2.52	3.20	3.62	3.92	4.16	4.35	4.51	4.65	4.78	4.89	4.99	5.08	5.16	5.24	5.31	5.37	5.44	5.49	5.55	5.74
13	2.50	3.18	3.59	3.88	4.12	4.30	4.46	4.60	4.72	4.83	4.93	5.02	5.10	5.18	5.25	5.31	5.37	5.43	5.48	5.67
14	2.49	3.16	3.56	3.85	4.08	4.27	4.42	4.56	4.68	4.79	4.88	4.97	5.05	5.12	5.19	5.26	5.32	5.37	5.43	5.61
15	2.48	3.14	3.54	3.83	4.05	4.23	4.39	4.52	4.64	4.75	4.84	4.93	5.01	5.08	5.15	5.21	5.27	5.32	5.38	5.56
16	2.47	3.12	3.52	3.80	4.03	4.21	4.36	4.49	4.61	4.71	4.81	4.89	4.97	5.04	5.11	5.17	5.23	5.28	5.33	5.52
17	2.46	3.11	3.50	3.78	4.00	4.18	4.33	4.46	4.58	4.68	4.77	4.86	4.93	5.01	5.07	5.13	5.19	5.24	5.30	5.48
18	2.45	3.10	3.49	3.77	3.98	4.16	4.31	4.44	4.55	4.65	4.75	4.83	4.90	4.98	5.04	5.10	5.16	5.21	5.26	5.44
19	2.45	3.09	3.47	3.75	3.97	4.14	4.29	4.42	4.53	4.63	4.72	4.80	4.88	4.95	5.01	5.07	5.13	5.18	5.23	5.41
20	2.44	3.08	3.46	3.74	3.95	4.12	4.27	4.40	4.51	4.61	4.70	4.78	4.85	4.92	4.99	5.05	5.10	5.16	5.20	5.38
24	2.42	3.05	3.42	3.69	3.90	4.07	4.21	4.34	4.44	4.54	4.63	4.71	4.78	4.85	4.91	4.97	5.02	5.07	5.12	5.29
30	2.40	3.02	3.39	3.65	3.85	4.02	4.16	4.28	4.38	4.47	4.56	4.64	4.71	4.77	4.83	4.89	4.94	4.99	5.03	5.20
40	2.38	2.99	3.35	3.60	3.80	3.96	4.10	4.21	4.32	4.41	4.49	4.56	4.63	4.69	4.75	4.81	4.86	4.90	4.95	5.11
60	2.36	2.96	3.31	3.56	3.75	3.91	4.04	4.16	4.25	4.34	4.42	4.49	4.56	4.62	4.67	4.73	4.78	4.82	4.86	5.02
120	2.34	2.93	3.28	3.52	3.71	3.86	3.99	4.10	4.19	4.28	4.35	4.42	4.48	4.54	4.60	4.65	4.69	4.74	4.78	4.92
inf.	2.33	2.90	3.24	3.48	3.66	3.81	3.93	4.04	4.13	4.21	4.28	4.35	4.41	4.47	4.52	4.57	4.61	4.65	4.69	4.83

v = degrees of freedom for Mean Square Within.

Entries are q values for use in calculating Tukey's Honestly Significant Difference, with $\alpha = 0.10$.

$$\text{HSD} = q\sqrt{\frac{\text{MSW}}{n}}$$

where MSW comes from the ANOVA for k subgroups of size n. (See page 99 in text.)

Table E
Percentiles of the Studentized Range Distribution
95th Percentiles

	k = Number of Subgroup Averages Being Compared																			
ν	2	3	4	5	6	7	8	9	10	11	12	13	14	15	16	17	18	19	20	24
1	18.0	27.0	32.8	37.1	40.4	43.1	45.4	47.4	49.1	50.6	52.0	53.2	54.3	55.4	56.3	57.2	58.0	58.8	59.6	62.1
2	6.08	8.33	9.80	10.9	11.7	12.4	13.0	13.5	14.0	14.4	14.7	15.1	15.4	15.7	15.9	16.1	16.4	16.6	16.8	17.5
3	4.50	5.91	6.82	7.50	8.04	8.48	8.85	9.18	9.46	9.72	9.95	10.2	10.3	10.5	10.7	10.8	11.0	11.1	11.2	11.7
4	3.93	5.04	5.76	6.29	6.71	7.05	7.35	7.60	7.83	8.03	8.21	8.37	8.52	8.66	8.79	8.91	9.03	9.13	9.23	9.58
5	3.64	4.60	5.22	5.67	6.03	6.33	6.58	6.80	6.99	7.17	7.32	7.47	7.60	7.72	7.83	7.93	8.03	8.12	8.21	8.55
6	3.46	4.34	4.90	5.30	5.63	5.90	6.12	6.32	6.49	6.65	6.79	6.92	7.03	7.14	7.24	7.34	7.43	7.51	7.59	7.86
7	3.34	4.16	4.68	5.06	5.36	5.61	5.82	6.00	6.16	6.30	6.43	6.55	6.66	6.76	6.85	6.94	7.02	7.10	7.17	7.43
8	3.26	4.04	4.53	4.89	5.17	5.40	5.60	5.77	5.92	6.05	6.18	6.29	6.39	6.48	6.57	6.65	6.73	6.80	6.87	7.11
9	3.20	3.95	4.41	4.76	5.02	5.24	5.43	5.59	5.74	5.87	5.98	6.09	6.19	6.28	6.36	6.44	6.51	6.58	6.64	6.88
10	3.15	3.88	4.33	4.65	4.91	5.12	5.30	5.46	5.60	5.72	5.83	5.93	6.03	6.11	6.19	6.27	6.34	6.40	6.47	6.69
11	3.11	3.82	4.26	4.57	4.82	5.03	5.20	5.35	5.49	5.61	5.71	5.81	5.90	5.98	6.06	6.13	6.20	6.27	6.33	6.54
12	3.08	3.77	4.20	4.51	4.75	4.95	5.12	5.27	5.39	5.51	5.61	5.71	5.80	5.88	5.95	6.02	6.09	6.15	6.21	6.41
13	3.06	3.73	4.15	4.45	4.69	4.88	5.05	5.19	5.32	5.43	5.53	5.63	5.71	5.79	5.86	5.93	5.99	6.05	6.11	6.31
14	3.03	3.70	4.11	4.41	4.64	4.83	4.99	5.13	5.25	5.36	5.46	5.55	5.64	5.71	5.79	5.85	5.91	5.97	6.03	6.22
15	3.01	3.67	4.08	4.37	4.59	4.78	4.94	5.08	5.20	5.31	5.40	5.49	5.57	5.65	5.72	5.78	5.85	5.90	5.96	6.15
16	3.00	3.65	4.05	4.33	4.56	4.74	4.90	5.03	5.15	5.26	5.35	5.44	5.52	5.59	5.66	5.73	5.79	5.84	5.90	6.08
17	2.98	3.63	4.02	4.30	4.52	4.70	4.86	4.99	5.11	5.21	5.31	5.39	5.47	5.54	5.61	5.67	5.73	5.79	5.84	6.02
18	2.97	3.61	4.00	4.28	4.49	4.67	4.82	4.96	5.07	5.17	5.27	5.35	5.43	5.50	5.57	5.63	5.69	5.74	5.79	5.97
19	2.96	3.59	3.98	4.25	4.47	4.65	4.79	4.92	5.04	5.14	5.23	5.31	5.39	5.46	5.53	5.59	5.65	5.70	5.75	5.93
20	2.95	3.58	3.96	4.23	4.45	4.62	4.77	4.90	5.01	5.11	5.20	5.28	5.36	5.43	5.49	5.55	5.61	5.66	5.71	5.89
24	2.92	3.53	3.90	4.17	4.37	4.54	4.68	4.81	4.92	5.01	5.10	5.18	5.25	5.32	5.38	5.44	5.49	5.55	5.47	5.76
30	2.89	3.49	3.85	4.10	4.30	4.46	4.60	4.72	4.82	4.92	5.00	5.08	5.15	5.21	5.27	5.33	5.38	5.43	5.47	5.64
40	2.86	3.44	3.79	4.04	4.23	4.39	4.52	4.63	4.73	4.82	4.90	4.98	5.04	5.11	5.16	5.22	5.27	5.31	5.36	5.51
60	2.83	3.40	3.74	3.98	4.16	4.31	4.44	4.55	4.65	4.73	4.81	4.88	4.94	5.00	5.06	5.11	5.15	5.20	5.24	5.39
120	2.80	3.36	3.68	3.92	4.10	4.24	4.36	4.47	4.56	4.64	4.71	4.78	4.84	4.90	4.95	5.00	5.04	5.09	5.13	5.27
inf.	2.77	3.31	3.63	3.86	4.03	4.17	4.29	4.39	4.47	4.55	4.62	4.68	4.74	4.80	4.85	4.89	4.93	4.97	5.01	5.14

ν = degrees of freedom for estimate of standard deviation.

Entries are q values for computing Tukey's Honestly Significant Difference with $\alpha = 0.05$.

$$HSD = q\sqrt{\frac{MSW}{n}}$$

where MSW comes from the ANOVA for k subgroups of size n.

Table E
Percentiles of the Studentized Range Distribution
99th Percentiles

k = Number of Subgroup Averages Being Compared

v	2	3	4	5	6	7	8	9	10	11	12	13	14	15	16	17	18	19	20	24
1	90.0	135.	164.	186.	202.	216.	227.	237.	246.	253.	260.	266.	272.	277.	282.	286.	290.	294.	298.	311.
2	14.0	19.0	22.3	24.7	26.6	28.2	29.5	30.7	31.7	32.6	33.4	34.1	34.8	35.4	36.0	36.5	37.0	37.5	37.9	39.5
3	8.26	10.6	12.2	13.3	14.2	15.0	15.6	16.2	16.7	17.1	17.5	17.9	18.2	18.5	18.8	19.1	19.3	19.5	19.8	20.5
4	6.51	8.12	9.17	9.96	10.6	11.1	11.5	11.9	12.3	12.6	12.8	13.1	13.3	13.5	13.7	13.9	14.1	14.2	14.4	14.9
5	5.70	6.97	7.80	8.42	8.91	9.32	9.67	9.97	10.2	10.5	10.7	10.9	11.1	11.2	11.4	11.6	11.7	11.8	11.9	12.5
6	5.24	6.33	7.03	7.56	7.97	8.32	8.61	8.87	9.10	9.30	9.49	9.65	9.81	9.95	10.1	10.2	10.3	10.4	10.5	10.9
7	4.95	5.92	6.54	7.01	7.37	7.68	7.94	8.17	8.37	8.55	8.71	8.86	9.00	9.12	9.24	9.35	9.46	9.55	9.65	10.0
8	4.74	5.63	6.20	6.63	6.96	7.24	7.47	7.68	7.87	8.03	8.18	8.31	8.44	8.55	8.66	8.76	8.85	8.94	9.03	9.32
9	4.60	5.43	5.96	6.35	6.66	6.91	7.13	7.32	7.49	7.65	7.78	7.91	8.03	8.13	8.23	8.32	8.41	8.49	8.57	8.85
10	4.48	5.27	5.77	6.14	6.43	6.67	6.87	7.05	7.21	7.36	7.48	7.60	7.71	7.81	7.91	7.99	8.07	8.15	8.22	8.48
11	4.39	5.14	5.62	5.97	6.25	6.48	6.67	6.84	6.99	7.13	7.25	7.36	7.46	7.56	7.65	7.73	7.81	7.88	7.95	8.20
12	4.32	5.04	5.50	5.84	6.10	6.32	6.51	6.67	6.81	6.94	7.06	7.17	7.26	7.36	7.44	7.52	7.59	7.66	7.73	7.96
13	4.26	4.96	5.40	5.73	5.98	6.19	6.37	6.53	6.67	6.79	6.90	7.01	7.10	7.19	7.27	7.34	7.42	7.48	7.55	7.78
14	4.21	4.89	5.32	5.63	5.88	6.08	6.26	6.41	6.54	6.66	6.77	6.87	6.96	7.05	7.12	7.20	7.27	7.33	7.39	7.62
15	4.17	4.83	5.25	5.56	5.80	5.99	6.16	6.31	6.44	6.55	6.66	6.76	6.84	6.93	7.00	7.07	7.14	7.20	7.26	7.48
16	4.13	4.78	5.19	5.49	5.72	5.92	6.08	6.22	6.35	6.46	6.56	6.66	6.74	6.82	6.90	6.97	7.03	7.09	7.15	7.36
17	4.10	4.74	5.14	5.43	5.66	5.85	6.01	6.15	6.27	6.38	6.48	6.57	6.66	6.73	6.80	6.87	6.94	7.00	7.05	7.26
18	4.07	4.70	5.09	5.38	5.60	5.79	5.94	6.08	6.20	6.31	6.41	6.50	6.58	6.65	6.72	6.79	6.85	6.91	6.96	7.17
19	4.05	4.67	5.05	5.33	5.55	5.73	5.89	6.02	6.14	6.25	6.34	6.43	6.51	6.58	6.65	6.72	6.78	6.84	6.89	7.08
20	4.02	4.64	5.02	5.29	5.51	5.69	5.84	5.97	6.09	6.19	6.29	6.37	6.45	6.52	6.59	6.65	6.71	6.76	6.82	7.01
24	3.96	4.54	4.91	5.17	5.37	5.54	5.69	5.81	5.92	6.02	6.11	6.19	6.26	6.33	6.39	6.45	6.51	6.56	6.61	6.79
30	3.89	4.45	4.80	5.05	5.24	5.40	5.54	5.65	5.76	5.85	5.93	6.01	6.08	6.14	6.20	6.26	6.31	6.36	6.41	6.57
40	3.82	4.37	4.70	4.93	5.11	5.27	5.39	5.50	5.60	5.69	5.77	5.84	5.90	5.96	6.02	6.07	6.12	6.17	6.21	6.36
60	3.76	4.28	4.60	4.82	4.99	5.13	5.25	5.36	5.45	5.53	5.60	5.67	5.73	5.79	5.84	5.89	5.93	5.98	6.02	6.16
120	3.70	4.20	4.50	4.71	4.87	5.01	5.12	5.21	5.30	5.38	5.44	5.51	5.56	5.61	5.66	5.71	5.75	5.79	5.83	5.96
Inf.	3.64	4.12	4.40	4.60	4.76	4.88	4.99	5.08	5.16	5.23	5.29	5.35	5.40	5.45	5.49	5.54	5.57	5.61	5.65	5.77

v = degrees of freedom for estimate of standard deviation.

Entries are q values for computing Tukey's Honestly Significant Difference with $\alpha = 0.01$.

$$HSD = q\sqrt{\frac{MSW}{n}}$$

where MSW comes from the ANOVA for k subgroups of size n.

Table F

Percentiles for the F-Distribution

90th Percentiles

$\alpha = 0.10$

Tabled values are the 90th percentiles of the F-distribution. If the F-ratio follows an F-distribution, then the calculated value of the F-ratio will be less than the tabled value approximately 90 percent of the time, and it will exceed the tabled value approximately 10 percent of the time.

ν_1 = Degrees of Freedom for the Numerator of the F-ratio

ν_2	1	2	3	4	5	6	7	8	9	10	12	15	20	24	30	40	60	120	inf.
1	39.9	49.5	53.6	55.8	57.2	58.2	58.9	59.4	59.9	60.2	60.7	61.2	61.7	62.0	62.3	62.5	62.8	63.1	63.3
2	8.53	9.00	9.16	9.24	9.29	9.33	9.35	9.37	9.38	9.39	9.41	9.42	9.44	9.45	9.46	9.47	9.47	9.48	9.49
3	5.54	5.46	5.39	5.34	5.31	5.28	5.27	5.25	5.24	5.23	5.22	5.20	5.18	5.18	5.17	5.16	5.15	5.14	5.13
4	4.54	4.32	4.19	4.11	4.05	4.01	3.98	3.95	3.94	3.92	3.90	3.87	3.84	3.83	3.82	3.80	3.79	3.78	3.76
5	4.06	3.78	3.62	3.52	3.45	3.40	3.37	3.34	3.32	3.30	3.27	3.24	3.21	3.19	3.17	3.16	3.14	3.12	3.10
6	3.78	3.46	3.29	3.18	3.11	3.05	3.01	2.98	2.96	2.94	2.90	2.87	2.84	2.82	2.80	2.78	2.76	2.74	2.72
7	3.59	3.26	3.07	2.96	2.88	2.83	2.78	2.75	2.72	2.70	2.67	2.63	2.59	2.58	2.56	2.54	2.51	2.49	2.47
8	3.46	3.11	2.92	2.81	2.73	2.67	2.62	2.59	2.56	2.54	2.50	2.46	2.42	2.40	2.38	2.36	2.34	2.32	2.29
9	3.36	3.01	2.81	2.69	2.61	2.55	2.51	2.47	2.44	2.42	2.38	2.34	2.30	2.28	2.25	2.23	2.21	2.18	2.16
10	3.28	2.92	2.73	2.61	2.52	2.46	2.41	2.38	2.35	2.32	2.28	2.24	2.20	2.18	2.16	2.13	2.11	2.08	2.06
11	3.23	2.86	2.66	2.54	2.45	2.39	2.34	2.30	2.27	2.25	2.21	2.17	2.12	2.10	2.08	2.05	2.03	2.00	1.97
12	3.18	2.81	2.61	2.48	2.39	2.33	2.28	2.24	2.21	2.19	2.15	2.10	2.06	2.04	2.01	1.99	1.96	1.93	1.90
13	3.14	2.76	2.56	2.43	2.35	2.28	2.23	2.20	2.16	2.14	2.10	2.05	2.01	1.98	1.96	1.93	1.90	1.88	1.85
14	3.10	2.73	2.52	2.39	2.31	2.24	2.19	2.15	2.12	2.10	2.05	2.01	1.96	1.94	1.91	1.89	1.86	1.83	1.80
15	3.07	2.70	2.49	2.36	2.27	2.21	2.16	2.12	2.09	2.06	2.02	1.97	1.92	1.90	1.87	1.85	1.82	1.79	1.76
16	3.05	2.67	2.46	2.33	2.24	2.18	2.13	2.09	2.06	2.03	1.99	1.94	1.89	1.87	1.84	1.81	1.78	1.75	1.72
18	3.01	2.62	2.42	2.29	2.20	2.13	2.08	2.04	2.00	1.98	1.93	1.89	1.84	1.81	1.78	1.75	1.72	1.69	1.66
20	2.97	2.59	2.38	2.25	2.16	2.09	2.04	2.00	1.96	1.94	1.89	1.84	1.79	1.77	1.74	1.71	1.68	1.64	1.61
22	2.95	2.56	2.35	2.22	2.13	2.06	2.01	1.97	1.93	1.90	1.86	1.81	1.76	1.73	1.70	1.67	1.64	1.60	1.57
24	2.93	2.54	2.33	2.19	2.10	2.04	1.98	1.94	1.91	1.88	1.83	1.78	1.73	1.70	1.67	1.64	1.61	1.57	1.53
26	2.91	2.52	2.31	2.17	2.08	2.01	1.96	1.92	1.88	1.86	1.81	1.76	1.71	1.68	1.65	1.61	1.58	1.54	1.50
28	2.89	2.50	2.29	2.16	2.06	2.00	1.94	1.90	1.87	1.84	1.79	1.74	1.69	1.66	1.63	1.59	1.56	1.52	1.48
30	2.88	2.49	2.28	2.14	2.05	1.98	1.93	1.88	1.85	1.82	1.77	1.72	1.67	1.64	1.61	1.57	1.54	1.50	1.46
40	2.84	2.44	2.23	2.09	2.00	1.93	1.87	1.83	1.79	1.76	1.71	1.66	1.61	1.57	1.54	1.51	1.47	1.42	1.38
60	2.79	2.39	2.18	2.04	1.95	1.87	1.82	1.77	1.74	1.71	1.66	1.60	1.54	1.51	1.48	1.44	1.40	1.35	1.29
120	2.75	2.35	2.13	1.99	1.90	1.82	1.77	1.72	1.68	1.65	1.60	1.55	1.48	1.45	1.41	1.37	1.32	1.26	1.19
inf.	2.71	2.30	2.08	1.94	1.85	1.77	1.72	1.67	1.63	1.60	1.55	1.49	1.42	1.38	1.34	1.30	1.24	1.17	1.00

ν_2 = Degrees of Freedom for the denominator of the F-ratio.

Table F

Percentiles for the F-Distribution

95th Percentiles

$\alpha = 0.05$

Tabled values are the 95th percentiles of the F-distribution. If the F-ratio follows an F-distribution, then the calculated value of the F-ratio will be less than the tabled value approximately 95 percent of the time, and it will exceed the tabled value approximately 5 percent of the time.

v_1 = Degrees of Freedom for the Numerator of the F-ratio

v_2	1	2	3	4	5	6	7	8	9	10	12	15	20	24	30	40	60	120	inf.
1	161.4	199.5	215.7	224.6	230.2	234.0	236.8	238.9	240.5	241.9	243.9	245.9	248.0	249.1	250.1	251.1	252.2	253.3	254.3
2	18.51	19.00	19.16	19.25	19.30	19.33	19.35	19.37	19.38	19.40	19.41	19.43	19.45	19.45	19.46	19.47	19.48	19.49	19.50
3	10.13	9.55	9.28	9.12	9.01	8.94	8.89	8.85	8.81	8.79	8.74	8.70	8.66	8.64	8.62	8.59	8.57	8.55	8.53
4	7.71	6.94	6.59	6.39	6.26	6.16	6.09	6.04	6.00	5.96	5.91	5.86	5.80	5.77	5.75	5.72	5.69	5.66	5.63
5	6.61	5.79	5.41	5.19	5.05	4.95	4.88	4.82	4.77	4.74	4.68	4.62	4.56	4.53	4.50	4.46	4.43	4.40	4.36
6	5.99	5.14	4.76	4.53	4.39	4.28	4.21	4.15	4.10	4.06	4.00	3.94	3.87	3.84	3.81	3.77	3.74	3.70	3.67
7	5.59	4.74	4.35	4.12	3.97	3.87	3.79	3.73	3.68	3.64	3.57	3.51	3.44	3.41	3.38	3.34	3.30	3.27	3.23
8	5.32	4.46	4.07	3.84	3.69	3.58	3.50	3.44	3.39	3.35	3.28	3.22	3.15	3.12	3.08	3.04	3.01	2.97	2.93
9	5.12	4.26	3.86	3.63	3.48	3.37	3.29	3.23	3.18	3.14	3.07	3.01	2.94	2.90	2.86	2.83	2.79	2.75	2.71
10	4.96	4.10	3.71	3.48	3.33	3.22	3.14	3.07	3.02	2.98	2.91	2.85	2.77	2.74	2.70	2.66	2.62	2.58	2.54
11	4.84	3.98	3.59	3.36	3.20	3.09	3.01	2.95	2.90	2.85	2.79	2.72	2.65	2.61	2.57	2.53	2.49	2.45	2.40
12	4.75	3.89	3.49	3.26	3.11	3.00	2.91	2.85	2.80	2.75	2.69	2.62	2.54	2.51	2.47	2.43	2.38	2.34	2.30
13	4.67	3.81	3.41	3.18	3.03	2.92	2.83	2.77	2.71	2.67	2.60	2.53	2.46	2.42	2.38	2.34	2.30	2.25	2.21
14	4.60	3.74	3.34	3.11	2.96	2.85	2.76	2.70	2.65	2.60	2.53	2.46	2.39	2.35	2.31	2.27	2.22	2.18	2.13
15	4.54	3.68	3.29	3.06	2.90	2.79	2.71	2.64	2.59	2.54	2.48	2.40	2.33	2.29	2.25	2.20	2.16	2.11	2.07
16	4.49	3.63	3.24	3.01	2.85	2.74	2.66	2.59	2.54	2.49	2.42	2.35	2.28	2.24	2.19	2.15	2.11	2.06	2.01
18	4.41	3.55	3.16	2.93	2.77	2.66	2.58	2.51	2.46	2.41	2.34	2.27	2.19	2.15	2.11	2.06	2.02	1.97	1.92
20	4.35	3.49	3.10	2.87	2.71	2.60	2.51	2.45	2.39	2.35	2.28	2.20	2.12	2.08	2.04	1.99	1.95	1.90	1.84
22	4.30	3.44	3.05	2.82	2.66	2.55	2.46	2.40	2.34	2.30	2.23	2.15	2.07	2.03	1.98	1.94	1.89	1.84	1.78
24	4.26	3.40	3.01	2.78	2.62	2.51	2.42	2.36	2.30	2.25	2.18	2.11	2.03	1.98	1.94	1.89	1.84	1.79	1.73
26	4.23	3.37	2.98	2.74	2.59	2.47	2.39	2.32	2.27	2.22	2.15	2.07	1.99	1.95	1.90	1.85	1.80	1.75	1.69
28	4.20	3.34	2.95	2.71	2.56	2.45	2.36	2.29	2.24	2.19	2.12	2.04	1.96	1.91	1.87	1.82	1.77	1.71	1.65
30	4.17	3.32	2.92	2.69	2.53	2.42	2.33	2.27	2.21	2.16	2.09	2.01	1.93	1.89	1.84	1.79	1.74	1.68	1.62
40	4.08	3.23	2.84	2.61	2.45	2.34	2.25	2.18	2.12	2.08	2.00	1.92	1.84	1.79	1.74	1.69	1.64	1.58	1.51
60	4.00	3.15	2.76	2.53	2.37	2.25	2.17	2.10	2.04	1.99	1.92	1.84	1.75	1.70	1.65	1.59	1.53	1.47	1.39
120	3.92	3.07	2.68	2.45	2.29	2.17	2.09	2.02	1.96	1.91	1.83	1.75	1.66	1.61	1.55	1.50	1.43	1.35	1.25
inf.	3.84	3.00	2.60	2.37	2.21	2.10	2.01	1.94	1.88	1.83	1.75	1.67	1.57	1.52	1.46	1.39	1.32	1.22	1.00

v_2 = Degrees of Freedom for the denominator of the F-ratio.

Table F

Percentiles for the F-Distribution

99th Percentiles
$\alpha = 0.01$

Tabled values are the 99th percentiles of the F-distribution. If the F-ratio follows an F-distribution, then the calculated value of the F-ratio will be less than the tabled value approximately 99 percent of the time, and it will exceed the tabled value approximately 1 percent of the time.

ν_1 = Degrees of Freedom for the Numerator of the F-ratio

ν_2	1	2	3	4	5	6	7	8	9	10	12	15	20	24	30	40	60	120	inf.
1	4052.	5000.	5403.	5625.	5764.	5859.	5928.	5981.	6022.	6056.	6106.	6157.	6209.	6235.	6261.	6287.	6313.	6339.	6366.
2	98.5	99.0	99.2	99.3	99.3	99.3	99.4	99.4	99.4	99.4	99.4	99.4	99.4	99.5	99.5	99.5	99.5	99.5	99.5
3	34.1	30.8	29.5	28.7	28.2	27.9	27.7	27.5	27.4	27.2	27.1	26.9	26.7	26.6	26.5	26.4	26.3	26.2	26.1
4	21.2	18.0	16.7	16.0	15.5	15.2	15.0	14.8	14.7	14.6	14.4	14.2	14.0	13.9	13.8	13.8	13.7	13.6	13.5
5	16.3	13.3	12.1	11.4	11.0	10.7	10.5	10.3	10.2	10.1	9.89	9.72	9.55	9.47	9.38	9.29	9.20	9.11	9.02
6	13.8	10.9	9.78	9.15	8.75	8.47	8.26	8.10	7.98	7.87	7.72	7.56	7.40	7.31	7.23	7.14	7.06	6.97	6.88
7	12.3	9.55	8.45	7.85	7.46	7.19	6.99	6.84	6.72	6.62	6.47	6.31	6.16	6.07	5.99	5.91	5.82	5.74	5.65
8	11.3	8.65	7.59	7.01	6.63	6.37	6.18	6.03	5.91	5.81	5.67	5.52	5.36	5.28	5.20	5.12	5.03	4.95	4.86
9	10.6	8.02	6.99	6.42	6.06	5.80	5.61	5.47	5.35	5.26	5.11	4.96	4.81	4.73	4.65	4.57	4.48	4.40	4.31
10	10.0	7.56	6.55	5.99	5.64	5.39	5.20	5.06	4.94	4.85	4.71	4.56	4.41	4.33	4.25	4.17	4.08	4.00	3.91
11	9.65	7.21	6.22	5.67	5.32	5.07	4.89	4.74	4.63	4.54	4.40	4.25	4.10	4.02	3.94	3.86	3.78	3.69	3.60
12	9.33	6.93	5.95	5.41	5.06	4.82	4.64	4.50	4.39	4.30	4.16	4.01	3.86	3.78	3.70	3.62	3.54	3.45	3.36
13	9.07	6.70	5.74	5.21	4.86	4.62	4.44	4.30	4.19	4.10	3.96	3.82	3.66	3.59	3.51	3.43	3.34	3.25	3.17
14	8.86	6.51	5.56	5.04	4.69	4.46	4.28	4.14	4.03	3.94	3.80	3.66	3.51	3.43	3.35	3.27	3.18	3.09	3.00
15	8.68	6.36	5.42	4.89	4.56	4.32	4.14	4.00	3.89	3.80	3.67	3.52	3.37	3.29	3.21	3.13	3.05	2.96	2.87
16	8.53	6.23	5.29	4.77	4.44	4.20	4.03	3.89	3.78	3.69	3.55	3.41	3.26	3.18	3.10	3.02	2.93	2.84	2.75
18	8.29	6.01	5.09	4.58	4.25	4.01	3.84	3.71	3.60	3.51	3.37	3.23	3.08	3.00	2.92	2.84	2.75	2.66	2.57
20	8.10	5.85	4.94	4.43	4.10	3.87	3.70	3.56	3.46	3.37	3.23	3.09	2.94	2.86	2.78	2.69	2.61	2.52	2.42
22	7.95	5.72	4.82	4.31	3.99	3.76	3.59	3.45	3.35	3.26	3.12	2.98	2.83	2.75	2.67	2.58	2.50	2.40	2.31
24	7.82	5.61	4.72	4.22	3.90	3.67	3.50	3.36	3.26	3.17	3.03	2.89	2.74	2.66	2.58	2.49	2.40	2.31	2.21
26	7.72	5.53	4.64	4.14	3.82	3.59	3.42	3.29	3.18	3.09	2.96	2.81	2.66	2.58	2.50	2.42	2.33	2.23	2.13
28	7.64	5.45	4.57	4.07	3.75	3.53	3.36	3.23	3.12	3.03	2.90	2.75	2.60	2.52	2.44	2.35	2.26	2.17	2.06
30	7.56	5.39	4.51	4.02	3.70	3.47	3.30	3.17	3.07	2.98	2.84	2.70	2.55	2.47	2.39	2.30	2.21	2.11	2.01
40	7.31	5.18	4.31	3.83	3.51	3.29	3.12	2.99	2.89	2.80	2.66	2.52	2.37	2.29	2.20	2.11	2.02	1.92	1.80
60	7.08	4.98	4.13	3.65	3.34	3.12	2.95	2.82	2.72	2.63	2.50	2.35	2.20	2.12	2.03	1.94	1.84	1.73	1.60
120	6.85	4.79	3.95	3.48	3.17	2.96	2.79	2.66	2.56	2.47	2.34	2.19	2.03	1.95	1.86	1.76	1.66	1.53	1.38
inf.	6.63	4.61	3.78	3.32	3.02	2.80	2.64	2.51	2.41	2.32	2.18	2.04	1.88	1.79	1.70	1.59	1.47	1.32	1.00

ν_2 = Degrees of Freedom for the denominator of the F-ratio.

Table G
Estimated Percentiles of a Modulus Ratio Statistic: Mod F (j, p, k-1)

Given the sums of squares for (k-1) orthogonal contrasts,
arrange the (k-1) SS values in ascending order of magnitude and label them as

$$\{ SS_1 \leq SS_2 \leq SS_3 \leq \ldots \leq SS_{k-1} \}.$$

Also let MSE_p denote the average of the p smallest of these (k-1) Sums of Squares:

$$MSE_p = \frac{SS_1 + SS_2 + \ldots + SS_p}{p}.$$

Then, under the assumption that all of the Sums of Squares are estimates of the same quantity, the Distribution of the Mod F Ratio:

$$\text{Mod } F(j, p, k-1) = \frac{SS_j}{MSE_p} \qquad \text{for } j = k-1, k-2, \ldots, p+1.$$

will have (approximately) the percentiles shown in the following tables. If one or more of the Sums of Squares is inflated due to some nonrandom differences between the subgroups, then the Mod F Ratio involving this inflated Sum of Squares will tend to be inflated. (The exception comes when some of the pooled Sums of Squares are also inflated by nonrandom differences between the subgroups.)

The tables are given for 7, 11, 15, 19, 23, 27, and 31 contrasts.
{ (k-1) = 7, 11, 15, 19, 23, 27, and 31. }

The number of Sums of Squares pooled to obtain the MSE term ranges from 2 to 10.
{ p = 2 to 10. }

For each combination of (k-1) and p, the rank of the numerator of the Mod F Ratio ranges from (k-1) down to (p+1). { j = (k-1) down to (p+1). }

Percentiles estimated are:

$\text{Mod } F_{.50}$ = the Median of the Distribution of the Mod F Ratio.
$\text{Mod } F_{.75}$ = the Upper Quartile of the Dist. of the Mod F Ratio.
$\text{Mod } F_{.90}$ = the Upper Decile of the Dist. of the Mod F Ratio.
$\text{Mod } F_{.95}$ = the Upper Five Percent of the Dist. of the Mod F Ratio.

The estimates include an uncertainty figure which is equal to ± 3 estimated standard errors of the estimate.

The 99th Percentiles of these Distributions were not obtained because the uncertainties for these estimated percentiles were so large that the values were useless.

Table G
Estimated Percentiles of a Modulus Ratio Statistic: Mod F (j, p, k-1)

k-1 = total number of orthogonal contrasts in the set of contrasts
p = number of contrasts pooled to form the MSE term
j = rank of the numerator Sum of Squares in the Mod F Ratio
1-α = percentile of the Mod F Distribution

(k-1) = 7

p	1-α	j=7	j=6	j=5	j=4	j=3
2	.50	49.1 ± 2.1	24.4 ± 0.9	13.5 ± 0.5	7.27 ± .20	3.38 ± .08
	.75	137 ± 8	64.5 ± 4.3	33.8 ± 1.9	16.8 ± 0.7	6.72 ± .21
	.90	413 ± 43	195 ± 19	96.2 ± 13.5	44.6 ± 4.0	16.1 ± 1.6
	.95	876 ± 120	428 ± 76	217 ± 44	93.0 ± 11.2	31.7 ± 4.2
3	.50	23.1 ± 0.7	11.6 ± 0.4	6.29 ± .18	3.43 ± .08	
	.75	50.5 ± 2.5	23.5 ± 0.9	12.2 ± 0.5	5.58 ± .23	
	.90	117 ± 9	50.6 ± 2.8	25.2 ± 1.5	10.8 ± 0.7	
	.95	208 ± 25	90.0 ± 9.2	42.5 ± 3.3	17.5 ± 1.3	
4	.50	12.6 ± 0.6	6.24 ± .20	3.56 ± .05		
	.75	23.9 ± 1.0	10.9 ± 0.4	5.31 ± .17		
	.90	46.4 ± 2.8	20.2 ± 1.4	8.94 ± .41		
	.95	71.8 ± 5.5	30.6 ± 2.5	12.8 ± 0.8		
5	.50	7.69 ± .28	3.89 ± .09			
	.75	13.2 ± 0.5	5.81 ± .21			
	.90	23.1 ± 1.1	9.16 ± .35			
	.95	33.3 ± 2.2	12.8 ± 0.8			
6	.50	4.82 ± .08				
	.75	7.39 ± .23				
	.90	11.9 ± 0.4				
	.95	16.1 ± 0.9				

Table G
Estimated Percentiles of a Modulus Ratio Statistic: Mod F (j, p, k-1)

(k-1) = 11

p	1-α	j=11	j=10	j=9	j=8	j=7	j=6	j=5	j=4	j=3
2	.50	154 ± 7	91.7 ± 4.5	59.6 ± 3.0	40.3 ± 2.0	28.2 ± 1.3	18.7 ± 0.9	11.6 ± 0.5	6.70 ± .24	3.28 ± .10
	.75	465 ± 30	260 ± 15	170 ± 9	116 ± 8	75.9 ± 4.4	48.6 ± 3.3	29.1 ± 1.8	15.3 ± 1.0	6.41 ± .30
	.90	1432 ± 130	801 ± 90	509 ± 53	343 ± 38	229 ± 20	142 ± 12	82.5 ± 7.6	42.7 ± 3.3	16.4 ± 1.1
	.95	2993 ± 369	1663 ± 230	1087 ± 162	731 ± 113	476 ± 69	306 ± 50	180 ± 23	90.6 ± 11.3	32.7 ± 3.7
3	.50	72.7 ± 3.4	42.4 ± 1.9	28.3 ± 1.3	19.1 ± 0.8	13.0 ± 0.5	8.65 ± .26	5.51 ± .14	3.25 ± .06	
	.75	170 ± 10	96.1 ± 5.0	61.3 ± 2.8	41.6 ± 1.7	27.6 ± 1.2	17.4 ± 0.7	10.5 ± 0.4	5.26 ± .17	
	.90	395 ± 29	230 ± 20	148 ± 12	97.2 ± 8.5	63.4 ± 6.6	38.5 ± 3.4	22.1 ± 1.8	10.2 ± 0.7	
	.95	735 ± 78	412 ± 50	268 ± 28	176 ± 22	116 ± 16	67.6 ± 9.6	39.0 ± 4.9	16.0 ± 1.7	
4	.50	41.8 ± 1.9	24.4 ± 1.0	16.3 ± 0.6	11.2 ± 0.5	7.68 ± .29	5.06 ± .15	3.31 ± .06		
	.75	84.9 ± 3.6	47.8 ± 2.2	30.7 ± 1.2	20.3 ± 0.9	13.3 ± 0.5	8.37 ± .31	4.85 ± .14		
	.90	180 ± 11	98.1 ± 6.6	59.8 ± 3.9	39.6 ± 2.6	24.8 ± 1.2	14.6 ± 0.9	7.80 ± .34		
	.95	286 ± 19	156 ± 11	95.5 ± 7.9	60.2 ± 4.4	37.5 ± 2.6	22.1 ± 2.1	11.3 ± 1.0		
5	.50	27.1 ± 1.0	15.8 ± 0.5	10.4 ± 0.3	7.17 ± .18	4.95 ±.08	3.37 ±.05			
	.75	50.1 ± 1.8	28.0 ± 0.9	17.7 ± 0.5	11.7 ± 0.3	7.66 ± .22	4.70 ±.10			
	.90	94.5 ± 4.9	49.4 ± 2.2	31.0 ± 1.5	19.6 ± 1.0	12.4 ± 0.6	6.91 ± .23			
	.95	140 ± 12	71.1 ± 4.3	44.2 ± 2.7	27.9 ± 1.9	17.1 ± 1.1	9.20 ± .39			
6	.50	18.8 ± 0.5	11.0 ± 0.2	7.23 ± .12	4.99 ± .11	3.47 ±.05				
	.75	32.4 ± 1.1	18.1 ± 0.6	11.4 ± 0.3	7.35 ± .19	4.72 ± .10				
	.90	56.2 ± 2.7	30.2 ± 1.6	18.4 ± 0.9	11.4 ± 0.4	6.73 ± .16				
	.95	79.5 ± 5.1	42.8 ± 2.5	24.9 ± 1.4	15.2 ± 0.9	8.76 ± .35				
7	.50	13.2 ± 0.3	7.83 ± .16	5.14 ± .11	3.58 ±.04					
	.75	21.7 ± 0.6	12.0 ± 0.3	7.57 ± .16	4.81 ±.08					
	.90	36.0 ± 1.4	18.7 ± 0.8	11.1 ± 0.4	6.57 ± .13					
	.95	48.9 ± 2.7	24.9 ± 1.3	14.7 ± 0.7	8.14 ± .18					
8	.50	9.70 ± .23	5.64 ±.09	3.80 ±.04						
	.75	15.5 ± 0.4	8.23 ± .16	5.11 ±.09						
	.90	23.6 ± 0.8	12.1 ± 0.4	7.00 ± .17						
	.95	31.2 ± 1.4	15.5 ± 0.5	8.78 ± .26						
9	.50	7.05 ± .14	4.21 ±.06							
	.75	10.6 ± 0.2	5.67 ±.09							
	.90	15.9 ± 0.5	7.71 ± .14							
	.95	20.4 ± 1.0	9.48 ± .32							
10	.50	5.18 ± .11								
	.75	7.25 ± .22								
	.90	10.4 ± 0.4								
	.95	13.2 ± 0.6								

Table G
Estimated Percentiles of a Modulus Ratio Statistic: Mod F (j, p, k-1)

(k-1) = 15

p	1-α	j=15	j=14	j=13	j=12	j=11	j=10	j=9	j=8
2	.50	316 ± 17	196 ± 9	140 ± 7	105 ± 6	78.6 ± 4.7	60.2 ± 3.3	46.0 ± 2.5	33.6 ± 1.8
	.75	939 ± 64	584 ± 41	408 ± 30	302 ± 21	232 ± 17	173 ± 14	128 ± 8	91.9 ± 5.6
	.90	3039 ± 356	1884 ± 181	1284 ± 143	961 ± 90	700 ± 69	516 ± 45	394 ± 36	281 ± 25
	.95	6759 ± 1111	4368 ± 861	3015 ± 532	2183 ± 353	1634 ± 222	1181 ± 176	880 ± 142	630 ± 88
3	.50	154 ± 8	97.0 ± 4.9	69.5 ± 3.4	51.6 ± 2.3	39.0 ± 1.6	29.6 ± 1.2	22.2 ± 0.9	16.4 ± 0.6
	.75	367 ± 24	228 ± 14	159 ± 10	116 ± 7	86.4 ± 5.3	65.0 ± 3.3	48.2 ± 2.5	34.6 ± 1.8
	.90	880 ± 68	543 ± 40	377 ± 26	272 ± 21	205 ± 17	152 ± 11	112 ± 9	77.2 ± 4.5
	.95	1553 ± 169	949 ± 98	670 ± 74	481 ± 40	359 ± 31	262 ± 19	189 ± 12	137 ± 11
4	.50	91.3 ± 3.9	57.4 ± 2.1	41.5 ± 1.5	30.6 ± 1.1	23.1 ± 0.8	17.3 ± 0.5	13.0 ± 0.4	9.58 ± .20
	.75	192 ± 10	116 ± 6	82.0 ± 4.1	60.7 ± 2.4	45.4 ± 2.4	33.8 ± 1.5	25.0 ± 1.1	17.7 ± 0.8
	.90	399 ± 32	241 ± 20	164 ± 10	118 ± 8	87.4 ± 5.1	63.6 ± 4.1	47.6 ± 2.8	33.2 ± 2.0
	.95	624 ± 57	371 ± 37	259 ± 27	187 ± 17	136 ± 11	102 ± 9	72.5 ± 6.2	50.4 ± 3.9
5	.50	58.5 ± 1.3	37.2 ± 0.9	26.6 ± 0.7	19.7 ± 0.5	15.0 ± 0.4	11.3 ± 0.2	8.53 ± .23	6.32 ± .14
	.75	112 ± 5	68.0 ± 3.3	47.7 ± 2.1	34.9 ± 1.5	26.0 ± 1.0	19.3 ± 0.7	14.2 ± 0.4	10.0 ± 0.3
	.90	206 ± 10	125 ± 7	86.7 ± 5.0	62.9 ± 3.9	46.2 ± 2.6	33.9 ± 1.9	24.4 ± 1.3	16.8 ± 0.9
	.95	326 ± 22	185 ± 14	133 ± 11	93.7 ± 6.9	67.4 ± 5.4	48.1 ± 3.6	34.7 ± 2.5	24.1 ± 1.8
6	.50	41.2 ± 1.3	25.8 ± 0.6	18.5 ± 0.5	13.6 ± 0.4	10.4 ± 0.3	7.89 ± .20	5.96 ± .12	4.43 ± .09
	.75	73.8 ± 2.7	45.4 ± 1.8	31.4 ± 1.0	22.7 ± 0.8	17.0 ± 0.5	12.5 ± 0.3	8.93 ± .27	6.40 ± .21
	.90	136 ± 7	78.5 ± 4.2	54.3 ± 3.5	38.6 ± 2.0	27.9 ± 1.7	20.1 ± 0.9	14.0 ± 0.6	9.63 ± .41
	.95	197 ± 15	112 ± 7	75.8 ± 6.6	54.3 ± 3.2	38.8 ± 2.9	27.8 ± 2.1	19.5 ± 1.3	12.8 ± 0.8
7	.50	30.4 ± 0.9	19.2 ± 0.4	13.5 ± 0.3	10.1 ± 0.2	7.62 ± .17	5.84 ± .11	4.44 ± .06	3.33 ± .03
	.75	51.4 ± 1.2	31.5 ± 0.7	21.9 ± 0.6	15.8 ± 0.5	11.6 ± 0.3	8.56 ± .20	6.15 ± .11	4.38 ± .06
	.90	85.6 ± 2.5	50.9 ± 1.5	34.7 ± 1.3	24.5 ± 1.0	17.7 ± 0.5	12.8 ± 0.5	8.83 ± .21	5.85 ± .13
	.95	121 ± 5	69.2 ± 3.4	47.0 ± 2.1	33.1 ± 1.7	23.7 ± 1.0	16.8 ± 0.7	11.4 ± 0.5	7.24 ± .31
8	.50	22.9 ± 0.7	14.3 ± 0.4	10.2 ± 0.2	7.61 ± .17	5.78 ± .11	4.42 ± .07	3.39 ± .04	
	.75	37.4 ± 0.9	22.6 ± 0.6	15.6 ± 0.4	11.3 ± 0.3	8.31 ± .22	6.07 ± .15	4.40 ± .07	
	.90	60.9 ± 2.4	35.5 ± 1.5	24.0 ± 0.9	16.8 ± 0.6	12.1 ± 0.4	8.48 ± .22	5.76 ± .13	
	.95	83.0 ± 4.3	47.0 ± 2.6	31.8 ± 2.1	22.4 ± 1.2	15.4 ± 0.7	10.6 ± 0.3	6.87 ± .27	
9	.50	17.6 ± 0.4	11.1 ± 0.2	7.91 ± .16	5.93 ± .12	4.51 ± .07	3.49 ± .04		
	.75	28.0 ± 0.9	16.9 ± 0.4	11.6 ± 0.2	8.35 ± .19	6.12 ± .13	4.47 ± .07		
	.90	44.2 ± 1.7	25.2 ± 0.8	17.2 ± 0.8	12.0 ± 0.4	8.44 ± .21	5.82 ± .13		
	.95	58.6 ± 3.0	32.8 ± 2.2	21.6 ± 1.1	15.1 ± 0.7	10.5 ± 0.4	6.94 ± .23		
10	.50	13.8 ± 0.3	8.74 ± .12	6.24 ± .08	4.68 ± .04	3.61 ± .04			
	.75	21.1 ± 0.5	12.8 ± 0.3	8.85 ± .17	6.35 ± .11	4.64 ± .07			
	.90	32.2 ± 1.2	18.6 ± 0.7	12.3 ± 0.4	8.57 ± .23	5.98 ± .13			
	.95	42.7 ± 2.5	23.6 ± 0.9	15.3 ± 0.6	10.3 ± 0.4	7.13 ± .26			

Table G
Estimated Percentiles of a Modulus Ratio Statistic: Mod F (j, p, k-1)

(k-1) = 15

p	1-α	j=7	j=6	j=5	j=4	j=3
2	.50	24.3 ± 1.0	16.5 ± 0.8	10.6 ± 0.5	6.35 ± .22	3.19 ± .07
	.75	64.7 ± 3.9	43.3 ± 2.7	26.5 ± 1.3	14.7 ± 0.9	6.20 ± .27
	.90	201 ± 19	133 ± 14	80.2 ± 8.3	40.9 ± 4.8	15.9 ± 1.7
	.95	432 ± 55	279 ± 35	174 ± 26	90 ± 14	31.9 ± 5.5
3	.50	11.7 ± 0.3	7.98 ± .22	5.24 ± .12	3.19 ±.07	
	.75	24.2 ± 1.1	16.0 ± 0.6	9.66 ± .42	5.07 ± .15	
	.90	53.4 ± 3.2	34.3 ± 2.5	19.4 ± 1.2	9.31 ± .61	
	.95	92.6 ± 7.9	56.2 ± 4.7	31.6 ± 2.7	14.6 ± 1.2	
4	.50	6.90 ± .16	4.82 ± .09	3.20 ± .06		
	.75	12.3 ± 0.5	7.90 ± .27	4.64 ± .11		
	.90	22.0 ± 1.1	14.1 ± 0.8	7.48 ± .37		
	.95	33.7 ± 2.5	20.0 ± 0.9	10.6 ± 0.7		
5	.50	4.60 ± .09	3.23 ± .07			
	.75	6.83 ± .17	4.46 ± .10			
	.90	10.9 ± 0.5	6.39 ± .21			
	.95	14.9 ± 0.9	8.46 ± .42			
6	.50	3.27 ± .05				
	.75	4.42 ± .10				
	.90	6.06 ± .19				
	.95	7.54 ± .34				

Table G
Estimated Percentiles of a Modulus Ratio Statistic: Mod F (j, p, k-1)

(k-1) = 19

p	1-α	j=19	j=18	j=17	j=16	j=15	j=14	j=13	j=12
3	.50	325 ± 30	199 ± 17	140 ± 6	110 ± 10	84.7 ± 7.3	67.2 ± 3.7	53.5 ± 2.9	44.0 ± 3.1
	.75	757 ± 69	463 ± 72	329 ± 30	248 ± 32	184 ± 18	145 ± 13	117 ± 9	94.3 ± 9.3
	.90	1727 ± 392	1132 ± 320	767 ± 154	576 ± 140	439 ± 87	338 ± 78	271 ± 61	211 ± 45
	.95	2938 ± 733	1743 ± 477	1312 ± 457	1013 ± 283	776 ± 312	575 ± 177	451 ± 138	361 ± 127
4	.50	201 ± 12	125 ± 9	90 ± 10	69.9 ± 8.3	55.0 ± 4.9	42.7 ± 2.7	33.9 ± 2.9	27.4 ± 2.7
	.75	439 ± 35	265 ± 31	182 ± 12	135 ± 9	108 ± 8	85 ± 10	67.1 ± 8.8	51.3 ± 3.8
	.90	935 ± 183	564 ± 133	377 ± 57	284 ± 37	223 ± 31	179 ± 24	136 ± 19	106 ± 12
	.95	1497 ± 290	924 ± 153	651 ± 99	469 ± 93	381 ± 82	296 ± 53	214 ± 31	169 ± 13
5	.50	130 ± 6	78.7 ± 8.6	57.3 ± 4.3	43.9 ± 3.1	34.1 ± 3.1	26.8 ± 1.9	21.7 ± 1.2	17.3 ± 1.2
	.75	261 ± 25	161 ± 21	108 ± 16	80 ± 14	61.9 ± 8.3	50.1 ± 7.2	38.5 ± 5.0	29.6 ± 2.6
	.90	535 ± 45	311 ± 45	214 ± 19	157 ± 18	118 ± 20	91 ± 11	70 ± 11	54.5 ± 6.7
	.95	804 ± 99	470 ± 84	318 ± 50	229 ± 47	175 ± 19	130 ± 12	99 ± 13	79.2 ± 9.3
6	.50	89.0 ± 5.4	56.0 ± 4.5	40.1 ± 2.5	30.8 ± 2.0	23.9 ± 1.1	19.2 ± 0.9	15.1 ± 0.4	11.9 ± 0.6
	.75	171 ± 10	101 ± 8	69.7 ± 5.3	52.2 ± 3.5	40.8 ± 1.9	32.2 ± 1.2	25.2 ± 1.0	19.8 ± 1.3
	.90	291 ± 33	178 ± 14	129 ± 9	94.1 ± 3.7	72.6 ± 4.9	55.3 ± 4.8	43.9 ± 3.8	32.9 ± 2.9
	.95	411 ± 68	238 ± 19	175 ± 25	133 ± 23	105 ± 21	78.6 ± 5.8	60.3 ± 8.1	45.8 ± 4.1
7	.50	66.7 ± 6.7	40.5 ± 2.8	29.3 ± 0.9	22.2 ± 1.0	17.5 ± 0.6	14.0 ± 1.0	11.0 ± 0.8	8.85 ± .43
	.75	119 ± 14	70.6 ± 8.1	49.0 ± 5.2	36.1 ± 3.7	27.8 ± 2.9	22.2 ± 1.5	17.3 ± 1.1	13.6 ± 0.8
	.90	207 ± 44	121 ± 17	81 ± 13	61.8 ± 9.2	46.5 ± 6.4	35.3 ± 4.1	26.8 ± 3.7	21.2 ± 2.8
	.95	288 ± 79	170 ± 28	117 ± 24	82.1 ± 9.3	65 ± 10	49.0 ± 7.3	37.6 ± 4.5	29.0 ± 3.1
8	.50	50.2 ± 4.5	30.4 ± 1.4	22.1 ± 0.9	16.8 ± 0.7	13.1 ± 0.3	10.4 ± 0.4	8.35 ± .34	6.69 ± .12
	.75	86.8 ± 4.1	52.9 ± 4.8	36.0 ± 1.0	26.3 ± 1.1	20.2 ± 0.4	15.9 ± 0.6	12.3 ± 0.8	9.74 ± .32
	.90	153 ± 2	83.5 ± 5.5	57.6 ± 2.6	41.7 ± 2.7	31.9 ± 1.4	24.0 ± 2.1	18.8 ± 0.5	14.4 ± 0.6
	.95	212 ± 33	116 ± 3	79.1 ± 3.7	56.1 ± 3.0	42.9 ± 3.8	31.3 ± 4.4	24.8 ± 0.8	17.9 ± 1.0
9	.50	40.3 ± 3.3	24.2 ± 1.4	17.2 ± 0.9	13.3 ± 0.6	10.5 ± 0.1	8.19 ± .32	6.55 ± .19	5.28 ± .10
	.75	66.5 ± 6.1	37.3 ± 0.5	26.1 ± 1.7	19.5 ± 1.1	15.0 ± 0.5	11.8 ± 0.4	9.38 ± .28	7.35 ± .16
	.90	108 ± 11	58.9 ± 5.6	40.9 ± 1.5	29.5 ± 0.9	21.7 ± 0.7	17.1 ± 0.7	12.8 ± 0.7	10.1 ± 0.5
	.95	143 ± 9	78.5 ± 8.5	52.9 ± 5.1	39.3 ± 2.6	29.4 ± 0.8	21.7 ± 1.5	16.4 ± 1.1	12.5 ± 0.3
10	.50	32.1 ± 1.2	19.3 ± 0.6	13.6 ± 0.3	10.5 ± 0.5	8.33 ± .13	6.61 ± .28	5.22 ± .16	4.20 ± .11
	.75	52.1 ± 2.9	30.0 ± 1.7	20.4 ± 0.7	15.5 ± 1.0	12.1 ± 0.8	9.26 ± .49	7.10 ± .23	5.52 ± .06
	.90	80.5 ± 2.5	46.0 ± 5.5	31.1 ± 2.7	22.7 ± 2.2	17.3 ± 2.0	13.4 ± 1.0	9.60 ± .59	7.29 ± .22
	.95	106 ± 7	60.8 ± 8.1	39.9 ± 4.0	29.4 ± 3.7	22.1 ± 3.2	16.5 ± 1.5	11.9 ± 1.3	8.81 ± .14

Table G
Estimated Percentiles of a Modulus Ratio Statistic: Mod F (j, p, k-1)

(k-1) = 19

p	1-α	j=11	j=10	j=9	j=8	j=7	j=6	j=5	j=4
3	.50	34.8 ± 3.8	27.1 ± 2.8	20.3 ± 1.7	15.0 ± 1.0	11.2 ± 1.1	7.79 ± .41	5.01 ± .18	3.18 ± .14
	.75	72.2 ± 6.8	56.3 ± 3.5	43.0 ± 3.3	32.2 ± 1.9	22.5 ± 1.1	15.0 ± 1.2	9.39 ± .33	4.94 ± .22
	.90	167 ± 37	128 ± 25	95 ± 26	66 ± 13	44.9 ± 4.1	30.5 ± 1.7	17.8 ± 1.1	8.59 ± .83
	.95	285 ± 76	209 ± 53	163 ± 45	116 ± 34	82 ± 21	52 ± 13	27.8 ± 2.8	12.4 ± 0.7
4	.50	21.2 ± 1.1	16.4 ± 0.4	12.7 ± 0.3	9.35 ± .47	6.90 ± .31	4.86 ± .20	3.22 ± .14	
	.75	39.3 ± 4.0	30.6 ± 3.3	23.5 ± 3.1	17.0 ± 1.1	12.2 ± 0.6	8.00 ± .47	4.73 ± .14	
	.90	82.7 ± 7.8	62.3 ± 6.5	45.4 ± 4.2	32.4 ± 3.3	22.8 ± 3.4	14.4 ± 1.6	7.57 ± .65	
	.95	132 ± 12	105 ± 19	66.4 ± 8.2	47.3 ± 2.6	33.5 ± 4.8	21.8 ± 4.2	10.8 ± 2.0	
5	.50	13.5 ± 0.8	10.6 ± 0.5	8.21 ± .18	6.18 ± .12	4.55 ± .06	3.27 ± .04		
	.75	22.9 ± 2.2	17.5 ± 1.4	13.2 ± 1.1	9.87 ± .93	6.90 ± .53	4.45 ± .10		
	.90	42.0 ± 7.5	31.7 ± 4.9	22.6 ± 3.2	16.1 ± 2.3	11.2 ± 1.6	6.56 ± .72		
	.95	65.9 ± 8.0	44.9 ± 4.5	32.0 ± 2.8	23.0 ± 4.6	15.3 ± 2.7	8.4 ± 1.1		
6	.50	9.41 ± .58	7.38 ± .47	5.74 ± .27	4.39 ± .20	3.24 ± .11			
	.75	15.6 ± 0.9	11.6 ± 0.4	8.68 ± .52	6.17 ± .41	4.32 ± .21			
	.90	24.6 ± 1.7	18.5 ± 0.9	13.7 ± 0.6	9.29 ± .22	5.85 ± .25			
	.95	35.1 ± 3.1	25.9 ± 1.3	19.4 ± 0.9	12.4 ± 0.8	7.22 ± .29			
7	.50	6.88 ± .21	5.42 ± .16	4.18 ± .24	3.20 ± .11				
	.75	10.5 ± 0.7	7.83 ± .29	5.86 ± .34	4.28 ± .29				
	.90	16.4 ± 2.4	12.0 ± 1.4	8.51 ± .51	5.72 ± .42				
	.95	22.1 ± 3.6	15.9 ± 3.4	11.0 ± 0.6	7.08 ± .39				
8	.50	5.28 ± .13	4.14 ± .13	3.26 ± .06					
	.75	7.52 ± .31	5.78 ± .28	4.22 ± .09					
	.90	11.0 ± 0.5	8.11 ± .51	5.62 ± .15					
	.95	13.8 ± 0.9	10.1 ± 1.3	6.80 ± .46					
9	.50	4.23 ± .11	3.37 ± .11						
	.75	5.58 ± .10	4.27 ± .11						
	.90	7.53 ± .36	5.50 ± .09						
	.95	9.11 ± .57	6.46 ± .34						
10	.50	3.39 ± .06							
	.75	4.18 ± .14							
	.90	5.37 ± .16							
	.95	6.42 ± .41							

Table G
Estimated Percentiles of a Modulus Ratio Statistic: Mod F (j, p, k-1)

(k-1) = 23

p	1-α	j = 23	j = 22	j = 21	j = 20	j = 19	j = 18	j = 17	j = 16
3	.50	446 ± 17	302 ± 10	237 ± 6	187 ± 4	155 ± 6	128 ± 1	108 ± 5	90.1 ± 5.1
	.75	1080 ± 68	698 ± 68	537 ± 36	440 ± 31	356 ± 23	290 ± 23	247 ± 15	206 ± 12
	.90	2675 ± 355	1688 ± 276	1287 ± 163	1027 ± 148	829 ± 143	679 ± 94	570 ± 87	455 ± 31
	.95	4775 ± 1906	3271 ± 1281	2491 ± 682	1993 ± 517	1558 ± 435	1293 ± 301	1051 ± 223	871 ± 162
4	.50	258 ± 11	169 ± 11	131 ± 10	106 ± 9	85.6 ± 7.7	70.5 ± 3.8	58.7 ± 3.8	49.7 ± 4.2
	.75	522 ± 58	362 ± 45	277 ± 16	213 ± 12	174 ± 14	142 ± 11	119 ± 10	99 ± 13
	.90	1158 ± 80	734 ± 57	566 ± 82	446 ± 94	361 ± 89	295 ± 62	249 ± 40	210 ± 18
	.95	1900 ± 424	1135 ± 249	963 ± 201	730 ± 167	600 ± 151	505 ± 125	411 ± 96	350 ± 72
5	.50	161 ± 8	109 ± 8	83.2 ± 5.5	66.6 ± 6.7	54.5 ± 6.0	44.9 ± 2.7	37.6 ± 2.9	30.8 ± 2.9
	.75	315 ± 21	205 ± 13	155 ± 11	122 ± 8	100 ± 6	84.0 ± 5.5	67.6 ± 3.1	56.2 ± 3.5
	.90	683 ± 62	426 ±20	308 ± 25	253 ± 24	200 ± 22	162 ± 11	134 ± 17	110 ± 18
	.95	1165 ± 99	733 ± 103	515 ± 87	420 ± 75	318 ± 47	262 ± 22	223 ± 8	175 ± 6
6	.50	114 ± 11	77.4 ± 3.7	59.6 ± 4.9	48.1 ± 4.1	39.1 ± 2.1	32.6 ± 0.9	27.4 ± 0.8	23.0 ± 1.1
	.75	211 ± 28	140 ± 19	107 ± 12	83.1 ± 6.2	68.6 ± 6.5	55.4 ± 3.9	45.1 ± 3.8	37.8 ± 3.8
	.90	369 ± 53	240 ± 39	183 ± 27	143 ± 18	116 ± 16	96 ± 14	77.4 ± 9.6	66.1 ± 6.9
	.95	532 ± 62	323 ± 51	254 ± 35	206 ± 32	167 ± 30	135 ± 27	113 ± 23	92 ± 17
7	.50	85.7 ± 5.9	59.8 ± 4.8	45.0 ± 4.4	36.8 ± 4.1	30.2 ± 3.3	24.9 ± 2.2	20.6 ± 1.9	17.3 ± 1.5
	.75	156 ± 12	103 ± 7	79.9 ± 7.1	61.3 ± 2.7	49.7 ± 2.5	40.9 ± 2.4	33.7 ± 1.8	27.7 ± 1.4
	.90	267 ± 43	172 ± 20	129 ± 15	102 ± 11	82.4 ± 13	68.7 ± 9.7	55.9 ± 7.4	45.9 ± 5.5
	.95	365 ± 82	238 ± 50	169 ± 26	138 ± 12	108 ± 14	90.5 ± 9.2	75.0 ± 9.6	62 ± 14
8	.50	68.9 ± 5.5	45.7 ± 3.5	34.4 ± 1.7	27.7 ± 1.8	22.7 ± 1.7	18.7 ± 1.4	15.6 ± 1.2	13.1 ± 1.0
	.75	113 ± 9	73.6 ± 3.1	55.5 ± 3.1	44.1 ± 3.6	35.9 ± 2.6	29.2 ± 1.4	23.9 ± 1.2	19.7 ± 1.0
	.90	174 ± 11	115 ± 7	85.3 ± 8.7	66.4 ± 3.6	53.9 ± 3.3	44.6 ± 3.2	36.4 ± 2.8	29.0 ± 2.4
	.95	243 ± 31	155 ± 14	114 ± 15	90.5 ± 8.7	72.6 ± 7.3	58.1 ± 3.4	47.2 ± 3.1	39.5 ± 4.2
9	.50	51.5 ± 2.5	34.9 ± 1.3	26.9 ± 1.3	21.9 ± 1.1	17.6 ± 0.8	14.5 ± 0.7	12.4 ± 0.4	10.3 ± 0.4
	.75	84.3 ± 4.3	55.8 ± 5.0	41.5 ± 2.2	33.5 ± 2.2	27.6 ± 1.2	22.3 ± 0.6	18.3 ± 0.4	15.3 ± 0.9
	.90	146 ± 9	92.6 ± 5.9	67.7 ± 3.5	55.0 ± 4.0	43.6 ± 1.7	35.4 ± 1.1	28.6 ± 0.4	23.0 ± 0.7
	.95	199 ± 21	128 ± 10	94.1 ± 4.9	70.3 ± 4.1	56.8 ± 5.3	46.2 ± 6.5	36.7 ± 2.2	30.1 ± 0.3
10	.50	43.6 ± 1.7	29.4 ± 1.4	22.3 ± 1.0	17.8 ± 0.8	14.7 ± 0.7	12.3 ± 0.7	10.2 ± 0.6	8.43 ± .34
	.75	68.7 ± 5.7	46.3 ± 3.8	33.5 ± 2.5	27.0 ± 1.4	21.7 ± 1.3	17.7 ± 0.4	14.7 ± 0.6	11.8 ± 0.5
	.90	107 ± 8	66.4 ± 6.1	49.6 ± 4.3	40.0 ± 5.7	31.2 ± 2.7	25.1 ± 2.1	20.6 ± 1.6	17.2 ± 1.7
	.95	143 ± 13	85.6 ± 7.7	60.8 ± 5.7	48.4 ± 7.3	38.3 ± 3.3	30.9 ± 2.3	25.9 ± 3.4	20.7 ± 2.2

Table G
Estimated Percentiles of a Modulus Ratio Statistic: Mod F (j, p, k-1)

(k-1) = 23

p	1-α	j=15	j=14	j=13	j=12	j=11	j=10	j=9	j=8
3	.50	75.5 ± 3.7	61.2 ± 3.6	50.6 ± 3.9	40.6 ± 2.3	33.2 ± 1.0	26.3 ± 0.8	20.4 ± 0.5	15.4 ± 1.0
	.75	164 ± 14	133 ± 5	110 ± 1	89.1 ± 0.9	72.3 ± 1.3	57.3 ± 3.0	42.6 ± 1.6	31.7 ± 1.1
	.90	385 ± 34	319 ± 50	256 ± 31	194 ± 12	154 ± 17	120 ± 17	90 ± 14	67 ± 11
	.95	715 ± 178	610 ± 134	463 ± 73	338 ± 98	288 ± 78	213 ± 65	156 ± 40	112 ± 30
4	.50	41.5 ± 3.2	34.1 ± 3.0	28.5 ± 2.3	23.4 ± 2.3	18.8 ± 1.7	15.1 ± 1.3	11.7 ± 1.0	8.90 ± .61
	.75	83 ± 11	66.9 ± 9.1	56.1 ± 9.1	45.6 ± 5.6	36.0 ± 3.2	29.4 ± 2.9	22.1 ± 1.8	16.1 ± 1.3
	.90	171 ± 12	143 ± 12	113 ± 15	93.1 ± 8.6	72.1 ± 5.6	56.1 ± 5.4	42.5 ± 7.0	31.8 ± 5.1
	.95	270 ± 47	231 ± 47	194 ± 49	157 ± 38	116 ± 30	90 ± 15	69 ± 15	49 ± 12
5	.50	25.8 ± 2.0	21.6 ± 1.6	17.8 ± 1.4	14.6 ± 0.7	11.8 ± 0.9	9.55 ± .69	7.43 ± .59	5.75 ± .46
	.75	46.4 ± 3.1	37.5 ± 1.9	30.9 ± 1.2	25.3 ± 1.4	20.4 ± 2.2	16.2 ± 1.1	12.3 ± 1.2	9.0 ± 1.0
	.90	89 ± 13	76 ± 12	57.1 ± 7.5	48.0 ± 5.0	35.9 ± 5.6	27.6 ± 3.5	21.9 ± 2.6	14.9 ± 1.2
	.95	138 ± 11	112 ± 12	90 ± 11	74 ± 12	55.3 ± 9.8	43.2 ± 8.7	31.9 ± 6.2	22.4 ± 4.1
6	.50	19.3 ± 0.9	16.1 ± 0.9	13.3 ± 0.5	10.9 ± 0.2	8.75 ± .35	6.89 ± .11	5.41 ± .08	4.13 ± .19
	.75	31.3 ± 2.6	25.9 ± 3.0	21.4 ± 2.0	17.3 ± 1.6	14.0 ± 1.2	10.6 ± 0.5	8.09 ± .39	5.78 ± .22
	.90	54.2 ± 4.9	44.6 ± 5.2	35.3 ± 5.2	27.7 ± 2.8	22.0 ± 2.1	16.4 ± 1.3	12.3 ± 0.5	9.06 ± .68
	.95	73 ± 10	61.5 ± 5.0	47.7 ± 7.1	38.8 ± 4.6	31.0 ± 3.2	23.0 ± 2.1	16.7 ± 0.9	11.5 ± 1.1
7	.50	14.5 ± 1.2	12.2 ± 1.1	10.0 ± 0.9	8.18 ± .61	6.66 ± .32	5.27 ± .18	4.16 ± .17	3.26 ± .07
	.75	22.5 ± 1.3	18.9 ± 1.5	15.1 ± 1.1	12.2 ± 0.9	9.80 ± .30	7.73 ± .49	5.77 ± .21	4.25 ± .06
	.90	36.2 ± 4.6	29.4 ± 2.0	23.9 ± 1.9	18.8 ± 1.5	14.5 ± 1.3	11.3 ± 0.6	8.12 ± .25	5.75 ± .16
	.95	49.0 ± 7.7	40.1 ± 4.2	32.1 ± 3.9	25.6 ± 3.0	19.2 ± 2.0	14.1 ± 1.7	9.94 ± .51	6.89 ± .29
8	.50	11.0 ± 0.7	9.20 ± .59	7.49 ± .52	6.14 ± .52	5.02 ± .38	3.99 ± .26	3.17 ± .18	
	.75	16.3 ± 0.9	13.3 ± 0.7	10.8 ± 0.5	8.70 ± .45	6.95 ± .52	5.40 ± .43	3.99 ± .19	
	.90	24.4 ± 2.2	19.8 ± 1.2	15.7 ± 0.7	12.3 ± 0.4	9.86 ± .41	7.29 ± .24	5.23 ± .34	
	.95	31.4 ± 4.0	25.5 ± 2.1	20.8 ± 1.5	16.2 ± 1.5	12.1 ± 0.9	8.88 ± .46	6.09 ± .27	
9	.50	8.55 ± .29	7.15 ± .20	5.96 ± .15	4.95 ± .11	3.96 ± .06	3.22 ± .10		
	.75	12.6 ± 0.6	10.3 ± 0.2	8.34 ± .28	6.70 ± .14	5.27 ± .19	4.04 ± .07		
	.90	18.7 ± 0.3	15.1 ± 0.4	11.5 ± 0.6	9.36 ± .36	7.00 ± .21	5.30 ± .07		
	.95	24.4 ± 1.4	20.1 ± 1.3	15.6 ± 1.6	11.6 ± 0.1	8.46 ± .15	6.29 ± .30		
10	.50	7.13 ± .31	5.84 ± .19	4.81 ± .10	3.95 ± .10	3.26 ± .07			
	.75	9.92 ± .11	8.00 ± .21	6.50 ± .15	5.11 ± .09	4.02 ± .06			
	.90	14.0 ± 1.0	11.0 ± 0.5	8.78 ± .17	6.68 ± .23	5.05 ± .15			
	.95	16.9 ± 1.9	13.5 ± 0.9	10.3 ± 0.4	7.91 ± .21	5.94 ± .15			

Table G
Estimated Percentiles of a Modulus Ratio Statistic: Mod F (j, p, k-1)

(k-1) = 23

p	1-α	j=7	j=6	j=5	j=4
3	.50	11.2 ± 0.4	8.08 ± .32	5.28 ± .18	3.27 ± .14
	.75	22.9 ± 1.6	15.6 ± 1.4	9.61 ± .69	4.98 ± .11
	.90	46.3 ± 5.7	31.2 ± 5.6	18.8 ± 2.5	8.61 ± .27
	.95	78 ± 21	55 ± 16	29.4 ± 5.1	13.5 ± 2.1
4	.50	6.59 ± .39	4.76 ± .13	3.11 ± .13	
	.75	11.5 ± .84	7.62 ± .41	4.64 ± .31	
	.90	21.8 ± 4.3	13.7 ± 1.1	7.54 ± .93	
	.95	33.2 ± 7.4	20.2 ± 2.2	10.5 ± 2.0	
5	.50	4.34 ± .32	3.13 ± .21		
	.75	6.46 ± .64	4.32 ± .52		
	.90	10.3 ± 1.5	6.2 ± 1.1		
	.95	14.4 ± 2.2	8.3 ± 1.5		
6	.50	3.10 ± .09			
	.75	4.07 ± .17			
	.90	5.66 ± .42			
	.95	7.35 ± .67			

Table G
Estimated Percentiles of a Modulus Ratio Statistic: Mod F (j, p, k-1)

(k-1) = 27

p	1-α	j=27	j=26	j=25	j=24	j=23	j=22	j=21	j=20
4	.50	375 ± 42	255 ± 39	200 ± 30	162 ± 22	136 ± 16	115 ± 14	99 ± 14	81.8 ± 8.9
	.75	754 ± 127	510 ± 86	397 ± 48	322 ± 46	269 ± 51	222 ± 35	189 ± 37	161 ± 24
	.90	1725 ± 523	1171 ± 314	880 ± 295	699 ± 202	583 ± 144	478 ± 120	393 ± 70	342 ± 77
	.95	2838 ± 1024	1958 ± 613	1475 ± 525	1150 ± 427	951 ± 348	814 ± 322	685 ± 272	578 ± 224
5	.50	245 ± 26	170 ± 16	130 ± 7	109 ± 5	89 ± 6	75.2 ± 6.0	63.5 ± 6.5	54.2 ± 4.7
	.75	484 ± 66	322 ± 40	254 ± 26	207 ± 22	171 ± 21	143 ± 24	121 ± 16	103 ± 14
	.90	986 ± 94	650 ± 51	500 ± 46	403 ± 26	342 ± 43	274 ± 35	234 ± 27	197 ± 21
	.95	1512 ± 189	984 ± 160	760 ± 114	621 ± 123	476 ± 67	398 ± 49	339 ± 34	289 ± 39
6	.50	171 ± 16	120 ± 7	93.6 ± 4.5	77.1 ± 5.8	64.6 ± 4.0	53.9 ± 3.4	46.4 ± 2.8	39.9 ± 2.8
	.75	317 ± 36	214 ± 16	169 ± 16	140 ± 11	112 ± 9	92.7 ± 9.0	80.6 ± 6.4	69.3 ± 6.5
	.90	607 ± 62	395 ± 26	302 ± 37	244 ± 22	196 ± 14	166 ± 17	142 ± 15	122 ± 13
	.95	925 ± 62	593 ± 44	460 ± 51	365 ± 24	296 ± 20	243 ± 27	202 ± 20	177 ± 7
7	.50	124 ± 8	87.2 ± 3.6	68.9 ± 5.5	55.2 ± 3.6	46.5 ± 3.5	39.3 ± 1.9	33.8 ± 1.5	29.0 ± 1.1
	.75	214 ± 15	149 ± 7	114 ± 10	92.8 ± 5.9	76.2 ± 4.5	65.3 ± 5.7	55.1 ± 3.2	47.5 ± 1.9
	.90	398 ± 43	264 ± 12	204 ± 13	164 ± 12	131 ± 16	107 ± 10	91.3 ± 5.5	77.1 ± 3.4
	.95	588 ± 45	384 ± 35	298 ± 26	238 ± 22	199 ± 23	166 ± 7	139 ± 7	111 ± 7
8	.50	96.3 ± 4.0	66.2 ± 3.0	51.6 ± 2.0	42.3 ± 0.9	35.1 ± 0.6	29.5 ± 1.1	25.5 ± 1.1	22.0 ± 1.0
	.75	161 ± 6	112 ± 4	84.7 ± 3.0	68.0 ± 3.5	55.6 ± 1.9	46.6 ± 0.4	40.0 ± 1.2	33.8 ± 0.8
	.90	266 ± 14	175 ± 17	135 ± 6	112 ± 4	91.3 ± 7.4	74.2 ± 2.5	62.0 ± 4.7	52.8 ± 2.1
	.95	358 ± 19	233 ± 23	178 ± 12	145 ± 16	119 ± 8	101 ± 8	81.8 ± 3.5	70.6 ± 4.4
9	.50	78.9 ± 4.5	53.8 ± 2.4	42.1 ± 1.3	35.1 ± 1.0	29.4 ± 0.2	24.7 ± 0.7	21.0 ± 0.7	18.0 ± 0.7
	.75	131 ± 8	88.3 ± 1.9	68.2 ± 1.5	54.1 ± 1.5	45.3 ± 1.7	38.1 ± 0.7	32.1 ± 0.4	27.3 ± 1.2
	.90	213 ± 9	141 ± 8	109 ± 4	85.1 ± 5.6	72.4 ± 7.4	57.8 ± 4.8	50.2 ± 6.6	42.2 ± 3.6
	.95	290 ± 16	190 ± 23	144 ± 14	113 ± 9	94.7 ± 9.7	79.3 ± 8.3	66 ± 11	55.5 ± 9.3
10	.50	64.2 ± 6.5	43.6 ± 4.4	34.3 ± 2.5	27.6 ± 1.8	23.0 ± 1.4	19.3 ± 1.0	16.6 ± 0.8	14.2 ± 0.7
	.75	99.3 ± 6.9	67.7 ± 5.0	52.5 ± 2.5	42.2 ± 1.6	34.1 ± 1.4	28.4 ± 1.2	24.1 ± 0.8	20.9 ± 0.9
	.90	160 ± 5	102 ± 7	77.8 ± 1.4	62.2 ± 1.4	50.6 ± 1.1	42.3 ± 1.4	35.9 ± 1.8	30.7 ± 2.4
	.95	206 ± 9	137 ± 8	100 ± 13	81.3 ± 5.3	65.2 ± 4.3	53.3 ± 5.3	45.8 ± 5.5	38.3 ± 2.3

Table G
Estimated Percentiles of a Modulus Ratio Statistic: Mod F (j, p, k-1)

(k-1) = 27

p	1-α	j = 19	j = 18	j = 17	j = 16	j = 15	j = 14	j = 13	j = 12
4	.50	70.5 ± 8.8	60.8 ± 7.0	51.6 ± 4.9	44.4 ± 4.5	37.4 ± 3.4	31.2 ± 3.4	26.3 ± 3.2	21.8 ± 2.2
	.75	139 ± 23	120 ± 21	100 ± 19	86 ± 16	75 ± 13	60.9 ± 9.2	50.9 ± 7.1	41.8 ± 5.4
	.90	287 ± 74	244 ± 65	208 ± 57	179 ± 57	157 ± 51	128 ± 34	102 ± 25	82 ± 17
	.95	494 ± 212	434 ± 201	363 ± 156	299 ± 120	252 ± 104	213 ± 77	175 ± 75	132 ± 51
5	.50	47.0 ± 4.6	40.4 ± 2.6	33.8 ± 1.5	29.3 ± 1.6	24.6 ± 1.3	20.6 ± 1.3	17.3 ± 0.7	14.5 ± 0.6
	.75	88 ± 13	77 ± 14	63.7 ± 7.0	54.3 ± 7.0	46.8 ± 5.0	38.7 ± 2.9	31.9 ± 2.6	26.2 ± 1.3
	.90	168 ± 25	142 ± 25	120 ± 19	102 ± 18	83 ± 14	69 ± 13	53.6 ± 7.5	44.3 ± 7.1
	.95	244 ± 36	207 ± 29	173 ± 22	154 ± 19	123 ± 13	100 ± 13	83 ± 12	66 ± 10
6	.50	33.9 ± 2.3	28.8 ± 2.5	24.7 ± 2.0	21.1 ± 1.6	17.9 ± 1.1	15.0 ± 1.0	12.5 ± 0.4	10.4 ± 0.5
	.75	59.7 ± 8.4	49.9 ± 5.3	42.2 ± 5.0	35.9 ± 3.0	30.3 ± 3.2	25.3 ± 1.8	20.7 ± 1.7	16.9 ± 1.8
	.90	104 ± 10	84.7 ± 7.3	72.0 ± 7.5	63.3 ± 4.8	52.5 ± 4.1	43.5 ± 2.8	35.7 ± 1.3	29.4 ± 2.2
	.95	143 ± 4	122 ± 6	108 ± 8	86.6 ± 5.8	73.5 ± 3.6	60.4 ± 2.7	51.1 ± 1.0	39.4 ± 1.0
7	.50	24.5 ± .9	21.2 ± .4	18.0 ± .7	15.6 ± .9	13.2 ± 0.9	11.1 ± 0.6	9.29 ± .64	7.70 ± .43
	.75	40.2 ± 1.5	34.0 ± 1.0	29.5 ± 1.5	24.5 ± .7	20.4 ± 1.1	16.6 ± 1.1	14.0 ± 0.9	11.6 ± 0.6
	.90	64.2 ± 4.0	54.9 ± 2.1	47.9 ± 2.3	38.7 ± 4.1	33.0 ± 1.6	27.0 ± 2.0	22.3 ± 2.6	17.8 ± 1.6
	.95	93.8 ± 4.7	81.4 ± 6.3	70.5 ± 7.6	57.3 ± 7.5	48.5 ± 9.0	38.6 ± 4.1	31.9 ± 4.6	24.4 ± 3.1
8	.50	18.7 ± 0.5	16.0 ± 0.4	13.9 ± 0.4	11.8 ± 0.4	10.1 ± 0.5	8.45 ± .25	7.01 ± .17	5.83 ± .07
	.75	29.2 ± 1.5	24.7 ± 1.2	20.9 ± 1.2	17.6 ± 1.2	14.7 ± 0.9	12.6 ± 0.6	10.2 ± 0.1	8.28 ± .28
	.90	44.2 ± 2.7	37.9 ± 2.1	32.4 ± 1.6	27.1 ± 1.6	22.3 ± 1.5	18.2 ± 1.1	14.8 ± 1.1	12.2 ± 0.4
	.95	58.4 ± 5.3	50.1 ± 6.5	42.4 ± 3.3	35.6 ± 3.7	29.1 ± 1.3	23.7 ± 1.7	19.2 ± 2.2	15.5 ± 0.6
9	.50	15.5 ± 0.5	13.2 ± 0.3	11.4 ± 0.2	9.65 ± .32	8.21 ± .33	6.87 ± .12	5.79 ± .06	4.84 ± .11
	.75	23.5 ± 0.9	19.7 ± 0.7	16.8 ± 0.7	14.2 ± 0.8	11.8 ± 0.8	9.86 ± .86	8.00 ± .38	6.59 ± .25
	.90	34.6 ± 1.9	29.2 ± 2.5	24.9 ± 1.6	20.5 ± 2.4	17.2 ± 1.3	14.3 ± 1.0	11.5 ± 0.4	8.99 ± .32
	.95	46.7 ± 6.4	39.6 ± 3.9	32.8 ± 4.8	26.7 ± 3.5	22.9 ± 2.4	18.4 ± 1.7	14.9 ± 1.5	11.3 ± 0.8
10	.50	12.0 ± 0.8	10.3 ± 0.4	8.81 ± .31	7.55 ± .24	6.42 ± .20	5.39 ± .13	4.59 ± .09	3.82 ± .11
	.75	17.9 ± 0.7	15.0 ± 0.5	12.8 ± 0.2	10.6 ± 0.5	8.82 ± .27	7.35 ± .26	6.01 ± .22	4.90 ± .13
	.90	25.6 ± 1.8	21.2 ± 1.4	17.9 ± 1.2	15.1 ± 0.8	12.3 ± 0.7	10.2 ± 0.5	8.38 ± .63	6.51 ± .26
	.95	32.6 ± 1.6	28.2 ± 1.6	22.7 ± 2.3	19.3 ± 1.0	15.9 ± 0.7	12.9 ± 0.3	10.1 ± 0.9	8.00 ± .62

Table G
Estimated Percentiles of a Modulus Ratio Statistic: Mod F (j, p, k-1)

(k-1) = 27

p	1-α	j=11	j=10	j=9	j=8	j=7	j=6	j=5
4	.50	17.8 ± 1.7	14.4 ± 1.3	11.4 ± 0.6	8.80 ± .52	6.48 ± .29	4.59 ± .09	3.10 ± .05
	.75	33.7 ± 3.7	27.1 ± 3.0	20.9 ± 2.3	15.6 ± 1.4	11.3 ± 1.3	7.30 ± .55	4.53 ± .26
	.90	65 ± 15	50.7 ± 8.7	38.5 ± 5.5	28.3 ± 4.1	20.6 ± 2.8	12.8 ± 1.8	7.1 ± 1.2
	.95	104 ± 38	80 ± 23	62 ± 16	44 ± 16	29.4 ± 9.0	18.6 ± 3.4	10.3 ± 2.4
5	.50	12.2 ± 0.6	9.61 ± .72	7.58 ± .47	5.93 ± .30	4.39 ± .21	3.18 ± .18	
	.75	21.1 ± 2.1	16.8 ± 0.8	12.6 ± 0.8	9.32 ± .47	6.54 ± .44	4.33 ± .23	
	.90	35.8 ± 5.7	27.2 ± 3.6	21.0 ± 4.4	15.2 ± 2.2	10.5 ± 1.1	6.16 ± .26	
	.95	51.4 ± 9.8	41.6 ± 10.6	31.5 ± 4.4	21.8 ± 3.2	13.6 ± 2.0	8.17 ± .90	
6	.50	8.54 ± .47	6.83 ± .34	5.53 ± .31	4.18 ± .13	3.14 ± .09		
	.75	13.6 ± 0.8	10.9 ± 0.9	8.20 ± .61	6.07 ± .32	4.26 ± .18		
	.90	22.9 ± 1.0	17.7 ± 2.2	12.8 ± 0.6	9.02 ± .27	5.92 ± .11		
	.95	32.0 ± 1.0	24.5 ± 2.6	18.3 ± 1.4	12.4 ± 0.9	7.39 ± .38		
7	.50	6.33 ± .33	5.07 ± .16	4.07 ± .10	3.14 ± .13			
	.75	9.26 ± .41	7.30 ± .41	5.67 ± .31	4.11 ± .25			
	.90	13.7 ± 0.6	11.0 ± 0.5	8.09 ± .52	5.50 ± .23			
	.95	18.9 ± 2.1	14.1 ± 0.7	9.92 ± .58	6.68 ± .38			
8	.50	4.86 ± .14	3.92 ± .12	3.13 ± .11				
	.75	6.62 ± .13	5.22 ± .16	4.03 ± .19				
	.90	9.43 ± .29	7.08 ± .26	5.14 ± .31				
	.95	11.8 ± 0.8	8.89 ± .35	6.09 ± .23				
9	.50	3.93 ± .11	3.23 ± .10					
	.75	5.21 ± .18	4.02 ± .13					
	.90	7.11 ± .37	5.23 ± .37					
	.95	8.55 ± .61	6.24 ± .53					
10	.50	3.20 ± .06						
	.75	3.94 ± .06						
	.90	5.01 ± .24						
	.95	5.80 ± .22						

Table G
Estimated Percentiles of a Modulus Ratio Statistic: Mod F (j, p, k-1)

(k-1) = 31

p	1-α	j=31	j=30	j=29	j=28	j=27	j=26	j=25	j=24	j=23
4	.50	548 ± 18	367 ± 5	289 ± 6	241 ± 6	201 ± 6	171 ± 6	150 ± 7	128 ± 6	111 ± 5
	.75	1160 ± 78	794 ± 72	617 ± 31	503 ± 33	424 ± 16	355 ± 29	310 ± 27	267 ± 18	239 ± 19
	.90	2519 ± 349	1757 ± 273	1398 ± 251	1134 ± 273	907 ± 194	790 ± 171	690 ± 176	609 ± 161	532 ± 115
	.95	4217 ± 971	2742 ± 822	2145 ± 505	1748 ± 539	1465 ± 387	1297 ± 375	1105 ± 371	915 ± 274	803 ± 242
5	.50	348 ± 5	242 ± 2	188 ± 2	153 ± 1	130 ± 5	112 ± 3	97.3 ± 4.2	84.3 ± 4.5	74.0 ± 4.5
	.75	730 ± 65	478 ± 40	369 ± 35	303 ± 18	256 ± 6	216 ± 7	184 ± 6	161 ± 4	140 ± 4
	.90	1399 ± 175	902 ± 53	698 ± 59	570 ± 53	473 ± 28	406 ± 18	348 ± 33	306 ± 20	273 ± 23
	.95	2196 ± 639	1379 ± 225	1044 ± 157	848 ± 75	728 ± 140	592 ± 100	518 ± 49	448 ± 38	384 ± 46
6	.50	243 ± 23	167 ± 5	131 ± 5	107 ± 5	90.8 ± 3.8	77.9 ± 2.3	67.4 ± 2.3	58.4 ± 1.7	51.6 ± 1.1
	.75	449 ± 37	303 ± 20	234 ± 14	193 ± 7	164 ± 7	141 ± 4	122 ± 5	105 ± 5	91.1 ± 6.2
	.90	833 ± 64	553 ± 19	423 ± 26	343 ± 25	278 ± 11	237 ± 17	205 ± 14	179 ± 17	151 ± 10
	.95	1158 ± 175	786 ± 83	606 ± 60	497 ± 50	396 ± 11	327 ± 27	282 ± 17	247 ± 15	213 ± 18
7	.50	182 ± 16	126 ± 6	101 ± 7	83.0 ± 5.8	69.0 ± 4.0	59.6 ± 3.5	50.9 ± 3.9	44.9 ± 3.4	39.2 ± 2.3
	.75	334 ± 39	220 ± 22	171 ± 15	139 ± 9	117 ± 7	98.3 ± 7.5	84.0 ± 6.6	72.7 ± 5.6	64.6 ± 5.1
	.90	572 ± 50	383 ± 23	299 ± 55	240 ± 39	206 ± 35	171 ± 29	146 ± 20	126 ± 14	111 ± 10
	.95	790 ± 56	518 ± 53	400 ± 62	324 ± 64	267 ± 35	225 ± 39	197 ± 34	173 ± 29	153 ± 33
8	.50	140 ± 9	98.2 ± 8.1	78.3 ± 7.0	64.7 ± 6.2	54.0 ± 3.8	46.3 ± 2.5	39.8 ± 2.7	34.1 ± 2.0	29.8 ± 1.5
	.75	240 ± 29	162 ± 20	126 ± 13	101 ± 9	85.0 ± 6.8	73.1 ± 6.1	63.9 ± 5.1	54.7 ± 4.0	47.1 ± 4.3
	.90	405 ± 51	258 ± 31	209 ± 26	165 ± 18	134 ± 14	113 ± 13	95 ± 12	85 ± 13	73 ± 11
	.95	588 ± 64	382 ± 62	279 ± 39	236 ± 36	190 ± 23	161 ± 25	134 ± 24	117 ± 24	99 ± 19
9	.50	112 ± 7	77.8 ± 2.3	62.9 ± 3.9	51.1 ± 1.5	43.3 ± 2.7	37.3 ± 1.8	32.1 ± 0.9	27.8 ± 1.2	23.9 ± 0.7
	.75	191 ± 5	126 ± 5	97.6 ± 6.8	80.2 ± 4.2	68.5 ± 3.6	58.5 ± 2.5	50.3 ± 4.0	43.0 ± 2.3	36.8 ± 1.1
	.90	305 ± 26	197 ± 10	148 ± 4	123 ± 5	101 ± 4	87.8 ± 7.1	73.8 ± 4.2	63.7 ± 2.2	55.1 ± 2.1
	.95	421 ± 20	263 ± 29	199 ± 19	164 ± 18	135 ± 19	110 ± 10	97 ± 12	82 ± 11	71.0 ± 5.6
10	.50	93.2 ± 3.9	63.4 ± 3.3	50.5 ± 2.2	40.9 ± 1.8	34.2 ± 1.3	29.3 ± 1.3	25.1 ± 0.9	22.1 ± 0.9	19.1 ± 1.1
	.75	148 ± 5	100 ± 4	77.9 ± 4.1	63.1 ± 2.4	53.4 ± 3.5	46.2 ± 3.1	39.9 ± 2.1	34.2 ± 1.5	29.6 ± 1.0
	.90	241 ± 7	156 ± 6	120 ± 9	95.5 ± 4.2	80.3 ± 2.4	68.7 ± 1.7	59.0 ± 3.2	51.1 ± 3.4	43.8 ± 2.3
	.95	343 ± 30	210 ± 25	166 ± 15	132 ± 13	108 ± 20	87.3 ± 7.7	76.3 ± 5.5	64.3 ± 5.3	54.9 ± 6.3

Table G
Estimated Percentiles of a Modulus Ratio Statistic: Mod F (j, p, k-1)

(k-1) = 31

p	1-α	j = 22	j = 21	j = 20	j = 19	j = 18	j = 17	j = 16	j = 15	j = 14
4	.50	95.9 ± 4.4	84.8 ± 5.1	73.4 ± 5.8	64.6 ± 5.0	56.9 ± 4.5	48.1 ± 3.4	41.8 ± 3.2	35.4 ± 2.9	30.3 ± 2.6
	.75	207 ± 16	178 ± 8	153 ± 13	133 ± 9	115 ± 8	98.0 ± 5.0	84.8 ± 5.8	72.0 ± 3.3	59.8 ± 4.4
	.90	461 ± 91	400 ± 86	346 ± 90	298 ± 69	258 ± 67	222 ± 58	185 ± 45	157 ± 38	130 ± 25
	.95	698 ± 204	621 ± 212	535 ± 132	468 ± 130	425 ± 117	362 ± 96	306 ± 72	256 ± 43	223 ± 51
5	.50	63.6 ± 3.2	56.0 ± 2.8	49.0 ± 2.4	43.3 ± 2.6	37.4 ± 1.7	32.6 ± 1.8	27.9 ± 0.9	24.0 ± 1.5	20.2 ± 0.8
	.75	123 ± 4	107 ± 3	92.5 ± 5.6	80.5 ± 7.9	69.5 ± 5.6	60.1 ± 3.1	51.6 ± 3.5	43.2 ± 2.6	35.8 ± 1.3
	.90	235 ± 29	199 ± 22	176 ± 21	153 ± 27	132 ± 13	111 ± 11	95 ± 13	81 ± 13	66.3 ± 9.3
	.95	333 ± 60	283 ± 30	256 ± 29	226 ± 29	193 ± 30	162 ± 31	139 ± 27	117 ± 19	100 ± 16
6	.50	45.1 ± 1.9	39.4 ± 0.6	34.1 ± 0.6	30.0 ± 0.4	26.3 ± 0.1	22.9 ± 0.4	19.5 ± 0.4	16.6 ± 0.5	14.0 ± 0.3
	.75	80.1 ± 4.8	69.1 ± 3.7	59.7 ± 4.4	51.8 ± 3.5	45.1 ± 4.2	38.7 ± 3.7	33.9 ± 3.8	28.4 ± 1.7	23.8 ± 1.8
	.90	132 ± 14	117 ± 16	101 ± 12	90 ± 12	75.8 ± 9.4	64.4 ± 4.1	56.8 ± 5.2	46.9 ± 5.6	39.3 ± 2.4
	.95	186 ± 25	166 ± 26	142 ± 18	126 ± 17	109 ± 25	89 ± 17	76 ± 12	65 ± 10	53.4 ± 8.9
7	.50	34.0 ± 2.9	29.5 ± 1.7	25.7 ± 1.8	22.6 ± 1.4	19.5 ± 1.2	17.1 ± 0.9	14.7 ± 0.7	12.6 ± 0.3	10.9 ± 0.4
	.75	56.6 ± 3.4	49.5 ± 3.4	43.4 ± 3.2	37.5 ± 3.7	32.0 ± 2.7	27.3 ± 1.3	23.7 ± 1.0	19.8 ± 1.3	16.6 ± 0.9
	.90	96 ± 13	82 ± 12	71.8 ± 7.9	61.4 ± 7.3	53.1 ± 5.5	48.8 ± 4.8	38.7 ± 3.0	31.2 ± 2.3	25.7 ± 2.6
	.95	132 ± 13	116 ± 12	98 ± 10	82 ± 13	71 ± 11	62 ± 12	52.1 ± 7.7	42.6 ± 5.9	35.4 ± 5.4
8	.50	26.3 ± 1.5	22.8 ± 1.1	20.0 ± 1.2	17.4 ± 1.0	15.3 ± 1.1	13.2 ± 1.1	11.2 ± 0.9	9.71 ± .90	8.24 ± .70
	.75	40.8 ± 3.8	35.1 ± 2.8	30.5 ± 1.7	26.3 ± 1.8	22.7 ± 1.6	19.2 ± 1.2	16.7 ± 1.5	14.2 ± 1.4	12.0 ± 1.4
	.90	63.4 ± 8.5	55.1 ± 7.9	47.4 ± 7.7	41.5 ± 7.4	35.3 ± 5.9	30.0 ± 3.3	25.5 ± 2.7	21.7 ± 3.6	17.8 ± 2.8
	.95	86 ± 14	75 ± 17	63 ± 11	54 ± 11	47 ± 11	39.4 ± 7.8	33.3 ± 7.5	28.5 ± 6.6	22.9 ± 4.8
9	.50	21.1 ± 0.7	18.5 ± 0.7	16.1 ± 0.7	14.0 ± 0.5	12.2 ± 0.6	10.6 ± 0.4	9.16 ± .39	7.89 ± .31	6.60 ± .23
	.75	32.0 ± 1.0	28.2 ± 1.0	24.1 ± 1.1	20.8 ± 1.2	17.9 ± 0.6	15.4 ± 0.8	13.0 ± 0.4	11.1 ± 0.5	9.29 ± .41
	.90	48.6 ± 3.0	41.7 ± 1.9	35.6 ± 2.9	30.4 ± 3.0	25.8 ± 1.9	22.0 ± 2.0	18.8 ± 1.5	15.8 ± 0.9	12.8 ± 0.6
	.95	61.6 ± 3.2	52.6 ± 4.6	44.8 ± 2.4	38.2 ± 2.5	33.5 ± 2.5	28.8 ± 2.1	23.9 ± 1.1	20.0 ± 1.3	16.2 ± 1.5
10	.50	16.8 ± 0.6	14.7 ± 0.7	12.9 ± 0.3	11.4 ± 0.2	9.87 ± .29	8.53 ± .12	7.46 ± .20	6.39 ± .15	5.47 ± .07
	.75	25.9 ± 1.0	22.2 ± 0.3	19.1 ± 0.3	16.3 ± 0.2	14.1 ± 0.4	12.1 ± 0.3	10.3 ± 0.3	8.86 ± .20	7.48 ± .06
	.90	38.1 ± 2.8	33.5 ± 3.6	28.6 ± 3.2	24.7 ± 1.4	20.9 ± 1.7	17.7 ± 1.0	14.6 ± 1.1	11.9 ± 0.5	9.86 ± .13
	.95	47.3 ± 6.0	42.1 ± 6.7	36.3 ± 5.4	32.4 ± 6.0	26.8 ± 3.3	22.5 ± 2.6	18.6 ± 1.7	15.5 ± 2.1	11.9 ± 0.6

Table G
Estimated Percentiles of a Modulus Ratio Statistic: Mod F (j, p, k-1)

(k-1) = 31

p	1-α	j=13	j=12	j=11	j=10	j=9	j=8	j=7	j=6	j=5
4	.50	25.8 ± 1.5	21.6 ± 1.3	17.6 ± 1.2	14.2 ± 0.6	11.3 ± 0.3	8.60 ± .28	6.44 ± .19	4.75 ± .12	3.13 ± .09
	.75	49.7 ± 2.8	41.5 ± 4.2	34.1 ± 3.0	26.5 ± 1.5	20.5 ± 1.9	15.5 ± 1.5	11.2 ± 1.1	7.44 ± .63	4.66 ± .07
	.90	106 ± 15	87 ± 12	69 ± 14	54.0 ± 8.0	41.1 ± 6.5	31.0 ± 5.3	20.4 ± 4.1	13.3 ± 1.2	7.32 ± .49
	.95	175 ± 34	138 ± 20	108 ± 17	87 ± 11	66.7 ± 5.6	44.5 ± 5.1	31.5 ± 5.2	19.8 ± 2.5	11.0 ± 1.1
5	.50	17.2 ± 0.7	14.3 ± 0.6	11.7 ± 0.4	9.49 ± .59	7.50 ± .27	5.80 ± .21	4.41 ± .09	3.13 ± .07	
	.75	29.5 ± 1.3	24.9 ± 1.6	20.3 ± 2.0	16.1 ± 0.7	12.2 ± 0.8	9.23 ± .67	6.55 ± .44	4.30 ± .13	
	.90	57.5 ± 6.8	47.2 ± 5.5	35.9 ± 4.4	28.3 ± 3.8	21.4 ± 2.2	15.4 ± 2.1	10.5 ± 0.7	6.20 ± .35	
	.95	82 ± 14	69.3 ± 7.2	54.9 ± 6.1	43.6 ± 6.3	31.4 ± 3.7	21.8 ± 1.0	14.6 ± 1.1	8.30 ± .59	
6	.50	11.9 ± 0.7	9.89 ± .39	8.22 ± .17	6.63 ± .37	5.23 ± .20	4.05 ± .15	3.11 ± .09		
	.75	19.8 ± 0.9	16.4 ± 0.7	13.3 ± 0.4	10.3 ± 0.4	8.01 ± .09	5.84 ± .07	4.16 ± .08		
	.90	32.4 ± 1.7	26.7 ± 2.5	21.0 ± 1.5	16.5 ± 0.5	12.5 ± 0.7	8.63 ± .16	5.75 ± .31		
	.95	43.8 ± 6.7	36.0 ± 3.9	28.4 ± 3.2	22.8 ± 2.3	16.1 ± 1.1	11.3 ± 0.7	7.21 ± .57		
7	.50	9.14 ± .44	7.56 ± .25	6.22 ± .17	5.02 ± .08	3.93 ± .10	3.07 ± .13			
	.75	13.8 ± 0.6	11.3 ± 0.6	9.14 ± .37	7.15 ± .32	5.63 ± .13	4.06 ± .17			
	.90	21.1 ± 1.6	17.3 ± 1.3	13.5 ± 0.2	10.2 ± 0.5	7.98 ± .44	5.51 ± .16			
	.95	28.5 ± 5.6	22.6 ± 2.8	17.3 ± 2.3	13.3 ± 1.5	10.0 ± 0.9	6.84 ± .44			
8	.50	6.93 ± .41	5.80 ± .33	4.91 ± .22	3.91 ± .19	3.14 ± .09				
	.75	9.9 ± 1.0	8.35 ± .58	6.70 ± .37	5.20 ± .32	3.99 ± .19				
	.90	14.7 ± 1.4	11.9 ± 1.2	9.23 ± .51	7.05 ± .42	5.11 ± .48				
	.95	18.5 ± 3.7	14.9 ± 3.2	11.2 ± 1.0	8.59 ± .47	6.08 ± .54				
9	.50	5.51 ± .34	4.66 ± .24	3.81 ± .13	3.12 ± .06					
	.75	7.62 ± .44	6.22 ± .34	4.94 ± .28	3.93 ± .19					
	.90	10.4 ± 0.3	8.42 ± .66	6.62 ± .51	4.94 ± .30					
	.95	12.9 ± 0.6	10.4 ± 0.7	7.90 ± .74	5.84 ± .50					
10	.50	4.59 ± .05	3.82 ± .10	3.16 ± .06						
	.75	6.15 ± .15	4.95 ± .03	3.91 ± .04						
	.90	8.14 ± .33	6.50 ± .20	4.98 ± .14						
	.95	9.67 ± .30	7.54 ± .12	5.81 ± .08						

Table H.1
Some Total or Overall (Method One) Estimators for Dispersion Parameters

Given k subgroups of size n (with a total of N = nk observations) and
given one of the following dispersion statistics which has been computed using all N = nk values:

Dispersion Statistic	Estimators for SD(X) Biased	Estimators for SD(X) Unbiased	Estimators for V(X) Biased	Estimators for V(X) Unbiased	Degrees of Freedom	notes
s	s	$\frac{s}{c_4}$	----	s^2	(nk -1)	[1] [3] [4]
s_n	s_n	$\frac{s_n}{c_2}$	s_n^2	$\frac{nk}{nk-1} s_n^2$	(nk -1)	[2] [3] [4]
R	$\frac{R}{d_2^*}$	$\frac{R}{d_2}$	$\left(\frac{R}{d_2}\right)^2$	$\left(\frac{R}{d_2^*}\right)^2$	$\nu_{B(1,N)}$	[4] [5]

[1] s denotes the statistic with the divisor of (N-1) = (nk-1).
[2] s_n denotes the statistic with the divisor of N = nk.
[3] The unbiased estimators in these rows yield identical values.
[4] Use c_2, c_4, and d_2 values for subgroups of size N = nk: d_2^* value for ONE subgroup of size N = nk.
[5] $\nu_{B(1,N)}$ = degrees of freedom from Table B for 1 subgroup of size N = nk.

All Method One Estimators are subject to being inflated whenever the data display a lack of statistical control.

This method of estimation is therefore discouraged except in circumstances where the user knows, beyond any reasonable doubt, that the data have come from a physical process which displays statistical control. Whenever a lack of control may exist, Method Two Estimators should be used instead of Method One or Method Three Estimators.

Table H.2:
Some Within-Subgroup (Method Two) Estimators for Dispersion Parameters

Given k subgroups of size n, and a dispersion statistic computed for each subgroup, the following estimators are based upon the AVERAGE of the k dispersion statistics:

Name of Estimator	Estimators for SD(X) Biased	Unbiased	Estimators for V(X) Biased	Unbiased	Degrees of Freedom	notes
Pooled Variance	$\sqrt{\overline{s^2}}$	$\frac{\sqrt{\overline{s^2}}}{c_4}$	----	$\overline{s^2}$	k(n-1)	[1]
Average Std. Dev.	$\bar{s}$	$\frac{\bar{s}}{c_4}$	$(\bar{s})^2$	$\left(\frac{\bar{s}}{c_4^*}\right)^2$	ν_a	[2] [3] [4]
Average RMS Dev.	$\bar{s}_n$	$\frac{\bar{s}_n}{c_2}$	$(\bar{s}_n)^2$	$\left(\frac{\bar{s}_n}{c_2^*}\right)^2$	ν_a	[2] [3] [4]
Average Range	$\frac{\bar{R}}{d_2^*}$	$\frac{\bar{R}}{d_2}$	$\left(\frac{\bar{R}}{d_2}\right)^2$	$\left(\frac{\bar{R}}{d_2^*}\right)^2$	$\nu_{B(k,n)}$	[5] [6]

[1] Use c_4 value for subgroups of size (1 + k(n-1)).

[2] Use c_2 and c_4 values for subgroups of size n. The c_2^* and c_4^* values are defined below.

[3] The unbiased estimators in these rows yield identical values.

[4] ν_a = degrees of freedom from formula below.

[5] Use d_2 value for subgroups of size n, and use d_2^* value for k subgroups of size n.

[6] $\nu_{B(k,n)}$ = degrees of freedom from Table B for k subgroups of size n.

The symbol ν_a will denote the "equivalent degrees of freedom" for the Average Standard Deviation Estimators. These degrees of freedom may be found according to the following formula:

If $n = 2$ or $n = 3$, then $\nu_a = \nu_{B(k,n)}$

If $n \geq 4$, then $\nu_a = [k(n-1) - 0.23(k-1)]$.

These degrees of freedom may be non-integral, but in use they will be rounded to the nearest integer.

Dr. James Maynard supplied the following formulas for c_2^* and c_4^*:

$$c_2^* = \sqrt{\left(\frac{n-1}{nk}\right) + \left(\frac{k-1}{k}\right)(c_2)^2} \quad \text{and} \quad c_4^* = \sqrt{\left(\frac{1}{k}\right) + \left(\frac{k-1}{k}\right)(c_4)^2}$$

Table H.3:
Some Between Subgroup (Method Three) Estimators for Dispersion Parameters

Given k subgroups of size n, and
a single dispersion statistic computed using the k subgroup averages:

Dispersion Statistic	Estimators for SD(X) Biased	Unbiased	Estimators for V(X) Biased	Unbiased	Degrees of Freedom	notes
$s_{\bar{x}}$	$s_{\bar{x}}\sqrt{n}$	$\frac{s_{\bar{x}}}{c_4}\sqrt{n}$	----	$n\, s_{\bar{x}}^2$	$(k-1)$	[1] [3]
$s_{k\bar{x}}$	$s_{k\bar{x}}\sqrt{n}$	$\frac{s_{k\bar{x}}}{c_2}\sqrt{n}$	$n\, s_{k\bar{x}}^2$	$\frac{nk}{k-1} s_{k\bar{x}}^2$	$(k-1)$	[2] [3]
$R_{\bar{x}}$	$\frac{R_{\bar{x}}}{d_2^*}\sqrt{n}$	$\frac{R_{\bar{x}}}{d_2}\sqrt{n}$	$n\left(\frac{R_{\bar{x}}}{d_2}\right)^2$	$n\left(\frac{R_{\bar{x}}}{d_2^*}\right)^2$	$\nu_{B(1,k)}$	[4] [5]

[1] $s_{\bar{x}}$ has a divisor of (k-1): use c_4 value for subgroups of size k.

[2] $s_{k\bar{x}}$ has divisor of (k): use c_2 value for subgroups of size k.

[3] The unbiased estimators in these rows yield identical values.

[4] Use d_2 value for subgroups of size k: use d_2^* value for 1 subgroup of size k.

[5] $\nu_{B(1,k)}$ = degrees of freedom from Table B for 1 subgroup of size k.

All Method Three Estimators are subject to severe inflation whenever the subgroup averages fail to display statistical control.

For this reason, Method Three estimates are rarely computed except when they are to be compared with Method Two estimates.

Analysis of Means Worksheet

A. BASIC INFORMATION:

Data Set Name ______________________________ k = ____ n = ____

Grand Average = $\bar{\bar{X}}$ = [ ] Average Range = $\bar{R}$ = ____________

B. FIND THE CONTROL LIMITS FOR THE RANGE CHART: (Table C)

$LCL_R = D_3\bar{R} =$ $UCL_R = D_4\bar{R} =$

Are any subgroup ranges outside these limits?

C. ESTIMATE THE STANDARD DEVIATION OF X: (Using k = _____ subgroups of size n = ____)

Use Table B to find: ν = _____ d_2^* = ________

$$\text{EST. SD(X)} = \frac{\bar{R}}{d_2^*} = \text{\_\_\_\_\_\_\_\_} = \text{\_\_\_\_\_\_\_\_\_\_\_\_}$$

D. ESTIMATE THE STANDARD DEVIATION OF THE SUBGROUP AVERAGES:

$$\text{EST. SD}(\bar{X}) = \frac{\text{EST. SD(X)}}{\sqrt{n}} = \text{\_\_\_\_\_\_\_\_\_\_} = \boxed{}$$

The degrees of freedom for this estimate is the same as in Step C above.

E. FIND THE VALUE FOR H:

Choose α = _____. From Step C above, ν = _____.

The number of subgroup averages being compared is k = _____.

From Table D, the value for H is = [ ]

F. FIND THE DECISION LIMITS FOR THE ANOM CHART:

The decision limits for the ANOM chart are $\bar{\bar{X}} \pm H\left[\text{EST. SD}(\bar{X})\right]$:

$UDL_{\bar{X}} =$

$LDL_{\bar{X}} =$

* Permission to reproduce this worksheet is hereby extended to the owner of this text.

Analysis of Variance Worksheet

Data Set Name ______________________________ k=_____ n=_____

The values that are enclosed in boxes will be inserted into the ANOVA table.

I. TOTAL or OVERALL VARIATION

A. Find s^2 using all nk observations = s^2 = ____________

B. Check Value = Grand Average = $\bar{\bar{X}}$ = ____________

C. Total Degrees of Freedom = (nk - 1) = [ ]

D. Total Sum of Squares = (nk-1) s^2 = TSS = [ ].

II. WITHIN-SUBGROUP VARIATION

A. Obtain the Subgroup Average and Subgroup Variance for each of the k subgroups:

Subgroup Averages, $\bar{X}$			Subgroup Variances, s^2		
1_______	6_______	11_______	1_______	6_______	11_______
2_______	7_______	12_______	2_______	7_______	12_______
3_______	8_______	13_______	3_______	8_______	13_______
4_______	9_______	14_______	4_______	9_______	14_______
5_______	10_______	15_______	5_______	10_______	15_______

B. Average these s^2 values to get the Mean Square Within = $\overline{s^2}$ =

MSW = [ ]

C. Within-Subgroup Degrees of Freedom =

k (n - 1) = [ ]

D. Within-Subgroup Sum of Squares =

k(n-1) MSW = SSW = [ ].

III. BETWEEN-SUBGROUP VARIATION

A. Calculate s^2 using the k subgroup averages = $s_{\bar{x}}^2$ = ____________

B. Check Value = Grand Average = $\bar{\bar{X}}$ = ____________

(Does this value match the value in I. B. above? If not, you have made a mistake.)

C. Find the Mean Square Between by multiplying the $s_{\bar{x}}^2$ value in III.A. by n =

n $s_{\bar{x}}^2$ = MSB = [ ]

D. Between-Subgroup Degrees of Freedom =

(k - 1) = [ ]

E. Between-Subgroup Sum of Squares =

(k-1) MSB = SSB = [ ]

(Pythagoras claims that (III.E.) + (II.D.) = (I.D.). If they don't, guess who's wrong.)

Working With Contrasts

For data consisting of k subgroups of size n:

DEFINING A CONTRAST:

- Each subgroup is assigned a coefficient... $\{c_1, c_2, c_3, ..., c_k\}$.
- The sum of the k coefficients MUST be zero.
- Subgroups with coefficients = 0 are excluded from the comparison.
- The contrast will compare those subgroups with positive coefficients with those having negative coefficients.

FINDING THE SUM OF SQUARES FOR A CONTRAST:

- The estimated contrast value is $\hat{C} = c_1\bar{X}_1 + c_2\bar{X}_2 + \ldots + c_k\bar{X}_k$.
- The adjustment term is $\frac{1}{n}\sum c_j^2 = \frac{\text{sum of squared coefficients}}{\text{subgroup size}}$.
- The sum of squares for a contrast has one degree of freedom and is given by:

$$SS(C) = \frac{(\hat{C})^2}{\frac{1}{n}\sum c_j^2} = \frac{\text{square of estimated contrast value}}{\text{adjustment term}}$$

- Orthogonal contrasts have independent sums of squares.
- The F-ratio for a contrast = $\frac{SS(C)}{MSW}$, and it has 1 and k(n-1) degrees of freedom.
- For sets of orthogonal contrasts, SS(C) values may be used in a scree plot.

FINDING THE ESTIMATED CONTRAST EFFECT:

- Denote the estimated contrast effect by $\hat{l}$

$$\hat{l} = \frac{\hat{C}}{\text{sum of positive } c_j} = \frac{\hat{C}}{\frac{1}{2}\sum |c_j|}$$

- Interpret $\hat{l}$ as the difference of the averages for the two groups being compared.

Data Set Eleven

Run	ABCDEFG	Y_1	Y_2
1	+ + + + + + +	56.94	56.89
2	+ + + + + + -	56.00	57.25
3	+ + + + + - +	56.26	53.90
4	+ + + + + - -	55.42	54.10
5	+ + + + - + +	56.14	58.71
6	+ + + + - + -	57.50	56.70
7	+ + + + - - +	57.08	54.48
8	+ + + + - - -	56.56	53.26
9	+ + + - + + +	50.90	49.69
10	+ + + - + + -	50.30	51.28
11	+ + + - + - +	47.65	49.02
12	+ + + - + - -	49.27	49.74
13	+ + + - - + +	50.75	51.26
14	+ + + - - + -	50.27	50.71
15	+ + + - - - +	50.20	48.25
16	+ + + - - - -	48.34	49.15
17	+ + - + + + +	49.62	49.53
18	+ + - + + + -	49.37	49.18
19	+ + - + + - +	47.81	47.67
20	+ + - + + - -	47.41	45.60
21	+ + - + - + +	47.81	47.61
22	+ + - + - + -	46.39	47.62
23	+ + - + - - +	45.90	46.44
24	+ + - + - - -	45.96	46.18
25	+ + - - + + +	47.01	47.29
26	+ + - - + + -	47.29	48.81
27	+ + - - + - +	44.93	42.48
28	+ + - - + - -	45.48	45.72
29	+ + - - - + +	47.70	46.31
30	+ + - - - + -	46.40	48.29
31	+ + - - - - +	45.02	43.68
32	+ + - - - - -	45.67	45.42
33	+ - + + + + +	56.81	56.63
34	+ - + + + + -	56.09	58.57
35	+ - + + + - +	55.56	54.87
36	+ - + + + - -	56.14	54.09
37	+ - + + - + +	55.80	57.87
38	+ - + + - + -	57.91	57.50
39	+ - + + - - +	55.16	56.82
40	+ - + + - - -	54.71	56.20
41	+ - + - + + +	53.45	50.76
42	+ - + - + + -	51.54	51.66
43	+ - + - + - +	48.64	47.83
44	+ - + - + - -	47.08	49.21
45	+ - + - - + +	50.17	51.99
46	+ - + - - + -	50.30	52.19
47	+ - + - - - +	49.64	49.04
48	+ - + - - - -	50.02	48.16
49	+ - - + + + +	48.56	48.16
50	+ - - + + + -	47.71	49.04
51	+ - - + + - +	46.16	47.75
52	+ - - + + - -	46.00	47.48
53	+ - - + - + +	49.01	49.06
54	+ - - + - + -	50.49	48.93
55	+ - - + - - +	47.58	46.94
56	+ - - + - - -	45.14	46.17
57	+ - - - + + +	47.79	47.59
58	+ - - - + + -	45.79	45.57
59	+ - - - + - +	46.67	45.09
60	+ - - - + - -	47.05	44.62
61	+ - - - - + +	46.11	48.47
62	+ - - - - + -	48.32	45.34
63	+ - - - - - +	44.71	46.46
64	+ - - - - - -	46.25	45.50
65	- + + + + + +	55.82	57.89
66	- + + + + + -	57.00	56.82
67	- + + + + - +	53.85	55.36
68	- + + + + - -	53.75	54.83
69	- + + + - + +	57.98	57.33
70	- + + + - + -	56.20	58.15
71	- + + + - - +	54.77	55.79
72	- + + + - - -	54.38	54.18
73	- + + - + + +	50.94	51.48
74	- + + - + + -	49.29	50.23
75	- + + - + - +	48.64	47.83
76	- + + - + - -	47.08	49.21
77	- + + - - + +	50.81	51.01
78	- + + - - + -	51.95	50.01
79	- + + - - - +	49.64	49.04
80	- + + - - - -	50.02	48.16
81	- + - + + + +	50.59	49.03
82	- + - + + + -	49.57	49.17
83	- + - + + - +	46.75	46.56
84	- + - + + - -	46.94	47.80
85	- + - + - + +	49.99	51.08
86	- + - + - + -	49.47	47.89
87	- + - + - - +	44.99	44.39
88	- + - + - - -	47.75	47.69
89	- + - - + + +	47.36	46.88
90	- + - - + + -	46.67	47.12
91	- + - - + - +	45.10	44.66
92	- + - - + - -	44.42	44.49
93	- + - - - + +	45.95	46.85
94	- + - - - + -	46.40	45.28
95	- + - - - - +	43.60	43.93
96	- + - - - - -	45.34	44.88
97	- - + + + + +	56.81	56.91
98	- - + + + + -	57.89	57.07
99	- - + + + - +	55.68	55.39
100	- - + + + - -	55.44	54.96
101	- - + + - + +	57.00	56.26
102	- - + + - + -	57.15	56.38
103	- - + + - - +	56.60	56.03
104	- - + + - - -	54.17	54.59
105	- - + - + + +	53.51	51.87
106	- - + - + + -	50.39	51.21
107	- - + - + - +	48.41	50.07
108	- - + - + - -	48.01	48.20
109	- - + - - + +	54.10	50.12
110	- - + - - + -	51.22	51.07
111	- - + - - - +	49.19	48.76
112	- - + - - - -	49.90	48.83
113	- - - + + + +	47.53	50.29
114	- - - + + + -	47.33	48.19
115	- - - + + - +	48.83	47.34
116	- - - + + - -	47.77	46.11
117	- - - + - + +	48.65	47.29
118	- - - + - + -	47.75	49.40
119	- - - + - - +	47.32	46.88
120	- - - + - - -	46.67	47.65
121	- - - - + + +	47.09	47.50
122	- - - - + + -	47.02	46.42
123	- - - - + - +	43.49	43.95
124	- - - - + - -	43.71	46.06
125	- - - - - + +	46.92	46.79
126	- - - - - + -	48.58	47.40
127	- - - - - - +	44.64	44.80
128	- - - - - - -	46.31	44.07

The Basic 8-Run Plackett-Burman Design

Run	A	B	C	D	E	F	G
1	−	−	−	−	−	−	−
2	−	−	−	+	+	+	+
3	−	+	+	+	+	−	−
4	−	+	+	−	−	+	+
5	+	+	−	−	+	+	−
6	+	+	−	+	−	−	+
7	+	−	+	+	−	+	−
8	+	−	+	−	+	−	+

The Basic 12-Run Plackett-Burman Design

Run	A	B	C	D	E	F	G	H	I	J	K
1	−	−	−	−	−	−	−	−	−	−	−
2	−	−	−	+	+	+	−	−	+	+	+
3	−	−	+	−	+	+	+	+	−	−	+
4	−	+	+	−	+	−	−	+	+	+	−
5	−	+	+	+	−	−	+	−	+	−	+
6	−	+	−	+	−	+	+	+	−	+	−
7	+	+	−	+	+	−	−	+	−	−	+
8	+	+	−	−	+	+	+	−	+	−	−
9	+	+	+	−	−	+	−	−	−	+	+
10	+	−	+	+	−	+	−	+	+	−	−
11	+	−	+	+	+	−	+	−	−	+	−
12	+	−	−	−	−	−	+	+	+	+	+

The Basic 16-Run Plackett-Burman Design

Run	A	B	C	D	E	F	G	H	I	J	K	L	M	N	O
1	−	−	−	−	−	−	−	−	−	−	−	−	−	−	−
2	−	−	−	−	−	−	−	+	+	+	+	+	+	+	+
3	−	−	−	+	+	+	+	+	+	+	+	−	−	−	−
4	−	−	−	+	+	+	+	−	−	−	−	+	+	+	+
5	−	+	+	+	+	−	−	−	−	+	+	+	+	−	−
6	−	+	+	+	+	−	−	+	+	−	−	−	−	+	+
7	−	+	+	−	−	+	+	+	+	−	−	+	+	−	−
8	−	+	+	−	−	+	+	−	−	+	+	−	−	+	+
9	+	+	−	−	+	+	−	−	+	+	−	−	+	+	−
10	+	+	−	−	+	+	−	+	−	−	+	+	−	−	+
11	+	+	−	+	−	−	+	+	−	−	+	−	+	+	−
12	+	+	−	+	−	−	+	−	+	+	−	+	−	−	+
13	+	−	+	+	−	+	−	−	+	−	+	+	−	+	−
14	+	−	+	+	−	+	−	+	−	+	−	−	+	−	+
15	+	−	+	−	+	−	+	+	−	+	−	+	−	+	−
16	+	−	+	−	+	−	+	−	+	−	+	−	+	−	+

The Basic 20-Run Plackett-Burman Design

Contrast Labels

Run	A	B	C	D	E	F	G	H	I	J	K	L	M	N	O	P	Q	R	S
1	-	-	-	-	-	-	-	-	-	-	-	-	-	-	-	-	-	-	-
2	-	-	-	-	+	-	+	+	-	+	+	+	-	-	+	-	+	+	+
3	-	-	-	+	+	+	+	+	+	-	+	+	-	+	-	+	-	-	-
4	-	-	+	+	-	+	+	-	+	-	-	+	+	-	+	-	-	+	+
5	-	-	+	+	+	-	-	+	-	-	-	-	+	+	+	+	+	+	-
6	-	+	+	+	+	-	-	-	+	+	+	-	-	+	-	-	-	+	+
7	-	+	+	-	+	+	+	-	-	+	+	-	+	-	+	+	-	-	-
8	-	+	+	-	-	+	-	+	+	+	-	+	-	+	+	-	+	-	-
9	-	+	-	-	-	+	-	+	+	-	+	-	+	-	-	+	+	+	+
10	-	+	-	+	-	-	+	-	-	+	-	+	+	+	-	+	+	-	+
11	+	+	-	+	-	+	+	-	-	-	+	-	-	+	+	-	+	+	-
12	+	+	-	+	+	+	-	+	-	+	-	+	+	-	-	-	-	+	-
13	+	+	-	-	+	-	+	+	+	-	-	-	+	+	+	-	-	-	+
14	+	+	+	-	+	-	+	-	+	-	-	+	-	-	-	+	+	+	-
15	+	+	+	+	-	-	-	+	-	-	+	+	-	-	+	+	-	-	+
16	+	-	+	+	-	-	+	+	+	+	+	-	+	-	-	-	+	-	-
17	+	-	+	-	-	+	+	+	-	+	-	-	-	+	-	+	-	+	+
18	+	-	+	-	+	+	-	-	-	-	+	+	+	+	-	-	+	-	+
19	+	-	-	-	-	-	-	-	+	+	+	+	+	+	+	+	-	+	-
20	+	-	-	+	+	+	-	-	+	+	-	-	-	-	+	+	+	-	+
	A	B	C	D	E	F	G	H	I	J	K	L	M	N	O	P	Q	R	S

The Basic 24-Run Plackett-Burman Design

Contrast Labels

Run	A	B	C	D	E	F	G	H	I	J	K	L	M	N	O	P	Q	R	S	T	U	V	W
1	-	-	-	-	-	-	-	-	-	-	-	-	-	-	-	-	-	-	-	-	-	-	-
2	-	-	-	-	-	-	+	-	+	-	+	+	+	+	+	+	-	-	-	+	+	+	+
3	-	-	-	+	-	+	-	+	-	-	+	+	+	-	-	-	+	+	+	+	-	+	+
4	-	-	+	+	+	+	+	-	+	+	+	+	-	-	-	-	-	+	-	+	+	-	-
5	-	-	+	+	-	+	+	+	-	+	-	+	-	+	+	+	-	-	+	-	-	-	+
6	-	-	+	-	+	+	-	-	-	-	-	-	-	+	+	+	+	+	+	+	+	+	-
7	-	+	+	-	+	+	-	-	+	+	+	-	+	-	+	-	-	-	+	-	-	+	+
8	-	+	+	-	+	-	-	+	+	-	+	+	-	+	-	+	+	+	-	-	-	-	+
9	-	+	+	+	-	-	-	+	-	+	+	-	+	+	+	-	+	-	-	+	+	-	-
10	-	+	-	+	-	-	+	-	+	+	-	-	-	+	-	-	+	+	+	-	+	+	+
11	-	+	-	+	+	-	+	+	+	-	-	-	+	-	+	+	-	+	+	+	-	-	-
12	-	+	-	-	+	+	+	+	-	+	-	+	+	-	-	+	+	-	-	-	+	+	-
13	+	+	-	-	+	+	+	+	-	-	+	-	-	+	-	-	-	-	+	+	+	-	+
14	+	+	-	-	-	+	-	+	+	+	-	+	-	+	+	-	-	+	-	+	-	+	-
15	+	+	-	+	-	+	-	-	+	-	+	+	-	-	+	+	+	-	+	-	+	-	-
16	+	+	+	+	+	-	+	-	-	-	-	+	-	-	+	-	+	-	-	+	-	+	+
17	+	+	+	+	-	+	+	-	-	-	+	-	+	+	-	+	-	+	-	-	-	+	-
18	+	+	+	-	-	-	-	-	-	+	-	+	+	-	-	+	-	+	+	+	+	-	+
19	+	-	+	-	-	-	+	+	+	+	+	-	-	-	-	+	+	-	+	+	-	+	-
20	+	-	+	-	-	+	+	+	+	-	-	-	+	-	+	-	+	+	-	-	+	-	+
21	+	-	+	+	+	-	-	+	+	-	-	+	+	+	-	-	-	-	+	-	+	+	-
22	+	-	-	+	+	+	-	-	+	+	-	-	+	+	-	+	+	-	-	+	-	-	+
23	+	-	-	+	+	-	-	+	-	+	+	-	-	-	+	+	-	+	-	-	+	+	+
24	+	-	-	-	+	-	+	-	-	+	+	+	+	+	+	-	+	+	+	-	-	-	-
	A	B	C	D	E	F	G	H	I	J	K	L	M	N	O	P	Q	R	S	T	U	V	W

The Basic 28-Run Plackett-Burman Design

Contrast Labels

Run	A	B	C	D	E	F	G	H	I	J	K	L	M	N	O	P	Q	R	S	T	U	V	W	X	Y	Z	*1*
1	-	-	-	-	-	-	-	-	-	-	-	-	-	-	-	-	-	-	-	-	-	-	-	-	-	-	-
2	-	-	-	-	+	-	-	-	-	+	-	-	-	-	+	+	+	+	+	+	+	+	+	-	+	+	+
3	-	-	-	+	+	+	-	+	+	+	+	-	-	+	+	-	-	-	+	+	-	+	-	+	+	-	-
4	-	-	-	+	-	-	+	+	-	+	+	+	+	-	+	-	-	+	-	+	+	-	-	+	-	+	+
5	-	-	+	+	-	-	-	+	-	-	+	+	+	+	+	+	+	+	-	-	-	+	+	-	+	-	-
6	-	-	+	-	-	+	-	+	+	+	-	+	-	+	-	+	-	-	-	-	+	+	+	+	-	+	+
7	-	-	+	-	+	-	+	-	+	-	-	+	+	+	-	-	-	+	+	+	-	-	+	+	+	+	-
8	-	+	+	-	+	-	+	-	+	-	+	+	-	-	+	+	+	-	-	+	-	+	-	+	-	-	+
9	-	+	+	+	+	-	-	+	+	-	+	-	+	-	-	+	-	-	+	-	+	-	-	-	+	+	+
10	-	+	+	+	+	+	-	-	+	+	-	-	+	-	+	-	+	+	-	-	+	-	+	+	-	-	-
11	-	+	+	+	-	+	+	-	-	+	-	-	+	+	-	+	-	+	+	+	-	+	-	-	-	-	+
12	-	+	-	+	-	+	+	-	-	+	+	+	-	-	-	+	+	-	+	-	-	-	+	+	+	+	-
13	-	+	-	-	-	+	+	+	+	-	+	-	-	+	-	-	+	+	-	+	+	-	+	-	+	-	+
14	-	+	-	-	+	+	+	+	-	-	-	+	+	+	+	-	+	-	+	-	+	+	-	-	-	+	-
15	+	+	-	-	+	-	+	+	-	+	-	-	+	+	+	+	-	-	-	-	-	-	+	+	+	-	+
16	+	+	-	+	+	-	-	+	+	+	-	+	-	+	-	+	+	+	-	+	-	-	-	-	-	+	-
17	+	+	-	+	+	-	-	-	-	-	+	+	-	+	-	-	-	+	+	-	+	+	+	+	-	-	+
18	+	+	-	+	-	+	-	-	+	-	-	+	+	-	+	-	-	-	-	+	-	+	+	-	+	+	+
19	+	+	+	-	-	+	-	+	-	-	-	+	-	-	+	+	-	+	+	+	+	-	-	+	+	-	-
20	+	+	+	-	-	-	+	+	+	+	+	-	-	-	+	-	-	+	+	-	-	+	+	-	-	+	-
21	+	+	+	-	-	-	-	-	-	+	+	-	+	+	-	-	+	-	-	+	+	+	-	+	+	+	-
22	+	-	+	-	+	+	-	+	-	+	+	+	+	-	-	-	+	-	+	+	-	-	+	-	-	-	+
23	+	-	+	+	+	+	+	+	-	-	-	-	-	-	-	-	+	+	-	-	-	+	-	+	+	+	+
24	+	-	+	+	+	+	+	-	-	-	+	-	-	+	+	+	-	-	-	+	+	-	+	-	-	+	-
25	+	-	+	+	-	-	+	-	+	+	-	+	-	+	+	-	+	-	+	-	+	-	-	-	+	-	+
26	+	-	-	+	-	-	+	+	+	-	-	-	+	-	-	+	+	-	+	+	+	+	+	+	-	-	-
27	+	-	-	-	-	+	-	-	+	-	+	-	+	+	+	+	+	+	+	-	-	-	-	+	-	+	+
28	+	-	-	-	+	+	+	-	+	+	+	+	+	-	-	+	-	+	-	-	+	+	-	-	+	-	-
	A	B	C	D	E	F	G	H	I	J	K	L	M	N	O	P	Q	R	S	T	U	V	W	X	Y	Z	*1*

The Basic 32-Run Plackett-Burman Design

Contrast Labels

Run	A	B	C	D	E	F	G	H	I	J	K	L	M	N	O	P	Q	R	S	T	U	V	W	X	Y	Z	*1*	2	*3*	*4*	*5*
1	-	-	-	-	-	-	-	-	-	-	-	-	-	-	-	-	-	-	-	-	-	-	-	-	-	-	-	-	-	-	-
2	-	-	-	-	-	-	-	-	-	-	-	-	-	-	-	+	+	+	+	+	+	+	+	+	+	+	+	+	+	+	+
3	-	-	-	-	-	-	-	+	+	+	+	+	+	+	+	+	+	+	+	+	+	+	+	-	-	-	-	-	-	-	-
4	-	-	-	-	-	-	-	+	+	+	+	+	+	+	+	-	-	-	-	-	-	-	-	+	+	+	+	+	+	+	+
5	-	-	-	+	+	+	+	+	+	+	+	-	-	-	-	-	-	-	-	+	+	+	+	+	+	+	+	-	-	-	-
6	-	-	-	+	+	+	+	+	+	+	+	-	-	-	-	+	+	+	+	-	-	-	-	-	-	-	-	+	+	+	+
7	-	-	-	+	+	+	+	-	-	-	-	+	+	+	+	+	+	+	+	-	-	-	-	+	+	+	+	-	-	-	-
8	-	-	-	+	+	+	+	-	-	-	-	+	+	+	+	-	-	-	-	+	+	+	+	-	-	-	-	+	+	+	+
9	-	+	+	+	+	-	-	-	-	+	+	+	+	-	-	-	-	+	+	+	+	-	-	-	-	+	+	+	+	-	-
10	-	+	+	+	+	-	-	-	-	+	+	+	+	-	-	+	+	-	-	-	-	+	+	+	+	-	-	-	-	+	+
11	-	+	+	+	+	-	-	+	+	-	-	-	-	+	+	+	+	-	-	-	-	+	+	-	-	+	+	+	+	-	-
12	-	+	+	+	+	-	-	+	+	-	-	-	-	+	+	-	-	+	+	+	+	-	-	+	+	-	-	-	-	+	+
13	-	+	+	-	-	+	+	+	+	-	-	+	+	-	-	-	-	+	+	-	-	+	+	+	+	-	-	+	+	-	-
14	-	+	+	-	-	+	+	+	+	-	-	+	+	-	-	+	+	-	-	+	+	-	-	-	-	+	+	-	-	+	+
15	-	+	+	-	-	+	+	-	-	+	+	-	-	+	+	+	+	-	-	+	+	-	-	+	+	-	-	+	+	-	-
16	-	+	+	-	-	+	+	-	-	+	+	-	-	+	+	-	-	+	+	-	-	+	+	-	-	+	+	-	-	+	+
17	+	+	-	-	+	+	-	-	+	+	-	-	+	+	-	-	+	+	-	-	+	+	-	-	+	+	-	-	+	+	-
18	+	+	-	-	+	+	-	-	+	+	-	-	+	+	-	+	-	-	+	+	-	-	+	+	-	-	+	+	-	-	+
19	+	+	-	-	+	+	-	+	-	-	+	+	-	-	+	+	-	-	+	+	-	-	+	-	+	+	-	-	+	+	-
20	+	+	-	-	+	+	-	+	-	-	+	+	-	-	+	-	+	+	-	-	+	+	-	+	-	-	+	+	-	-	+
21	+	+	-	+	-	-	+	+	-	-	+	-	+	+	-	-	+	+	-	+	-	-	+	+	-	-	+	-	+	+	-
22	+	+	-	+	-	-	+	+	-	-	+	-	+	+	-	+	-	-	+	-	+	+	-	-	+	+	-	+	-	-	+
23	+	+	-	+	-	-	+	-	+	+	-	+	-	-	+	+	-	-	+	-	+	+	-	+	-	-	+	-	+	+	-
24	+	+	-	+	-	-	+	-	+	+	-	+	-	-	+	-	+	+	-	+	-	-	+	-	+	+	-	+	-	-	+
25	+	-	+	+	-	+	-	-	+	-	+	+	-	+	-	-	+	-	+	+	-	+	-	-	+	-	+	+	-	+	-
26	+	-	+	+	-	+	-	-	+	-	+	+	-	+	-	+	-	+	-	-	+	-	+	+	-	+	-	-	+	-	+
27	+	-	+	+	-	+	-	+	-	+	-	-	+	-	+	+	-	+	-	-	+	-	+	-	+	-	+	+	-	+	-
28	+	-	+	+	-	+	-	+	-	+	-	-	+	-	+	-	+	-	+	+	-	+	-	+	-	+	-	-	+	-	+
29	+	-	+	-	+	-	+	+	-	+	-	+	-	+	-	-	+	-	+	-	+	-	+	+	-	+	-	+	-	+	-
30	+	-	+	-	+	-	+	+	-	+	-	+	-	+	-	+	-	+	-	+	-	+	-	-	+	-	+	-	+	-	+
31	+	-	+	-	+	-	+	-	+	-	+	-	+	-	+	+	-	+	-	+	-	+	-	+	-	+	-	+	-	+	-
32	+	-	+	-	+	-	+	-	+	-	+	-	+	-	+	-	+	-	+	-	+	-	+	-	+	-	+	-	+	-	+
	A	B	C	D	E	F	G	H	I	J	K	L	M	N	O	P	Q	R	S	T	U	V	W	X	Y	Z	*1*	2	*3*	*4*	*5*

The Basic 64-Run Plackett-Burman Design: Part I

Contrast Labels:

Run	ABCD	EFGH	IJKL	MNOP	QRST	UVWX	YZab	cdef	ghij	klmn	opqr	stuv	wxyz	αβγδ	ηθμπ	ρσω
			111	1111	1112	2222	2222	2333	3333	3334	4444	4444	4555	5555	5556	666
	1234	5678	9012	3456	7890	1234	5678	9012	3456	7890	1234	5678	9012	3456	7890	123
1	----	----	----	----	----	----	----	----	----	----	----	----	----	----	----	---
2	----	----	----	----	----	----	----	---+	++++	++++	++++	++++	++++	++++	++++	+++
3	----	----	----	---+	++++	++++	++++	++++	++++	++++	++++	+++-	----	----	----	---
4	----	----	----	---+	++++	++++	++++	+++-	----	----	----	---+	++++	++++	++++	+++
5	----	---+	++++	++++	++++	+++-	----	----	----	---+	++++	++++	++++	+++-	----	---
6	----	---+	++++	++++	++++	+++-	----	---+	++++	+++-	----	----	----	---+	++++	+++
7	----	---+	++++	+++-	----	---+	++++	++++	++++	+++-	----	---+	++++	+++-	----	---
8	----	---+	++++	+++-	----	---+	++++	+++-	----	---+	++++	+++-	----	---+	++++	+++
9	---+	++++	+++-	----	---+	++++	+++-	----	---+	++++	+++-	----	---+	++++	+++-	---
10	---+	++++	+++-	----	---+	++++	+++-	---+	+++-	----	---+	++++	+++-	----	---+	+++
11	---+	++++	+++-	---+	+++-	----	---+	++++	+++-	----	---+	+++-	---+	++++	+++-	---
12	---+	++++	+++-	---+	+++-	----	---+	+++-	---+	++++	+++-	---+	+++-	----	---+	+++
13	---+	+++-	---+	++++	+++-	---+	+++-	----	---+	+++-	---+	++++	+++-	---+	+++-	---
14	---+	+++-	---+	++++	+++-	---+	+++-	---+	+++-	---+	+++-	----	---+	+++-	---+	+++
15	---+	+++-	---+	+++-	---+	+++-	---+	++++	+++-	---+	+++-	---+	+++-	---+	+++-	---
16	---+	+++-	---+	+++-	---+	+++-	---+	+++-	---+	+++-	---+	+++-	---+	+++-	---+	+++
17	-+++	+---	-+++	+---	-+++	+---	-+++	+---	-+++	+---	-+++	+---	-+++	+---	-+++	+--
18	-+++	+---	-+++	+---	-+++	+---	-+++	+--+	+---	-+++	+---	-+++	+---	-+++	+---	-++
19	-+++	+---	-+++	+--+	+---	-+++	+---	-+++	+---	-+++	+---	-++-	-+++	+---	-+++	+--
20	-+++	+---	-+++	+--+	+---	-+++	+---	-++-	-+++	+---	-+++	+--+	+---	-+++	+---	-++
21	-+++	+--+	+---	-+++	+---	-++-	-+++	+---	-+++	+--+	+---	-+++	+---	-++-	-+++	+--
22	-+++	+--+	+---	-+++	+---	-++-	-+++	+--+	+---	-++-	-+++	+---	-+++	+--+	+---	-++
23	-+++	+--+	+---	-++-	-+++	+--+	+---	-+++	+---	-++-	-+++	+--+	+---	-++-	-+++	+--
24	-+++	+--+	+---	-++-	-+++	+--+	+---	-++-	-+++	+--+	+---	-++-	-+++	+--+	+---	-++
25	-++-	-+++	+--+	+---	-++-	-+++	+--+	+---	-++-	-+++	+--+	+---	-++-	-+++	+--+	+--
26	-++-	-+++	+--+	+---	-++-	-+++	+--+	+--+	+--+	+---	-++-	-+++	+--+	+---	-++-	-++
27	-++-	-+++	+--+	+--+	+--+	+---	-++-	-+++	+--+	+---	-++-	-++-	-++-	-+++	+--+	+--
28	-++-	-+++	+--+	+--+	+--+	+---	-++-	-++-	-++-	-+++	+--+	+--+	+--+	+---	-++-	-++
29	-++-	-++-	-++-	-+++	+--+	+--+	+--+	+---	-++-	-++-	-++-	-+++	+--+	+--+	+--+	+--
30	-++-	-++-	-++-	-+++	+--+	+--+	+--+	+--+	+--+	+--+	+--+	+---	-++-	-++-	-++-	-++
31	-++-	-++-	-++-	-++-	-++-	-++-	-++-	-+++	+--+	+--+	+--+	+--+	+--+	+--+	+--+	+--
32	-++-	-++-	-++-	-++-	-++-	-++-	-++-	-++-	-++-	-++-	-++-	-++-	-++-	-++-	-++-	-++

The Basic 64-Run Plackett-Burman Design: Part II

Contrast Labels:

Run	ABCD	EFGH	IJKL	MNOP	QRST	UVWX	YZab	cdef	ghij	klmn	opqr	stuv	wxyz	αβγδ	ηθμπ	ρσω
			111	1111	1112	2222	2222	2333	3333	3334	4444	4444	4555	5555	5556	666
	1234	5678	9012	3456	7890	1234	5678	9012	3456	7890	1234	5678	9012	3456	7890	123
33	++--	++--	++--	++--	++--	++--	++--	++--	++--	++--	++--	++--	++--	++--	++--	++-
34	++--	++--	++--	++--	++--	++--	++--	++-+	--++	--++	--++	--++	--++	--++	--++	--+
35	++--	++--	++--	++-+	--++	--++	--++	--++	--++	--++	--++	--+-	++--	++--	++--	++-
36	++--	++--	++--	++-+	--++	--++	--++	--+-	++--	++--	++--	++-+	--++	--++	--++	--+
37	++--	++-+	--++	--++	--++	--+-	++--	++--	++--	++-+	--++	--++	--++	--+-	++--	++-
38	++--	++-+	--++	--++	--++	--+-	++--	++-+	--++	--+-	++--	++--	++--	++-+	--++	--+
39	++--	++-+	--++	--+-	++--	++-+	--++	--++	--++	--+-	++--	++-+	--++	--+-	++--	++-
40	++--	++-+	--++	--+-	++--	++-+	--++	--+-	++--	++-+	--++	--+-	++--	++-+	--++	--+
41	++-+	--++	--+-	++--	++-+	--++	--+-	++--	++-+	--++	--+-	++--	++-+	--++	--+-	++-
42	++-+	--++	--+-	++--	++-+	--++	--+-	++-+	--+-	++--	++-+	--++	--+-	++--	++-+	--+
43	++-+	--++	--+-	++-+	--+-	++--	++-+	--++	--+-	++--	++-+	--+-	++-+	--++	--+-	++-
44	++-+	--++	--+-	++-+	--+-	++--	++-+	--+-	++-+	--++	--+-	++-+	--+-	++--	++-+	--+
45	++-+	--+-	++-+	--++	--+-	++-+	--+-	++--	++-+	--+-	++-+	--++	--+-	++-+	--+-	++-
46	++-+	--+-	++-+	--++	--+-	++-+	--+-	++-+	--+-	++-+	--+-	++--	++-+	--+-	++-+	--+
47	++-+	--+-	++-+	--+-	++-+	--+-	++-+	--++	--+-	++-+	--+-	++-+	--+-	++-+	--+-	++-
48	++-+	--+-	++-+	--+-	++-+	--+-	++-+	--+-	++-+	--+-	++-+	--+-	++-+	--+-	++-+	--+
49	+-++	-+--	+-++	-+--	+-++	-+--	+-++	-+--	+-++	-+--	+-++	-+--	+-++	-+--	+-++	-+-
50	+-++	-+--	+-++	-+--	+-++	-+--	+-++	-+-+	-+--	+-++	-+--	+-++	-+--	+-++	-+--	+-+
51	+-++	-+--	+-++	-+-+	-+--	+-++	-+--	+-++	-+--	+-++	-+--	+-+-	+-++	-+--	+-++	-+-
52	+-++	-+--	+-++	-+-+	-+--	+-++	-+--	+-+-	+-++	-+--	+-++	-+-+	-+--	+-++	-+--	+-+
53	+-++	-+-+	-+--	+-++	-+--	+-+-	+-++	-+--	+-++	-+-+	-+--	+-++	-+--	+-+-	+-++	-+-
54	+-++	-+-+	-+--	+-++	-+--	+-+-	+-++	-+-+	-+--	+-+-	+-++	-+--	+-++	-+-+	-+--	+-+
55	+-++	-+-+	-+--	+-+-	+-++	-+-+	-+--	+-++	-+--	+-+-	+-++	-+-+	-+--	+-+-	+-++	-+-
56	+-++	-+-+	-+--	+-+-	+-++	-+-+	-+--	+-+-	+-++	-+-+	-+--	+-+-	+-++	-+-+	-+--	+-+
57	+-+-	+-++	-+-+	-+--	+-+-	+-++	-+-+	-+--	+-+-	+-++	-+-+	-+--	+-+-	+-++	-+-+	-+-
58	+-+-	+-++	-+-+	-+--	+-+-	+-++	-+-+	-+-+	-+-+	-+--	+-+-	+-++	-+-+	-+--	+-+-	+-+
59	+-+-	+-++	-+-+	-+-+	-+-+	-+--	+-+-	+-++	-+-+	-+--	+-+-	+-+-	+-+-	+-++	-+-+	-+-
60	+-+-	+-++	-+-+	-+-+	-+-+	-+--	+-+-	+-+-	+-+-	+-++	-+-+	-+-+	-+-+	-+--	+-+-	+-+
61	+-+-	+-+-	+-+-	+-++	-+-+	-+-+	-+-+	-+--	+-+-	+-+-	+-+-	+-++	-+-+	-+-+	-+-+	-+-
62	+-+-	+-+-	+-+-	+-++	-+-+	-+-+	-+-+	-+-+	-+-+	-+-+	-+-+	-+--	+-+-	+-+-	+-+-	+-+
63	+-+-	+-+-	+-+-	+-+-	+-+-	+-+-	+-+-	+-++	-+-+	-+-+	-+-+	-+-+	-+-+	-+-+	-+-+	-+-
64	+-+-	+-+-	+-+-	+-+-	+-+-	+-+-	+-+-	+-+-	+-+-	+-+-	+-+-	+-+-	+-+-	+-+-	+-+-	+-+

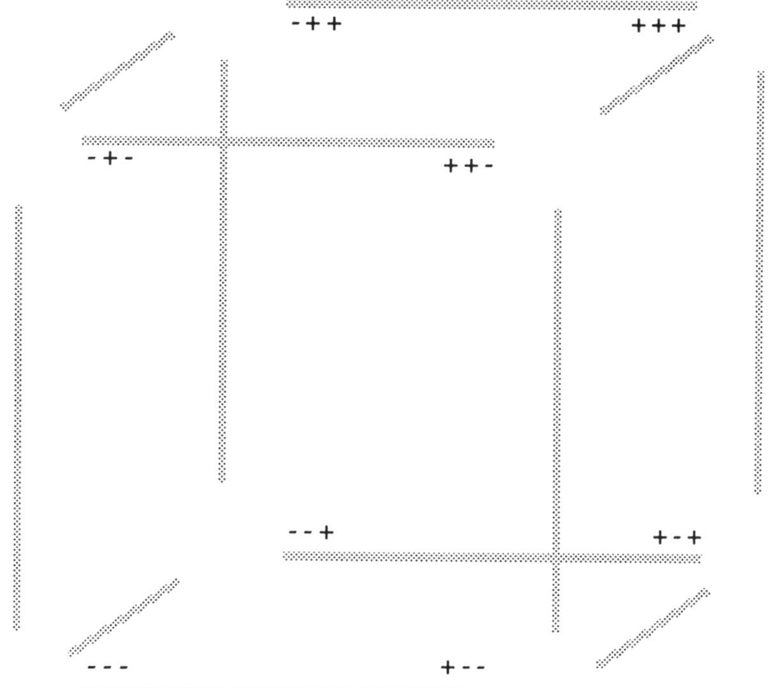

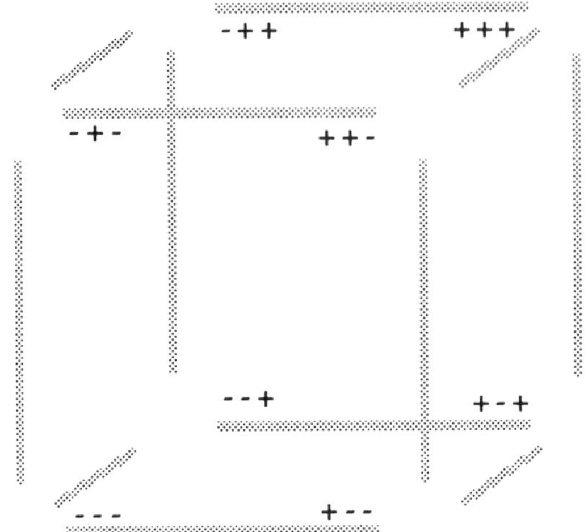

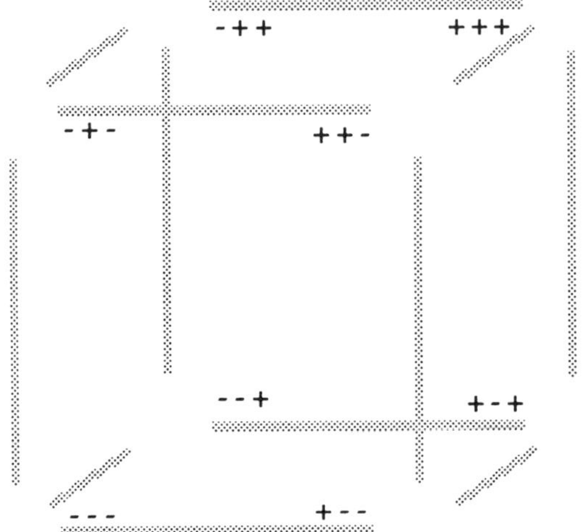

Scree Plot and Normal Probability Plot For 8-Run Design

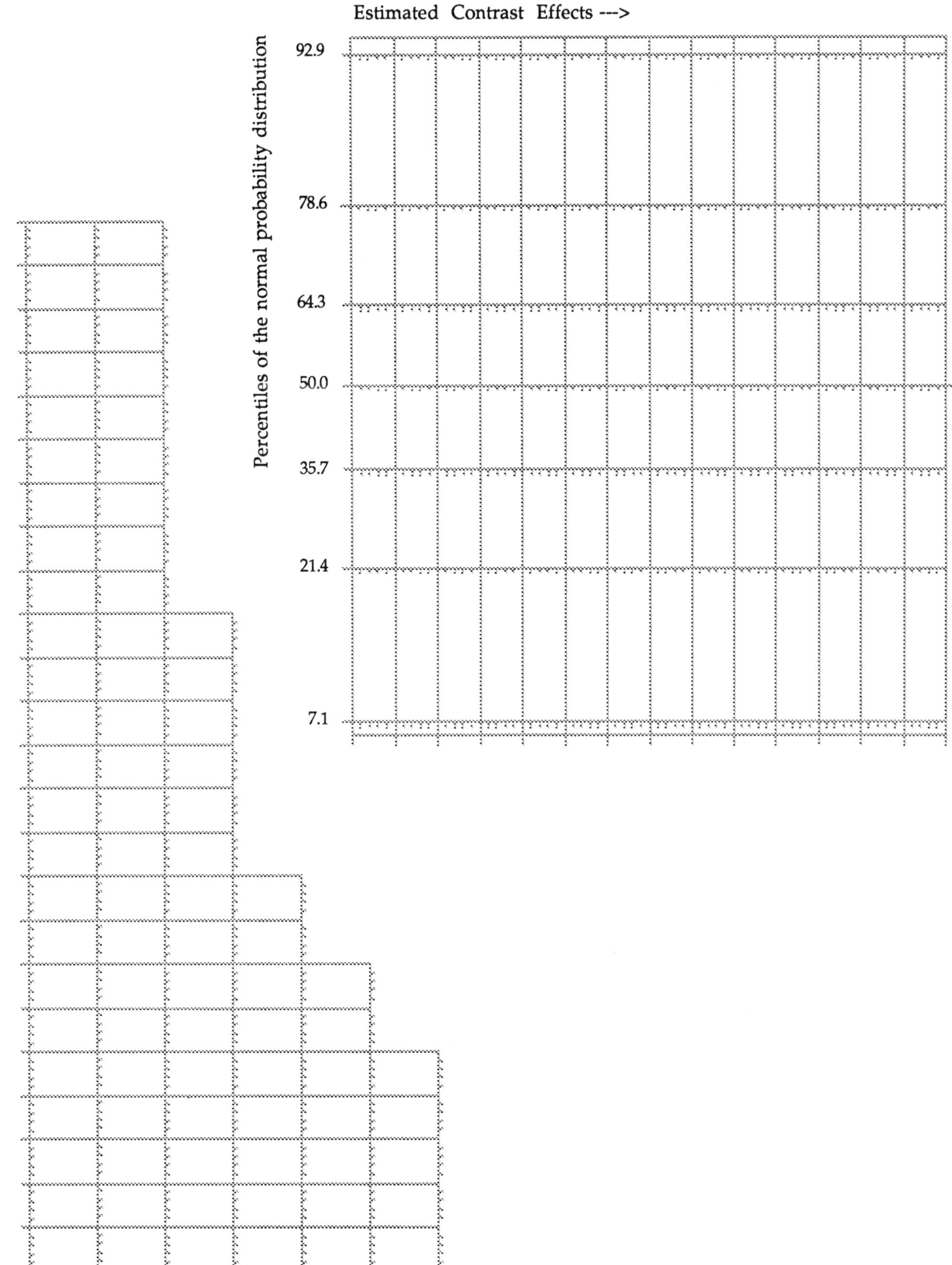

Scree Plot and Normal Probability Plot For 12-Run Design

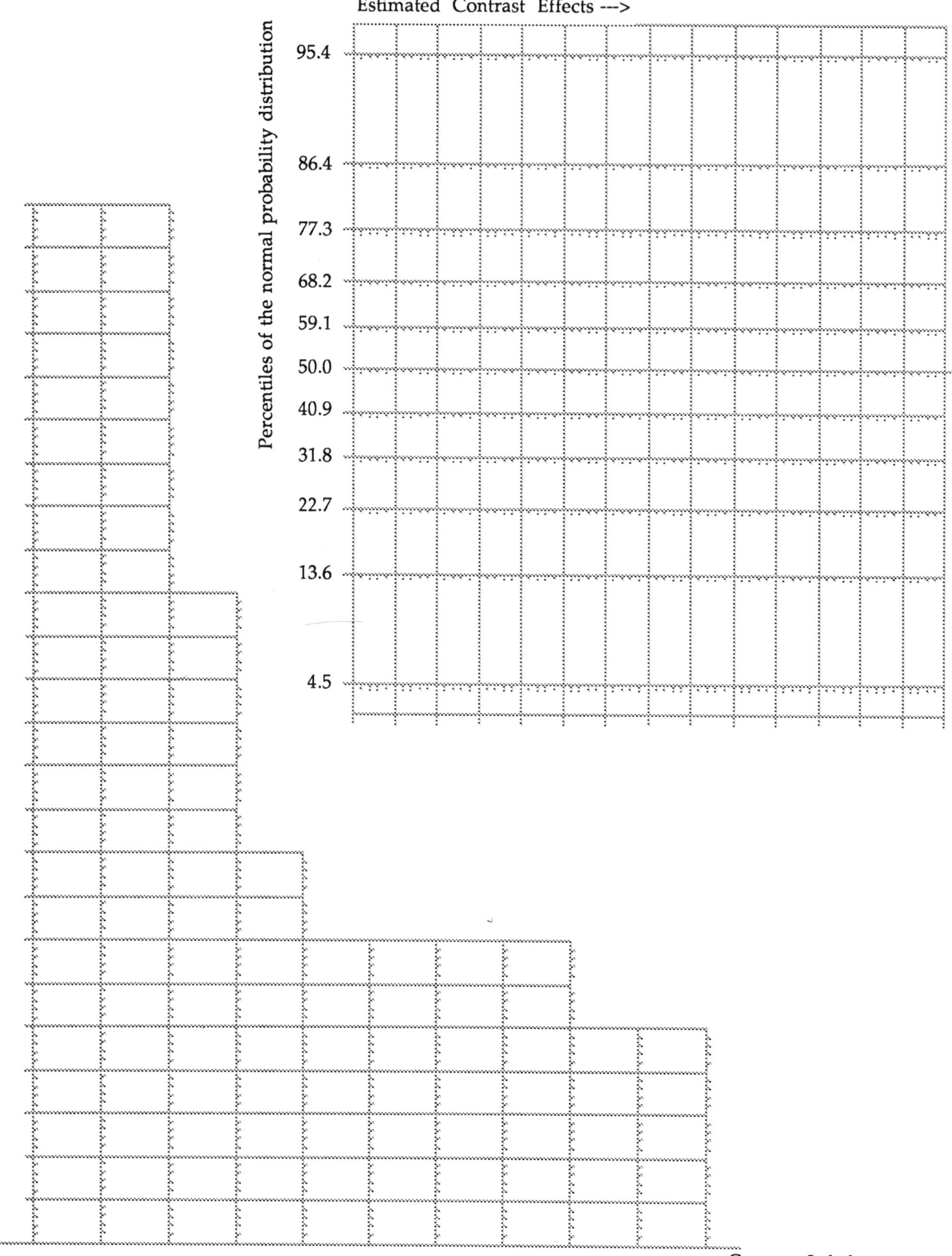

Scree Plot and Normal Probability Plot For 16-Run Design

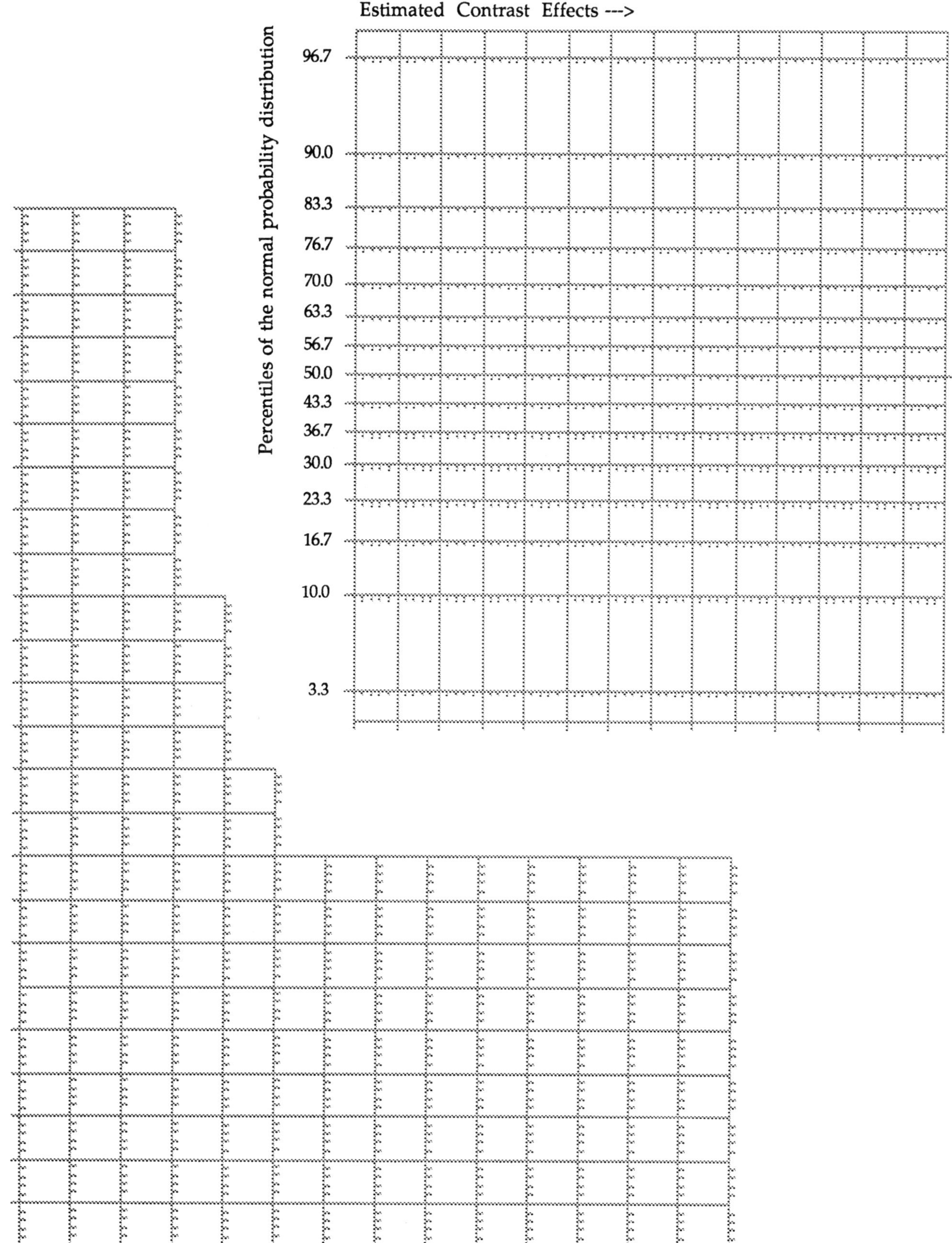

Scree Plot and Normal Probability Plot For 20-Run Design

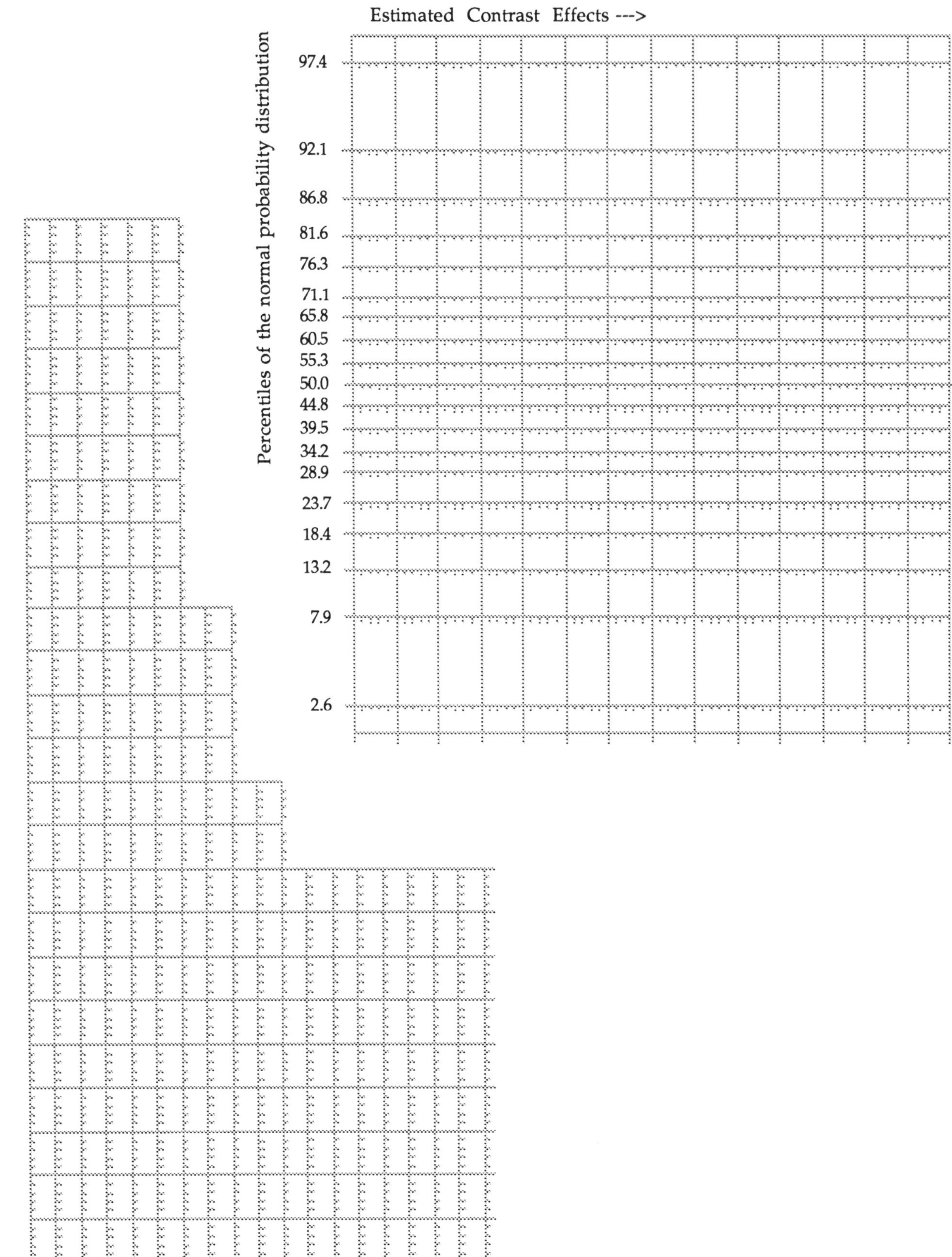

Scree Plot and Normal Probability Plot For 24-Run Design

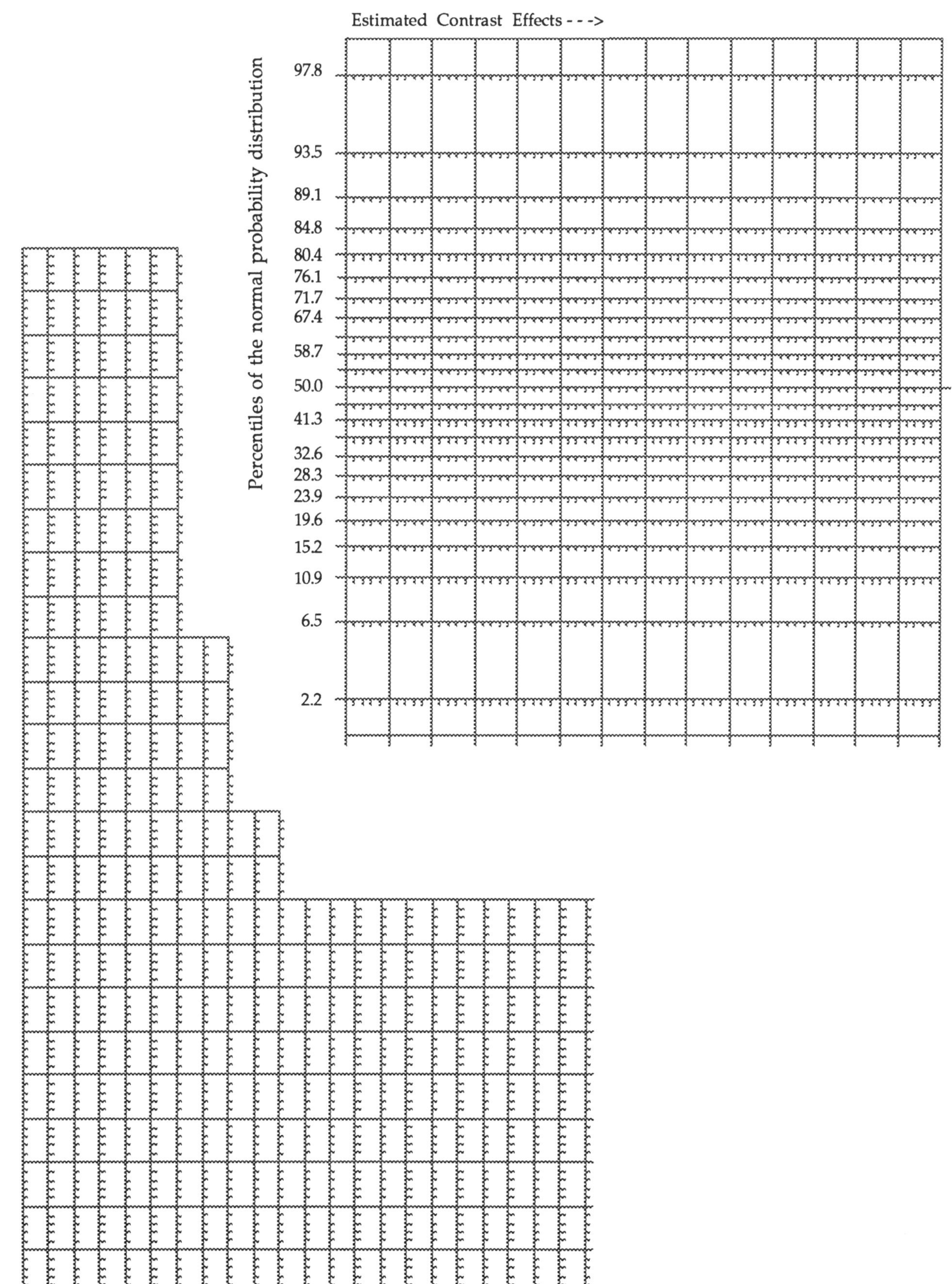

Scree Plot and Normal Probability Plot For 28-Run Design

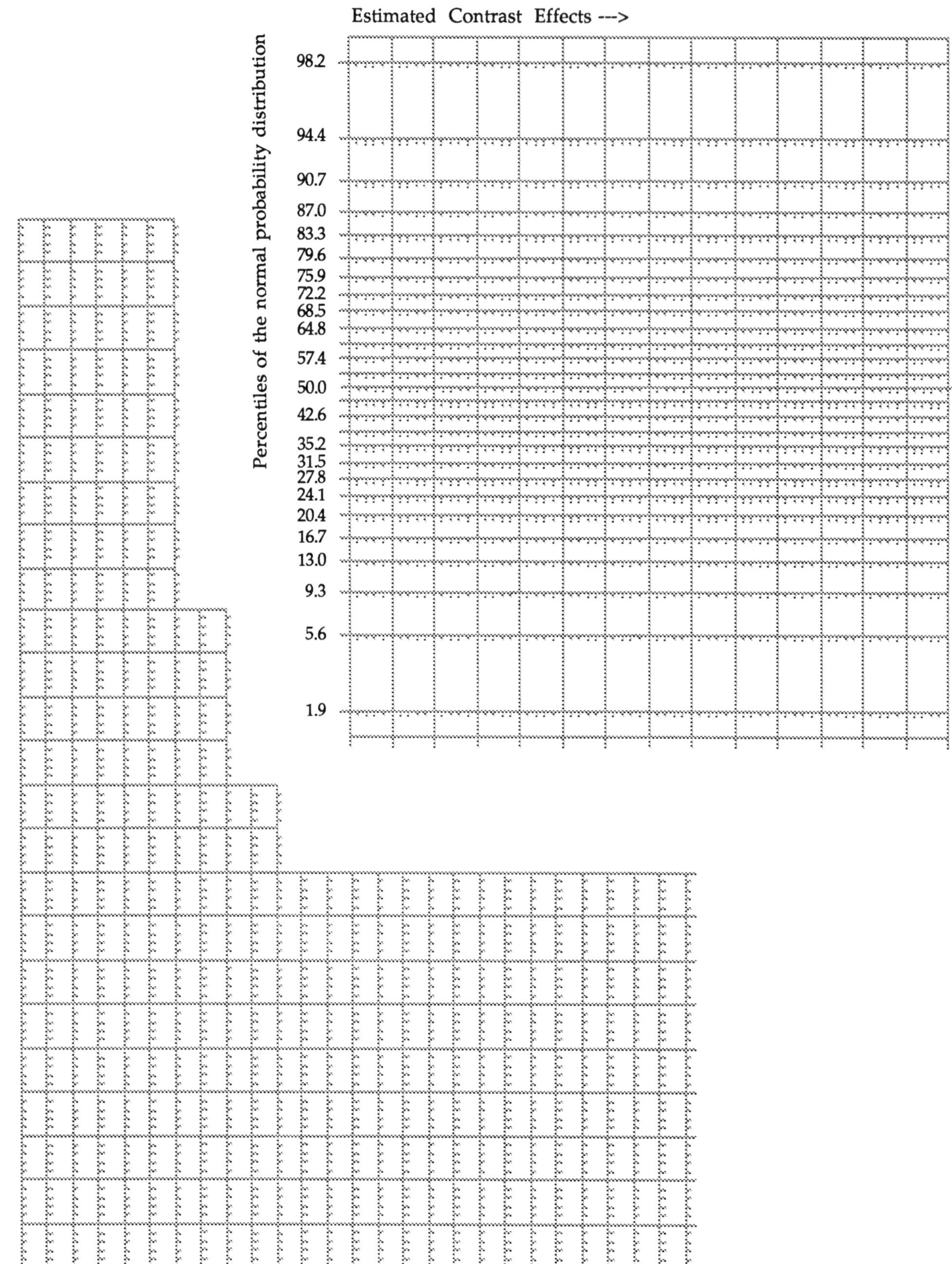

Scree Plot and Normal Probability Plot For 32-Run Design

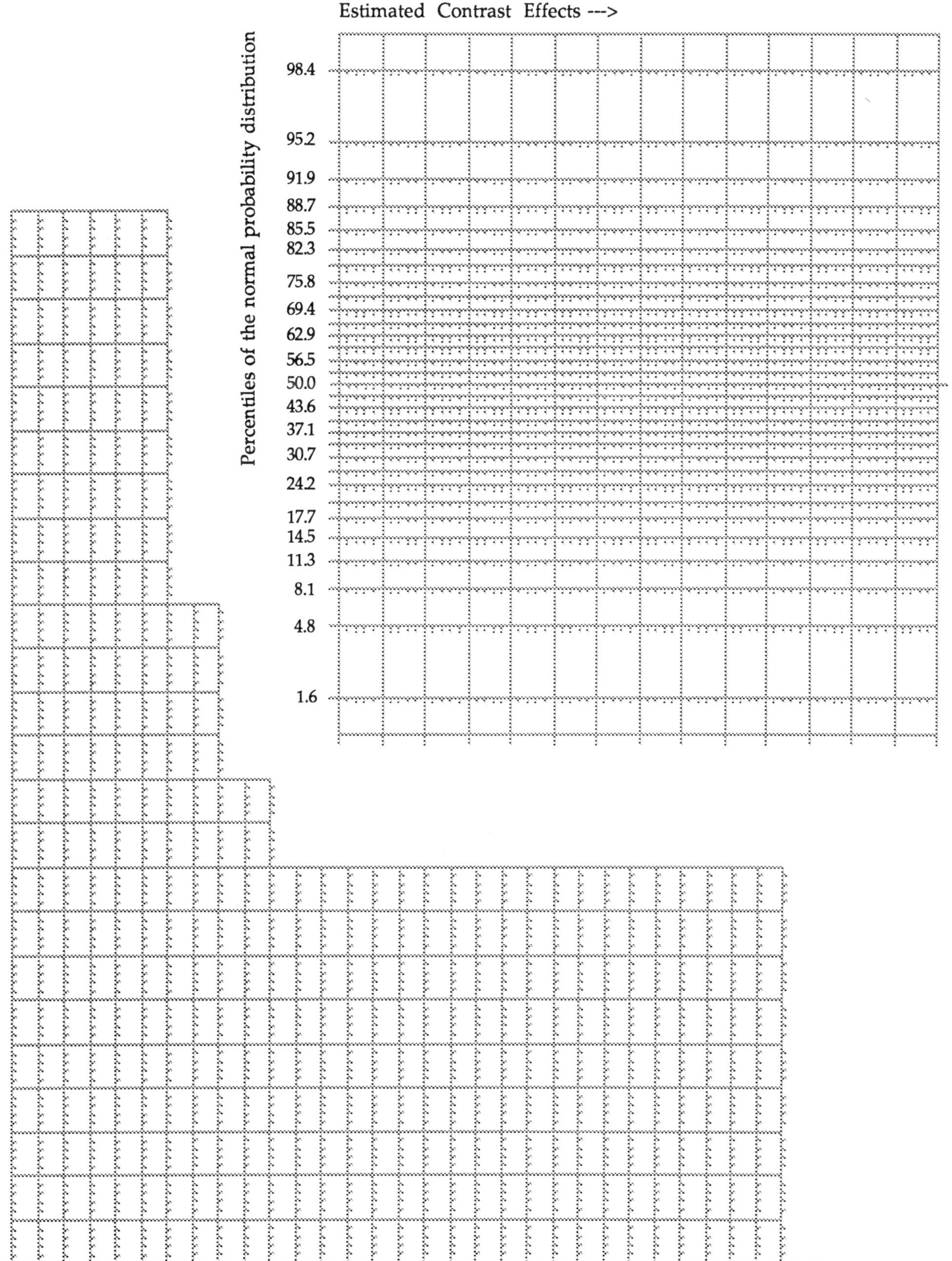

Bibliography and References

M.S. Bartlett and M.G. Kendall, "The Statistical Analysis of Variance—Heterogeneity and the Logarithmic Transformation," *Journal of the Royal Statistical Society, Series B, v.8,* pp.128–138, 1946.

G.E.P. Box and J.S. Hunter, (1961) "The 2^{k-p} Fractional Factorial Designs," *Technometrics, Vol.3,* pp.311-351, 449-458.

G.E.P. Box, W.G. Hunter and J.S. Hunter, (1978) *Statistics for Experimenters,* John Wiley and Sons, New York.

G.E.P. Box and K.B. Wilson, (1951) "On the Experimental Attainment of Optimum Conditions," *Journal of the Royal Statistical Society, Series B, Vol.XIII,* pp.1-35.

C. Daniel, (1962) "Sequences of Fractional Replicates in the 2^{p-q} Series," *Journal of American Statistical Association,* pp.403-429.

C. Daniel, (1976) *Applications of Statistics to Industrial Experimentation,* John Wiley and Sons, New York.

O.L. Davies and W. A. Hay, (1950) "The Construction and Uses of Fractional Factorial Designs in Industrial Research," *Biometrics,* pp.233-249.

W.J. Diamond, (1981) *Practical Experimental Designs,* Van Nostrand Reinhold, New York.

A.J. Duncan, (1965), *Quality Control and Industrial Statistics,* Richard D, Irwin, Inc. Homewood, Ill.

E.G. Johnson and J.W. Tukey, "Graphical Exploratory Analysis of Variance Illustrated on a Splitting of the Johnson and Tsao Data," ***Design, Data, & Analysis by Some Friends of Cuthbert Daniel,*** Colin L Mallows, Editor, John Wiley and Sons, New York, 1987, pp.178–180.

L.S. Nelson, (1983) "Exact Critical Values for the Analysis of Means," *Journal of Quality Technology, Vol.15,* pp.40-44.

P.R. Nelson, "Exact Critical Points for the Analysis of Means," *Communications in Statistics, v.11,* pp.699-709, 1982.

P.R. Nelson, "A Comparison of Sample Sizes for the Analysis of Means and the Analysis of Variance," *Journal of Quality Technology, v.15,* p.33-39, 1983.

P.R. Nelson, "Extended Tables and Additional Uses for Analysis of Means Critical Values," privately circulated manuscript, 1990.

E.R. Ott, (1967, 1983) "Analysis of Means--A Graphical Procedure," *Journal of Quality Technology, Vol.15,* pp.10-18.

Andrew C. Palm and Donald J. Wheeler, (1990) "Equivalent Degrees of Freedom for Estimates of the Process Standard Deviation Based on Shewhart Control Charts," unpublished manuscript.

P.B. Patnaik, (1950) "The Use of Mean Range as an Estimator of Variance in Statistical Tests," *Biometrika, Vol. 37,* pp.78–87.

R.L. Plackett and J. P. Burman, (1946), "Design of Optimum Multifactorial Experiments," *Biometrika, Vol.3,* p.305.

H. Scheffè, (1953) "A Method for Judging All Contrasts in the Analysis of Variance," *Biometrika, Vol.40,* pp.87–104.

H. Scheffè, (1959) *The Analysis of Variance,* John Wiley, New York.

E.G. Schilling, (1973) "A Systematic Approach to the Analysis ofd Means," *Journal of Quality Technology, v.5,* pp.93-108, pp.147-159.

Walter Shewhart, (1931) *Economic Control of Quality of Manufactured Product,* republished by American Society for Quality Control in1980.

Walter Shewhart, (1939) *Statistical Method from the Viewpoint of Quality Control,* republished by Dover Publications, New York, in1986.

John W. Tukey, (1953) *The Problem of Multiple Comparisons,* dittoed manuscript of 396 pages, Princeton University.

D.J. Wheeler, *Tables of Screening Designs, 2nd. Ed.,* SPC Press, Knoxville, 1989.

J.C. Whitwell and G. K. Morbey, (1961) "Reduced Designs of Resolution Five," *Technometrics, Vol.3,* pp.459-477.

Neil R. Ullman, (1989) "The Analysis of Means for Signal and Noise," *Journal of Quality Technology, Vol.21,* p.111-127.

Daniel T. Voss, (1988) "Generalized Modulus-Ratio Tests for Analysis of Factorial Designs with Zero Degrees of Freedom for Error," *Communications in Statistics—Theory and Methods, No.17(10),* p.3345–3359.

W.J. Youden, (1961) "Partial Confounding in Fractional Replication," *Technometrics, Vol.3,* pp.353-358.

Answers to Exercises

1.1 (a) 5.0 (b) 5.5 (c) 3.0 (d) 1.118034 (e) 1.195229

1.2 (a) 0.0 (b) $^1/_3$ (c) 0.577350
(d) –0.10, 0.20, 1.7, and 0.696419 (e) –0.122222, –0.2, 1.5, and 0.465773

1.3				
$s = 1.195229$	$s = 1.195$	$\frac{s}{c_4} = 1.239$		$s^2 = 1.4286$
$s_n = 1.118034$	$s_n = 1.118$	$\frac{s_n}{c_2} = 1.239$	$s_n^2 = 1.250$	
$R = 3.0$	$\frac{R}{d_2^*} = 1.013$	$\frac{R}{d_2} = 1.054$	$\left(\frac{R}{d_2}\right)^2 = 1.110$	$\left(\frac{R}{d_2^*}\right)^2 = 1.0265$

1.4 (a) 6 (b) 3.895 (c) 1.540

1.5 (a) 4.333 (b) 2.847 (c) 1.522

1.6 (a) 1.0 (b) 1.693 (c) 0.5907 (d) 1.671

1.7 (a) 10.0 (b) 3.895 (c) 2.567

1.8 (a) 4.333 (b) 2.847 (c) 1.522

1.9 (a) 7.0 (b) 1.693 (c) 4.135 (d) 11.695

1.10				
$\overline{s^2} = 628.33\overline{3}$	$\sqrt{\overline{s^2}} = 25.067$	----	----	$\overline{s^2} = 628.33$
$\bar{s} = 24.207$	$\bar{s} = 24.207$	$\frac{\bar{s}}{c_4} = 26.275$	$(\bar{s})^2 = 585.98$	----
$\bar{R} = 56$	$\frac{\bar{R}}{d_2^*} = 26.718$	$\frac{\bar{R}}{d_2} = 27.198$	$\left(\frac{\bar{R}}{d_2}\right)^2 = 739.71$	$\left(\frac{\bar{R}}{d_2^*}\right)^2 = 713.83$

1.11 (a) (i) 2.455 with $v = 9.0$ (ii) 2.990 with $v = 11$
(b) (i) 5.710 with $v = 9.0$ (ii) 8.545 with $v = 11$

1.12 (a) (i) 1.619 with $v = 8.4$ (ii) 1.589 with $v = 8.5$ (iii) 1.563 with $v = 9.0$
(b) (i) 2.472 with $v = 8.4$ (ii) $2.44\bar{4}$ with $v = 9.0$

Answers to Exercises

1.13 (a) (i) 7.088 with $v = 2.0$ (ii) 6.770 with $v = 2.0$
(b) (i) 39.638 with $v = 2.0$ (ii) 36.000 with $v = 2.0$

2.1 (a) See figure 2.11, p.74.
(b) UCL_R = 17.50 $\alpha = .10$ gives UDL = 39.82 and LDL = 27.65

2.2 UCL_R = 12.07 $\alpha = .10$ gives UDL = 7.64 and LDL = 0.27

2.3 For Concentrations: k = 2 and n = 15:

Averages are 29.20 and $38.2\bar{6}$ while UDL = 34.99 and LDL = 32.47

For Temperatures: k = 5 and n = 6:

Averages are $39.1\bar{6}$, $37.1\bar{6}$, $33.8\bar{3}$, $32.1\bar{6}$, $26.3\bar{3}$ while UDL = 37.31 and LDL = 30.16

2.4 $\bar{\bar{R}}$ = 6.8, $d_2 = 1.693$, $d_3 = 0.888$, and Est SD(R) = 3.567:

ANOMR for Concentration has L = 2 and m = 5;

Average Ranges are 5.0 and 8.6; and

UDL = 9.18 and LDL = 4.42 (Ullman: UDL = 8.97 and LDL = 4.63).

ANOMR for Temperatures has L = 5 and m = 2;

Average Ranges are 6.0, 7.0, 5.0, 10.0, 6.0

UDL = 13.18 and LDL = 0.42 (Ullman: UDL = 13.42 and LDL = 2.22)

Answers to Exercises

3.1

Source	SS	df	MS	F
Between	197.333	2	98.667	41.4
Within	50.000	21	2.381	
Total	247.333	23		

3.2

Source	SS	df	MS	F
Between	72.000	2	36.000	14.7
Within	22.000	9	2.444	
Total	94.000	11		

3.3

Source	SS	df	MS	F
Between	19830.0	4	4957.5	7.89
Within	9425.0	15	628.333	
Total	29255.0	19		

If you were not able to obtain these values return to page 92 and 93 and work through Example 3.3.

3.4

Source	SS	df	MS	F
Between	1403.8668	9	155.985	11.1
Within	280.0	20	14.0	
Total	1683.8666	29		

3.5 HSD = 9.74

5	4	3	10	2	8	1	7	9	6
19.7	23.7	31.3	33.0	34.7	36.3	36.7	39.7	40.7	41.7

so Subgroup 5 is detectably less than Subgroups 3, 10, 2, 8, 1, 7, 9 and 6,
and Subgroup 4 is detectably less than Subgroups 2, 8, 1, 7, 9 and 6,
and Subgroup 3 is detectably less than Subgroup 6

3.6

Source	SS	df	MS	F
Between	590.333	23	25.667	4.13
Within	447.500	72	6.2153	
Total	1037.833	95		

$F_{.90}(23,72) < F_{.90}(20,60) = 1.54$

Therefore, since the F-ratio exceeds the decision value, act as if there are signals within these data.

Answers to Exercises

3.7

ANOVA:

Source	S S	d f	M S	F
Between	234 607.5	7	33515.357	43.1
Within	24 870.0	32	777.1875	
Total	259 477.5	39		

TUKEY: HSD = 51.74

8	1	6	3	4	2	5	7
983	998	1104	1151	1152	1160	1184	1194

Subgroups 8 and 1 are detectably less than Subgroups 6, 3, 4, 2, 5, and 7,
and Subgroup 6 is detectably less than Subgroups 2, 5, and 7.

One-Way ANOM:
UCL for ranges = 137.4.
UDL = 1145.98,
LDL = 1085.52
Seven of the eight subgroup averages are detectably different from the Grand Averge.

Main Effect ANOMs:
for Catalyst: k = 2 and n = 20;
Averages are 1115.25 and 1116.25;
UDL = 1123.2, LDL = 1108.3.
for Time: k = 2 and n = 20;
Averages are 1111.5 and 1120.0;
UDL = 1123.2, LDL = 1108.3.
for Temperature: k = 2 and n = 20;
Averages are 1131.75 and 1099.75;
UDL = 1123.2, LDL = 1108.3.

4.1 See contrasts given in Exercise 4.2, p.116.

4.2 See values given in Exercise 4.3, p.120.

Answers to Exercises

4.3

Contrast	adjustment term	SS(C)	F
1	$^{10}/_3$	616.533	44.0
2	$^{4}/_3$	12.000	0.86
3	$^{12}/_3$	75.111	5.37
4	$^{24}/_3$	93.389	6.67
5	$^{40}/_3$	410.700	29.33

4.4

Contrast	$\hat{C}$	adjustment term	SS(C)	F	$\hat{l}$
1	24.5	6	100.04	16.1	2.04
2	15.25	3	77.52	12.5	2.54
3	1.25	3	0.52	0.08	
4	–43.5	6	315.38	50.7	–3.625
5	1.25	4	0.39	0.06	
6	–12.25	12	12.51	2.01	

Heat treatment W yields parts that average 2.04 units longer than those produced using Heat Treatment L.
Machine A parts average 2.54 units longer than those produced on Machine C.
Parts from Machines B and D average 3.625 units longer than those produced on Machines A and C.

5.1 See contrasts shown in Exercise 5.7, p.136.

5.2 Yes, Yes $\{C_1, C_2, C_3, C_4, C_5, C_6\}$

5.3 Yes

5.4 Yes, and No.

5.5/5.6

Contrast	$\hat{C}$	SS(C)	df	MS	F
6	0	0	1	0	0
7	0	0	1	0	0
8	36	162.0	1	162.0	11.57*
9	21.333	34.133	1	34.13	2.43
Within Subgroups		280.0	20	14.0	

$F_{.95}(1,20) = 4.35$

Answers to Exercises

5.7/5.8

Contrast	$\hat{C}$	SS(C)	df	F
1	4	10	1	0.01
2	34	722.5	1	0.9
3	-128	10240	1	13.1*
4	-256	40960	1	52.7*
5	-454	128822.5	1	165.7*
6	-292	53290	1	68.6*
7	30	562.5	1	0.7

$F_{.95}(1,32) < F_{.95}(1,30) = 4.17$

5.9 See Example 5.8, p.141 for correct values.

5.10 (a) MSE = 55,908.75

(b)

Contrast	F
1	0.00018
2	0.0129
3	0.183

(c) $F_{.90}(1,4) = 4.54$

5.11 (a) MSE = 431.666

(b) and (c)

Contrast	rank j	Mod F		
3	4	23.7	>	Mod $F_{.95}(4,3,7)$
4	5	94.9	>	Mod $F_{.95}(5,3,7)$
5	7	298.4	>	Mod $F_{.95}(7,3,7)$
6	6	123.5	>	Mod $F_{.95}(6,3,7)$

(d) Contrasts 3, 4, 5, and 6 all appear to represent potential signals.

5.12 Scree Plot has the following values (left to right):

129,000., 53,000., 41,000., 10,000., 1,000., 500., 0.

Normal Probability Plot has following points (left to right and bottom to top):
(-113.5, 7.1), (-73, 21.4), (-64, 35.7), (-32, 50.0), (1.0, 64.3), (7.5, 78.6), (8.5, 92.9)

Answers to Exercises

5.13 (a) Contrast coefficients for Contrast G are (top to bottom):

00-11 0000 0-11-1 1-110 000-1 1000

(b) MSW = 0.88768 with 11 d.f.

(c) Estimated value for Contrast D is 1.96

(d) Adjustment term is 2.5, see p.117.

(e) SS(D) = 1.536 and F-ratio = 1.73

(f) Using just the values in the Contrast for C, and rounding to the interger values:

C−	C+
47	57
48	55
47	49
48	55
47	54

One could also use all 35 values in this response plot.

6.1 B + AC + AD + BCD = 1.0

B + AC − AD − BCD = 2.5

B − AC + AD − BCD = 2.0

Therefore, if Contrast B represents a signal, it is most like to be the Main Effect for Factor B.

Estimated Effect for Main effect B = 1.833.

6.2

Contrast	$\hat{C}$	SS(C)	$\hat{l}$
A	1.43	0.256	0.36
B	−0.17	0.004	−0.04
C	22.77	64.809	5.69
D	20.09	50.451	5.02
E	−0.87	0.095	−0.22
F	10.41	13.546	2.60
G	12.23	18.697	3.06

Answers to Exercises

6.3 Contrast G = CD Interaction Effect

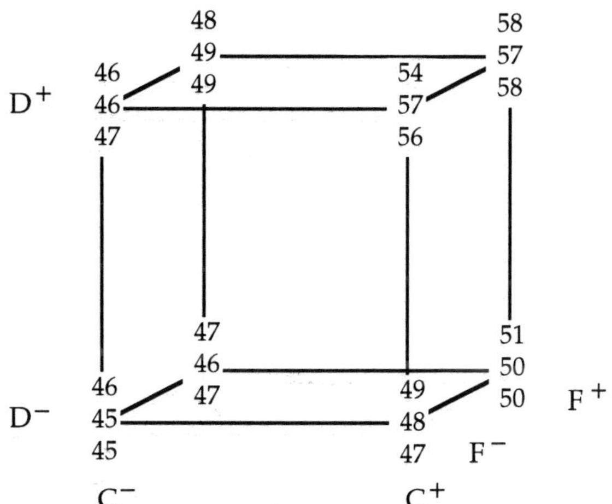

6.4
1. **I** = ABC and **I** = –ABD
2. **I** = ABC = –ABD = –CD
3. Resolution II

6.5
1. **I** = ABCD = BC = AD
2. A = BCD = ABC = D
3. AB = CD = AC = BD

Answers to Exercises

7.1/7.2

Contrast	$\hat{C}$	SS	df	F	$\hat{l}$
A	2.5	1.5625	1	0.58	
B	0.5	0.0625	1	0.02	
C	–1.5	0.5625	1	0.20	
D	3.5	3.0625	1	1.13	
E	–12.5	39.0625	1	14.53*	–3.125
F	–6.5	10.5625	1	3.92*	–1.625
G	5.5	7.5625	1	2.81	
Within		21.5	8		
Total		83.9375	15		

Contrast E represents E – AD – BG – CF
Contrast F represents F – AG – BD – CE

7.3/7.4

A	B	C	D	E	F	G
–	–	+	–	–	–	–
–	–	+	+	+	+	+
–	+	–	+	+	–	–
–	+	–	–	–	+	+
+	+	+	–	+	+	–
+	+	+	+	–	–	+
+	–	–	+	–	+	–
+	–	–	–	+	–	+

A	B	C	D	E	F	G
BC	AC	AB	–AE	–AD	–AG	–AF
–DE	–DF	DG	–BF	–BG	–BD	–BE
–FG	–EG	EF	CG	CF	CE	CD

Answers to Exercises

7.5/7.6

Contrast	$\hat{C}$	SS(C)	F
A	–8	4.0	1.52
B	6	2.25	0.86
C	12	9.0	3.43
D	-4	1.0	0.38
E	–10	6.25	2.38
F	32	64.0	24.38
G	6	2.25	0.86
H	–6	2.25	0.86
I	8	4.0	*pool*
J	6	2.25	0.86
K	0	0	0
L	–8	4.0	1.52
M	-2	0.25	*pool*
N	8	4.0	*pool*
B	–6	2.25	*pool*
Total		107.75	

MSE = 10.5/4 = 2.625 with 4 d.f. $F_{.90}(1,4) = 4.54$ and $\hat{l}$ for Contrast F = 4.0

The Response Plot for Main Effect F is:

F^-	F^+
0 1	8 3
0 0	6 5
0 0	2 0
0 0	4 5

So 60 seconds of mix time yields an average of 4.0 more voids per piece than does 30 seconds of mix time.

Index